Geographies of Globalization

D1315708

Critical Introductions to Geography

Critical Introductions to Geography is a series of textbooks for undergraduate courses covering the key geographical subdisciplines and providing broad and introductory treatment with a critical edge. They are designed for the North American and international market and take a lively and engaging approach with a distinct geographical voice that distinguishes them from more traditional and out-dated texts.

Prospective authors interested in the series should contact the series editor:
John Paul Jones III
Department of Geography and Regional Development
University of Arizona
jpjones@email.arizona.edu

Published
Cultural Geography
Don Mitchell

Political Ecology
Paul Robbins

Geographies of Globalization
Andrew Herod

Social Geography
Vincent J Del Casino Jr

Geographies of Media and Communication
Paul Adams

Forthcoming
Research Methods in Geography
John Paul Jones III and Basil Gomez

Mapping
Jeremy Crampton

Environment and Society
Paul Robbins

Geographic Thought
Tim Cresswell

Cultural Landscape
Donald Mitchell and Carolyn Breitbach

Geographies of Globalization

A Critical Introduction

Andrew Herod

WILEY-BLACKWELL

A John Wiley & Sons, Ltd., Publication

MARY M. STRIBLING LIBRARY

This edition first published 2009
© 2009 Andrew Herod

Blackwell Publishing was acquired by John Wiley & Sons in February 2007. Blackwell's publishing program has been merged with Wiley's global Scientific, Technical, and Medical business to form Wiley-Blackwell.

Registered Office
John Wiley & Sons Ltd, The Atrium, Southern Gate, Chichester, West Sussex, PO19 8SQ, United Kingdom

Editorial Offices
350 Main Street, Malden, MA 02148-5020, USA
9600 Garsington Road, Oxford, OX4 2DQ, UK
The Atrium, Southern Gate, Chichester, West Sussex, PO19 8SQ, UK

For details of our global editorial offices, for customer services, and for information about how to apply for permission to reuse the copyright material in this book please see our website at www.wiley.com/wiley-blackwell.

The right of Andrew Herod to be identified as the author of this work has been asserted in accordance with the Copyright, Designs and Patents Act 1988.

All rights reserved. No part of this publication may be reproduced, stored in a retrieval system, or transmitted, in any form or by any means, electronic, mechanical, photocopying, recording or otherwise, except as permitted by the UK Copyright, Designs and Patents Act 1988, without the prior permission of the publisher.

Wiley also publishes its books in a variety of electronic formats. Some content that appears in print may not be available in electronic books.

Designations used by companies to distinguish their products are often claimed as trademarks. All brand names and product names used in this book are trade names, service marks, trademarks or registered trademarks of their respective owners. The publisher is not associated with any product or vendor mentioned in this book. This publication is designed to provide accurate and authoritative information in regard to the subject matter covered. It is sold on the understanding that the publisher is not engaged in rendering professional services. If professional advice or other expert assistance is required, the services of a competent professional should be sought.

Library of Congress Cataloging-in-Publication Data

Herod, Andrew, 1964–
    Geographies of globalization : a critical introduction / Andrew Herod.
        p. cm.
    Includes bibliographical references and index.
    ISBN 978–1–4051–1052–5 (hardcover : alk. paper) — ISBN 978–1–4051–1091–4 (pbk. : alk. paper)   2. Globalization.   2. International economic integration.   3. Human geography.   I. Title.
    JZ1318.H477 2009
    303.48′2—dc22

                                                                    2008021456

A catalogue record for this book is available from the British Library.

Set in 10.5/13pt Minion by Graphicraft Limited, Hong Kong
Printed in Malaysia by Vivar Printing Sdn Bhd

[1]   [2009]

MARY M. STRIBLING LIBRARY

JZ
1318
.H47
2009

Contents

Figures

Tables

Preface

Issues of globalization are omnipresent today and "the global" has become the object of much analytical fascination. Wherever we turn, it appears, we cannot escape the "new global reality" of globalization. Indeed, if we were to be somewhat whimsical, we might even suggest that the term "globalization" has itself become a global one which has colonized our collective consciousness, for few things, we are often told, are untouched by it, either materially or rhetorically. From the ubiquitous claims of politicians that "we *must* implement such-and-such policy *because* of globalization," to global climate change, to fears of global pandemics of disease, to preparing students for life in a global economy, to fighting a global war on terrorism, references to things "global" appear to have gained greater currency than ever and globalization is perceived as a powerful catalyst for changing the way that things have "always been done." The contemporary transformation in the planet's political economy that we often take to be "globalization," it seems, has made the world so small that we can no longer use distance to shield us from events occurring in what were formerly far-off places, nor to hide us from the gaze of others located pretty much anywhere on Earth. The global, then, such discourses tell us, is the scale of social life from which there is no escape, the scale which denies us the geographical option of fleeing to a distant spatial sanctuary in some remote corner of the planet wherein we can insulate ourselves from the consequences of contemporary social and natural processes. As our world has shrunk, many of the out-of-the-way spaces into which people a hundred or two hundred years ago could slip and make themselves invisible have been opened up to inspection and regulation by external global forces. In the process, the planet has, arguably, been turned into a giant panopticon in which nothing is beyond our reach – a reality brought home by the ability of modern spy satellites to survey every square meter of the Earth's surface in some significant detail. Virtually nowhere, it seems, remains beyond either globalization's touch or its view.

I have written this book, then, to explore the debate around how the geography of the world economy is being transformed by contemporary processes. In so doing I am immediately, of course, confronted with the issue of what, exactly, is "globalization"? This is a difficult question because there is no clear-cut definition, and I certainly do not want to add to the cacophony by adding my own characterization. Instead, I am interested in exploring how the term "globalization" has been used and with what force. Whereas social scientists usually strive for clarity of conceptualization as the basis for explanation, I want to argue here that the ambiguity of the concept of globalization is actually what gives it such force as an idea, and it is this ambiguity that I will examine in some depth. Questions of whether we are experiencing "globalization" are inseparable from the issue of whether we believe we are doing so – if we think we live in a globalizing world we are likely to see globalization in virtually everything, and "globalization" will become something of a self-fulfilling prophecy as our actions conducted in the belief that we live in a globalizing world will be apt to bring about the very condition within which we believe we live. Hence, borrowing from Koselleck (1985), we might think of "globalization" both as a "space of experience" (i.e., as something which we may believe exists and so to which we respond) and as a "horizon of expectations" (i.e., as something which brings with it the implicit prospect that contemporary ways of thinking and living will be transformed by those actions we take in the present, actions which are based upon our contemporary space of experience concerning the reality of "globalization").

In terms of the book's specifics, the themes I explore herein have been developed within the context of a course on "Geographies of Globalization" which I have now taught for about a decade. The book is designed primarily as a text for upper-level undergraduates and graduate students who are beginning to explore this vast territory, together with others who may have an interest in the topic of globalization. Given my own disciplinary background, the core argument herein is that whatever else it may be, "globalization" is a geographical process, one which is about connecting different parts of the world together in different ways and creating various geographical imaginations about such connections. Consequently, in Chapter 1 I outline how we can understand globalization as a spatial process and understand economies as spatial entities. Chapter 2 examines representations of "the global" and how the image of the globe as a coherently connected whole has come about, at least in the West. In Chapter 3 I distinguish between two words that have frequently been used interchangeably – "globalization" and "internationalization." Chapter 4 examines discursive issues related to how we understand matters of geographical scale, for how we conceptualize the relationship between, say, "the global" and "the local" will dramatically affect how we comprehend globalization. Chapter 5 looks at how different parts of the planet were linked together through colonialism and mercantilism prior to the twentieth century. In Chapter 6 I explore how transnational corporations have served as powerful actors for reworking geographical linkages between places in the twentieth century. In Chapter 7 I outline the activities of several international regulatory bodies (the International Monetary Fund, the World Bank, the World Trade Organization) which have played significant roles in how the geography of contemporary global capitalism has been made. Following from this, in Chapter 8 I look at how workers and their organizations have sought to organize transnationally and how such actions have facilitated transnational connections between places. The concluding chapter considers how denoting certain eras as "pre-global" and "global" relies on constructing particular narratives about the past and organizing historical events in specific ways, together with what that means for our understanding of such events.

In writing this book I have accumulated a number of debts. I would particularly like to thank: J. P. Jones for encouraging me to write the book and being patient when I didn't; John Campbell, UGA reference librarian for tracking down statistics on historical business trends; Mark Lundeen, for bringing to my attention the data on US currency issuance and gold stocks used in Chapter 9; Wendy Giminski and Mark Dodson for drawing the diagrams used herein; the *New York Times* for permission to reproduce Figure 1.1; ProLogis for permission to reproduce Figure 1.4; the Bodleian Library, Oxford University, for permission to reproduce Figure 2.3; Hema Maps NZ for permission to reproduce Figure 2.6; the Alaska International Airport System for permission to reproduce Figure 2.7; and Sage Publications for permission to reproduce Figures 4.2–4.6 (which originally appeared in A. Herod [2003], "Scale: The local and the global," pp. 229–47 in *Key Concepts in Geography*, edited by S. Holloway, S. Rice, and G. Valentine). Figure 1.3 is redrawn from P. Dicken (2003), *Global Shift: Reshaping the Global Economic Map in the 21st Century*; Figures 1.5, 1.6, and 1.7 are freely available from www.telegeography.com; Figure 1.8 is redrawn from M. S. Monmonier (1977), *Maps, Distortion, and Meaning*, AAG Resource Paper 75-4; Figure 2.1 is freely available from NASA. Finally, I want to thank Jennifer for graciously living with this global project for so long.

Athens, Georgia, USA

A Note on Terminology, Naming, and the Calculation of Historical Monetary Values

Several times in this book the term "neoliberal" is used to refer to contemporary economic policies associated with how the global economy is regulated. In this expression "liberal" does not refer to ideas proposed by those towards the left of the political spectrum – that is to say, it is not used in contrast to "conservative" in a political sense. Rather, here the word "liberal" relates to regimes of global trade which seek to have as little government regulation as possible. Such use has its origins in the "liberal" (i.e., laissez-faire) free trade policies pursued by the British government in the nineteenth century. "Neoliberalism," then, is used to refer to the ideas of those who would like to pursue a renewed laissez-faire "free trade" trade and economic agenda. Put another way, "neoliberal" trade and economic policies are, in fact, politically conservative policies.

I use modern place names throughout the text, except when historical names are more appropriate. References to India prior to 1947 refer to what is today India, Pakistan, and Bangladesh. I use the terms "global South," "developing countries," and "Less Developed Countries" (LDCs) interchangeably, recognizing that each of these terms is problematic in its own way.

I have presented contemporary (2005) values for most monetary amounts listed in the text (the figures appear in square brackets after the original numbers). Unless otherwise specified, all calculations of the current value of past monetary measures follow Officer's (2006) methodology, who argues that calculating the contemporary value of an economy's total output of goods and services and other large-scale projects or expenditures is best done on the basis of their proportionate share of the Gross Domestic Product, then and now. Calculations of values in British pounds sterling (£) between 1830 and 2005 use Officer's (2006) GDP figures. Those between 1700 and 1829 use data on Britain's GDP given in Broadberry et al. (2006) and a figure of £1.209 trillion for Britain's 2005 GDP. The figures for the value of French colonial imports on p. 117 are based on French GDPs of 28.7 billion francs in 1892 and 346 billion in 1929 (Mitchell 2003b) and a French 2005 GDP of €1.7 trillion (the same

1929 and 2005 GDPs are used for the figures for French colonial stocks and bonds (on p. 119). Contemporary figures for France's exports to, and investment in, Central America (p. 127) are based upon estimates of French GDP of 25 144 million francs for 1887 and 28 758 million francs for 1896 (Mitchell 2003b). The figure for German investment is based upon Mitchell's (2003b) estimate of 40 643 million marks for German NNI (Net National Income) in 1906 and a figure of €2.2 trillion for Germany's 2005 GDP. The 2005 figures for Dutch investment in the US in 1914 and 1938 (p. 146) are based on Dutch NNIs of 2.730 billion guilders and 5.395 billion guilders respectively (Mitchell 2003b), exchange rates of $1 = 2.46 guilders (1914) and $1 = 1.82 guilders (1938) (Officer 2007), and a 2005 GDP of $595 billion. The 2005 figure for Japanese FDI in China in 1930 (p. 150) is based on a 1930 GNP of ¥14.7 billion (Mitchell 2003a), a 1930 average exchange rate of $1 = ¥2.025 (Officer 2007), and a 2005 GNP of $4 988 billion. That for Japanese FDI outflows to Asia (p. 160) is calculated using a Japanese GDP figure of ¥148 billion for 1975 and ¥503 billion for 2005 and multiplying the $9 billion FDI figure given by Mason by 3.4 (representing the factor by which Japanese GDP grew during this period). Occasionally, when figures for nations' GDPs are unavailable, I have calculated current figures on the basis of changes in the retail price index. This typically gives a lower value than does calculating current values as a proportion of GDP and is considered less appropriate by Officer. When a single contemporary figure is given for a range of past values, the 2005 figure is an average of the values for the beginning and ending of the series, unless GDP figures are available for only part of the range, in which case I have calculated the current value based upon the last year in the range (as on p. 115, where the figures of £6.4 million equaling £110.7 billion and £11.3 million equaling £205.8 billion in 2005 are calculated based upon a 1700 GDP of £69.9 million). In calculating the US dollar ($) equivalent of monetary amounts, the figure listed represents the amount it was at the time given for the original currency (based upon the average exchange rate for that year, per Officer [2007]). Generally, I have not given a 2005 figure when the original data year is after 1995. Year 2005 figures for FDI are calculated using the GDP of the originator country, not the host.

Abbreviations

AFL	American Federation of Labor
AFRO	African Regional Organisation
AIDS	Acquired Immune Deficiency Syndrome
AIFLD	American Institute for Free Labor Development
APRO	Asia and Pacific Regional Organisation
BATU	Brotherhood of Asian Trade Unionists
CFA franc	*franc des colonies françaises d'Afrique*
CGT	Confédération Générale du Travail
CIA	Central Intelligence Agency
CIO	Congress of Industrial Organizations
CISC	Confédération Internationale des Syndicats Chrétiens
COSATU	Congress of South African Trade Unions
CUT	Central Única dos Trabalhadores
EIC	British East India Company
EPZ	Export Processing Zone
ERO	European Regional Organisation
EU	European Union
EWCs	European Works Councils
FDI	foreign direct investment
FTUI	Free Trade Union Institute
GATS	General Agreement on Trade in Services
GATT	General Agreement on Tariffs and Trade
GDP	gross domestic product
GNP	gross national product
GUFs	Global Union Federations
IBRD	International Bank for Reconstruction and Development
ICEM	International Federation of Chemical, Energy and General Workers' Unions

ICFTU	International Confederation of Free Trade Unions
ICSID	International Centre for Settlement of Investment Disputes
IDA	International Development Association
IFC	International Finance Corporation
IFTU	International Federation of Trade Unions
ILO	International Labour Organization
IMF	International Monetary Fund
ITO	International Trade Organization
ITSs	international trade secretariats
ITUC	International Trade Union Confederation
IWW	Industrial Workers of the World
JIT	just-in-time
LDC	Less Developed Country
LMU	Latin Monetary Union
MIGA	Multilateral Investment Guarantee Agency
MITI	(Japanese) Ministry of International Trade and Industry
NICs	newly industrializing countries
OECD	Organization for Economic Cooperation and Development
PAFL	Pan-American Federation of Labor
SARS	Severe Acute Respiratory Syndrome
SEIU	Service Employees' International Union
SIGTUR	Southern Initiative on Globalisation and Trade Union Rights
SMU	Scandinavian Monetary Union
TINA	"There Is No Alternative"
TNC	transnational corporation
TRIM	Agreement on Trade-Related Aspects of Investment Measures
TRIP	Agreement on Trade-Related Aspects of Intellectual Property Rights
TUC	Trades Union Congress
TUIs	Trade Union Internationals
USWA	United Steelworkers of America
UTC	Coordinated Universal Time
VOC	*Vereenigde Oostindische Compagnie* (Dutch United East Indian Company)
VTM	Virtual Trade Mission
WCCs	world company councils
WCL	World Confederation of Labour
WFTU	World Federation of Trade Unions
WTO	World Trade Organization
WWCs	World Works Councils

Chapter 1

Introduction

Chapter summary: This chapter raises the question of what is meant by "globalization" and how we conceptualize economies as objects of analysis. It also explores how to think of economies as spatial entities, contrasting concepts of absolute and relative distance.

- The World Economy as a Spatial Entity
  - What is "the economy"?
  - How can we think of the economy as a spatial entity?
- Of Globalization, Shrinking Worlds, and the Paradoxes of Geography

What is "globalization" and how might we understand its geographies? Such an apparently simple pair of questions might demand an equally simple pair of answers. Thus, a popular internet commentary has suggested the phenomenon can be readily illustrated by the 1997 death of British Princess Diana in a Paris tunnel:

What is globalization? It's when an English princess with an Egyptian boyfriend crashes in a French tunnel in a German car with a Dutch engine driven by a Belgian who was high on Australian beer, Scottish whiskey, and Burmese dope whilst being followed closely by Spanish paparazzi on Italian motorcycles with Japanese cameras, and who was treated by an American doctor using Swiss drugs based on Brazilian medicines and medical technology that uses Bill Gates's software that he stole from the Taiwanese and which was loaded on hardware based on IBM clones that use Philippine-made chips and Malaysian-made monitors, clones which are assembled by Bangladeshi workers in a Singapore plant, transported by lorries made in Korea and driven by Indians, shipped by the Vietnamese crew of ships built in Northern Ireland but owned by Greeks and registered in Panama, ships hijacked by Indonesian pirates with guns made in Israel and smuggled by Africans and finally sold by an Arab salesman working through a Hong Kong front!

However, hidden within this explanation is a further question: is "globalization" merely the growth of "things international"? Put another way, does the mentioning of these many and diverse countries and people who were allegedly linked to Diana's death confirm the "globalized" nature of this event? Why is this a "global" rather than an "international" event, and is there any difference between these two terms? Certainly, the aftermath of the car accident that caused Diana's death was broadcast worldwide, virtually as it happened, and was taken to illustrate the "global problem" of paparazzi following famous people. But what does it mean to characterize her death as a "global" or even "international" event rather than as, say, a series of interlinked "local" events – an accident in a particular tunnel in Paris learned about by people in very particular locations (living rooms, cafés, offices, and street corners) across the planet? Moreover, given that the accident happened early in the morning in Paris, its timing meant that many people living in North America (where it was still evening) found out about the night's events before many Parisians, who were still sleeping. How, then, is this either a "global" or a "local" event when people living just a few streets from the accident may only have become aware of Diana's death several hours after people living across the Atlantic?

Such questions illustrate both that the term "globalization" is multifaceted and can mean quite different things to different people, and that the process of "globalization" has significant consequences for our lives and how we understand the world and our place in it. Although most generations probably imagine they live in times infinitely more complicated and worrisome than those of their forebears, the transformation of our contemporary world by "globalization" (whatever that may actually be) certainly seems to have augured tremendous consternation, concern, and confusion. Indeed, "globalization" appears to have made our lives very much more complicated as the world's old "commonsensical" geographical order seems to have been upset. For instance, disagreements between the French and US governments over going to war in Iraq led many Americans to boycott French goods. However, in a world of growing interconnectivity, in which transnational corporations (TNCs) criss-cross the planet with their investments, production chains, and lines of executive control and decision making, determining what, exactly, constituted a "French" product often proved difficult. Thus, one internet call urged Americans to boycott a long list of goods and companies, including such "obviously" French products and firms as Evian mineral water, Yoplait yogurt, Air France, Airbus, Cartier, and Yves St. Laurent. The only problem for the boycotters with such a strategy, though, was that many of the companies and products listed either were not actually French (Cartier is owned by a Swiss company, whilst Yves St. Laurent is Netherlands-based) or had significant ties with US companies. Hence, Evian's French parent company Groupe Danone SA uses Coca-Cola to distribute its mineral water in the US, whilst Yoplait is produced in the United States by General Mills and Air France has a strategic alliance with US carrier Delta, with which it code shares international routes and jointly owns a cargo-handling company.[1] Additionally, whilst urging US airlines not to purchase Airbuses was seen as a way to punish French companies and the French government (who together own almost 40 percent of Airbus), it would also punish firms from two US allies in the war – the UK and Spain, firms from which own, respectively, 20 percent and 6 percent of Airbus – as well as put out of work thousands of Americans employed by US-based companies which annually manufacture about $5.5 billion worth of Airbus components (McArtor 2003).

Furthermore, whilst many boycotters chose not to drink French wine, much "French" wine is actually blended and bottled in the United States – a process that provides employment for US winery, bottle manufacturing, and warehouse workers, not to mention, of course, the fact that it is distributed by US truck drivers. Equally, there arose the question of what to do about products like Waterman fountain pens, high quality pens that are made in both France and the UK by a company owned by the US giant Gillette. Should only the French-made pens be boycotted and, if so, how would they be distinguished by the average consumer from the UK-made ones, and what would this mean for the US investors who owned shares in Gillette? And should boycott supporters stop sleeping at Red Roof Inns across the USA and refuse to watch the *Jerry Springer Show*, read *Woman's Day* magazine, or download songs from MP3.com, all of which were owned at the time by French companies? What exactly, then, did a boycott of French goods mean when French firms were the fifth largest foreign investors in the US economy, owning businesses employing nearly 650 000 American workers (Hamilton & Quinlan 2004)? Moreover, what did it even mean to talk about a "French" firm – even if it were located in, and exporting from, France – when many of that firm's stockholders might be American, British, Brazilian, Korean, Nigerian, or Canadian?

The difficulty of determining whether Evian and Yoplait, Airbus and Air France are "really" French products and companies highlights the fact that whatever else it may be, globalization is a spatial process and phenomenon, one which seems to be blurring old geographical certainties in which products and people were easily and quite literally "put in their place." Thus, in the days when Japanese car makers manufactured their products in Japan and exported them to the United States it was relatively easy to determine what was and was not a Japanese car. Today, on the other hand, determining whether a Corolla assembled in Fremont, California (where Toyota operates a joint-venture assembly plant with General Motors) is really a Japanese car, an American car, both, or neither seems much more difficult.[2] Likewise, what does it mean to talk of Airbus being a "European" – much less a "French" – company when it sources billions of dollars' worth of parts from US plants and has an engineering facility in Wichita, Kansas (which designed the wing for its A380 aircraft), or of its major competitor Boeing being an "American" company when it annually buys some €4 billion in components from European suppliers (McArtor 2003)?

Seeing globalization as a geographical phenomenon might, at first glance, appear strange. However, with a little thought it quickly becomes apparent that globalization is reshaping in significant ways the spatial and temporal organization of contemporary capitalism worldwide. As production chains for many products literally stretch across the globe, as intercontinental air travel means that most places in the world are now no more than 24 or 48 hours away from each other, and as modern telecommunications technology means that we can sit in our living rooms and watch in real time events unfold half a world away, our planet seems to have become ever smaller and ever faster. Quite literally, we do not seem to have either the space or the time we used to have. Thus, whereas in the mid-seventeenth century it took four days for a letter to travel the 413 miles from London to Edinburgh (Picard 1997: 73), and in the mid-nineteenth century French author Jules Verne wrote a best-selling novel in which the characters were engaged in a race around the world to be undertaken in fewer than 80 days, today not only could one travel around the world in fewer than 80 hours, but one could also transmit several billion dollars around it in decidedly fewer than 80 seconds.

Furthermore, not only does the world appear smaller than it used to, but the relationship between people and places also seems to have changed, at least for many. Hence, whilst for much of recent history going to work meant having physically to cross geographic space to get from home to work, today it is not uncommon for millions of workers to "telecommute," working at home but connected to a central office via computer. Yet whereas a couple of decades ago telecommuting might allow workers to stay at home in the suburbs and avoid having to travel into a downtown office or perhaps allowed them to live in Idaho but work "in" Los Angeles, today it is increasingly common for workers to telecommute from one side of the planet to the other – as when Indian engineers in New Delhi log onto the mainframe computer of San Francisco-based Bechtel corporation and work, in real time, on projects with their US colleagues (Filkins 2000). Equally, in the past things like letters, business brochures, and government reports had to be physically transported (sometimes at great cost) from their point of origin to wherever they were going. Today, however, the ability to transmit such information via email in the form of PDF files or video clips means that nothing actually has to be physically transported from place to place – a phenomenon leading to the emergence in some sectors of what economist Danny Quah (1999) has called the "weightless economy."

Such spatial and temporal transformations have led some to argue that we live in a "shrinking globe" and an era of "fast capitalism" (Agger 1989) in which we are seeing the "acceleration of just about everything" (Gleick 1999); that we now exist in an "epoch of simultaneity" (Foucault 1986) in which, thanks to telecommunications technologies, the occurrence of events around the world and our knowledge of them are virtually simultaneous, regardless of how far we are physically removed from such events. Thus, whereas on November 22, 1883 the *New-York Tribune* reported that, "A Paris dispatch in *The Post* says that owing to the absence of a telegraph cable between Tonquin and Saigon the result of military operations in Tonquin can only be known a week hence," by the middle of the twentieth century electronic media had so "con-tract[ed] the world to a village or tribe where everything happens to everyone at the same time [that] everyone knows about, and therefore participates in, everything that is happening the minute it happens" (Carpenter & McLuhan 1960: xi). This speed with which information crosses space today can have significant and rapid economic impacts around the globe. Hence, on September 11, 2001, the London stock market began responding within minutes of the first plane hijacked by Al-Qaeda operatives hitting New York's World Trade Center (Figure 1.1). Indeed, by the end of that day's

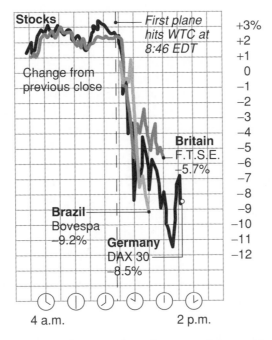

Figure 1.1   Impact of 9-11 attacks on stock markets around the world.

trading on London's FTSE 100 index, British Airways' stock had fallen more than 20 percent in value on fears that passenger numbers would decline as people chose not to fly. At the same time, the volume of trading on the FTSE shot up as investors around the world who were unable to trade in New York because its markets had closed used London as a proxy.

The perception that we are living in a faster and smaller world is, of course, the result of significant changes in technology and ways of social organization which allow distant places to be reached and/or interacted with in much shorter times than previously. It is also, however, partly a psychological aspect of what may seem to be unrelated elements of contemporary life. For instance, as life expectancies in the industrialized world have increased, society's "collective memory" has lengthened – whereas average US life expectancy in 1900 of only 47 years meant society's collective memory was about half a century into the past, today it is over three-quarters of a century. This is significant, because longer life expectancies (particularly in a time of rapid social change) mean that more people are likely to remember how things may have been radically different in the past – millions of people who were born into a world without automobiles lived to see humans walk on the Moon and satellites leave our solar system, for instance. Longer life expectancies, then, provide a background against which the speed of change appears to be ever faster – if one can remember a "larger" and "slower" world, contemporary changes often seem to be much more significant. This raises the question of how young people today, who have grown up in the age of the internet, transoceanic jet travel, and satellite television, will perceive the speed of future change – will they do so in ways similar to their parents or grandparents or does their youth make the world's smallness and fastness appear normal? Furthermore, such issues raise the question of how the speed of change is perceived in the so-called Less Developed Countries (LDCs), where life expectancies remain much lower but where they have been increasing significantly in some parts, even as they have decreased dramatically in others, like AIDS-ravished parts of Africa. Additionally, given that different cultures view time in different ways, such that the "Western concept of time [and hence of speed of change], which is abstract, external, linear, and quantitative, [often] makes little sense to members of other cultures where durations are measured not by the ticking of the clock, but by the unfolding of environmental events or the ordering of sacred rituals" (Rifkin 1987: 52), we are left with the question of how societally specific are perceptions that we are living in a smaller and faster world.[3] Even as we can empirically show that, for example, it takes less time today to get from New York to Dar es Salaam than it did 50 years ago, such realities may be understood in quite different ways across the planet.

If, for many of us, the geographical and temporal relationships between places appear to be changing rapidly, so, too, do the systems of spatial ordering which have helped us make sense of the world economy for much of the twentieth century. Thus, after the end of World War II it became increasingly common in many academic and government circles to allocate countries to one of three levels of development – First World countries were those seen to be "advanced" industrial capitalist nations, Second World countries were those which professed to be communist, whilst Third World countries were seen as those nations (virtually without exception, former European, Japanese, or US colonies) that were largely agrarian and which had failed to develop "modern" industrialized economies. Today, though, such a system of spatial ordering does not seem to make much sense. With the collapse of communism in the Soviet Union and Eastern Europe, most of the "Second World" has ceased to exist, whilst China's embrace of capitalism – even if it remains a communist state in official rhetoric – makes it increasingly difficult to place that country squarely within the Second World category. Equally, whereas some parts of the old "Third World" like Brazil have developed rapidly and emerged as "newly industrializing countries" (NICs) – in 2003 Brazil had the world's eleventh largest economy, ahead of the Netherlands, Australia, Russia, Switzerland, and Belgium – others, like Mali, the Central African Republic, and Zimbabwe, appear to be even worse off today than they were 30 or 40 years ago, a situation which has led some commentators to ascribe them "Fourth World" status. Likewise, the ability of suppliers to deliver components across the globe in 24 or fewer hours means we can really consider places like Manila or Taipei or São Paulo to be, effectively, manufacturing suburbs of Los Angeles or Pittsburgh or London – a consideration which confounds our generally taken-for-granted notions of the geographical relationships between places and what it means to talk about processes like the "suburbanization of manufacturing."

Putting all of this together, it is clear that something geographical is going on in terms of the contemporary economic restructuring of our world. How, then, do we make sense of these changes?

The World Economy as a Spatial Entity

Frequently, "the economy" is presented in rather non-geographical terms as an entity in which relationships of supply and demand play out amongst producers and consumers who are seen to exist largely in a spatial vacuum. Thus, much economic theory strips itself of geography by initially assuming that economic interactions take place over an undifferentiated "isotropic" plane or, alternatively, on the head of a pin – only after the "serious" work of theorizing the "purely economic" basics of supply and demand has been completed are the subsequent "complications" of geographic location and spatial interaction added to the economic model to make it more "realistic." However, as many geographers have been at pains to point out, such additive approaches, in which space is simply appended to economic models of the world almost as an afterthought, are profoundly unsatisfactory. Rather, given that social interactions must take place somewhere and are intimately shaped by spatial considerations (like "how far away is the market for my product?" and "will it be easy or difficult to access geographically resources located in a particular place?"), spatial relations must be at the heart of theorizing processes like globalization. But how should we think of economies as spatial entities?

In order to begin answering these questions we must, first, consider what we mean by the term "economy" and, second, must examine how economies are constituted spatially, how the production of economic landscapes is not just a reflection of social relationships between economic actors but also how spatial relations are themselves constitutive of economic relationships – that is to say, how space is not simply the stage upon which economic relations play out but is, rather, implicated and imbricated in the very construction of those economic relations.

What is "the economy"?

The question "what is 'the economy'?" may seem an odd one because "the economy" has typically been thought of as a "self-evident object of study" (Thrift 2000: 690), a patently obvious "thing" which encompasses the totality of the monetarized relations of production, distribution, and consumption of goods and services within a particular geographical area. Generally, this area is implicitly conceived to be the nation-state, such that references to economic matters at other geographical resolutions are commonly qualified by supplementary scalar markers, as in "the *international* economy" or "the *local* economy." In such a view, "the economy" is taken to be a largely autonomous and self-regulating entity encompassing those social relations that operate within a self-contained sphere ("the economic"), a sphere distinct from others like "the cultural" or "the political" (Mitchell 1998).

However, such a conception is a historically and geographically specific one. Thus, economists of the eighteenth and nineteenth centuries generally did not use the term "the economy" in its modern sense as an analytically coherent object of inquiry but, rather, used the designation to refer to the "proper husbanding of resources and the intelligent management of their circulation" (Mitchell 2002: 4). As Mitchell (1998: 84) argues, even Adam Smith, "dubiously claimed as the father of modern economics, never once refers in *The Wealth of Nations* to a structure or whole of this sort." In contrast to today's widespread understanding of "the economy" as an entity which covers the monetized relationships between people and places, economists like Smith viewed "the economy" in the quite different terms of a vast household under the control of a sovereign and, as such, its unifying theme was that of "labor," regardless of whether this labor was paid or not. Hence, for Smith "the economy [was] a collection of rational subjects, whose economising activity leads to greater productivity and thereby accumulation of the capital stock," which was itself taken to be the aggregate product of human labor (Tribe 1981: 145). The wealth of nations was measured not by the quantity of their money or goods but by their labor and its capacity to transform the world around them.

Perhaps surprisingly, given its ubiquity, the contemporary understanding and discursive representation of "the economy" as an entity which has a unity and exists as a general structure of economic relations in an autonomous domain called "the economic" has its origin only in the period from about the mid-1930s. Precipitated by the political and economic calamities of mass unemployment, the collapse of many nations' currencies, and their subsequent abandoning of the gold standard (and with it the belief that bank notes were inherently valuable because they represented gold, a belief at the core of nineteenth-century financial thinking), the 1920s and 1930s were marked by a momentous crisis in the systems of economic representation which had appeared as natural ways in which to describe economic relationships during much of the nineteenth century. As these erstwhile systems of representation "began to fall apart and [as] the social orders [they] underpinned lost their coherence ... the notion of the economy as a coherent structure came into circulation" (Mitchell 1998: 88). A key text in this regard is British economist John Maynard Keynes's 1936 work *The General Theory of Employment, Interest and Money*, a text often taken to mark the birth of what is today called "macro-economics" (Mitchell 1998, 2002). Whereas orthodox pre-Keynesian economics had conceived of the sphere of economic behavior as being the individual market for particular goods (an abstraction which had no geographical definition implicit within it), Keynes sought to theorize the "economic system as a whole."[4] Significantly, Keynes's ideas – and those of economists like Jan Tinbergen, who in 1936 published the first dynamic model claiming to represent an entire economy – had specific geographies implicated in them, for they were focused upon constituting a theoretical object of analysis ("the economy") within a particular territorial configuration: the nation-state. In the Keynesian approach, then, it was the *national* economy that was privileged theoretically and analytically (Radice 1984).

Certainly, economists before Keynes explored the essential elements of what would come to be called macro-economics, but it was Keynes who "put the pieces together by constructing a theory in which the aggregates of income, consumption, and investment were mathematically related to one another" within a system at whose heart lay the concept of "national income." Hence, rather than remaining a descriptive statistic to measure the economy's performance at any given time, for Keynes "national income" became "an object of policy that could be scientifically predicted and manipulated by . . . the state's economic engineers" (Adelstein 1991: 171). With the development of econometrics and a new language to talk about "the economy" (e.g., the "Gross National Product" and the "Gross Domestic Product," two terms which take as referents the nation-state's boundaries [Carson 1975]), the "world was [increasingly] pictured in the form of separate nation-states, with each state marking the boundary of a distinct economy" (Mitchell 1998: 90). The new concepts emerging in the mid-twentieth century, then, allowed nation-states to reimagine themselves as containers of "the economy."

Such understandings of "the economy" and "the economic" have been subjected recently to critique by, amongst others, feminist and post-colonial scholars.[5] Hence, Cameron and Gibson-Graham (2003) trace how unwaged laboring typically done by women (like house*work*) has often been conceptualized as "non-economic" because it is not *directly* remunerated for money, even though it may ultimately contribute to processes of capital accumulation. Such an approach to female labor, they show, has a history stretching back at least to the writings of nineteenth-century economists like Nassau William Senior, who included female labor market activity within his definition of "the economic" but excluded activity within the home (housework and child rearing, which was largely done by women) because this latter did not result directly in goods that could be exchanged in the market for money (Hewitson 1999). Others have suggested that only with the collapse of European empires after 1945 did it begin to make sense to talk of the "British economy" or the "French economy" in *national* terms, since previously these entities were generally considered to include colonies like India or Indo-China.[6] This was particularly so with the French empire, as overseas possessions like Réunion and Martinique constituted *départements* of France and enjoyed the same rights as any other *département* (like representation in the National Assembly).[7] All of this is not to say, of course, that economies did not exist prior to the mid-twentieth century nor that the term "the economy" does not refer to material things – clearly, they did and it does. But it *is* to say that the particular theoretical object that we call "the economy" is a relatively recent creation, one shaped by political and other considerations.

This discussion about "the economy" and how we conceive of it has at least two important implications for considering the world economy as a spatial entity and of how globalization is impacting its nature. First, the view of "the economy" and of "the economic" laid out by Keynes and others in the 1930s clearly implicates space in particular ways. As Keynesian approaches developed, economies increasingly came to be seen as territorially constituted at the national scale, with the boundaries of the nation-state serving as the spatial resolution at which statistics on production, employment, real wages, and so forth could be incorporated into the new econometric models. Significantly, contemporary processes like the stretching of production lines across national boundaries appear once again to be transforming the relationship between the nation-state and "the economy" and to be bringing with them real analytical problems of determining where one economy ends and another begins. In turn, this is giving rise to a sense of unease, as what have been taken during the past half-century to be commonsensical notions of the national economy no longer appear to work. Whereas the format of "the economy" implicit in Keynes's work represented a new discursive formation to replace that which had been shattered by the events of the 1920s and 1930s, contemporary restructuring appears to be highlighting the limits of this very mode of representation.

Second, the issue of discursive formations themselves brings into play geographical questions, for such formations – that is to say, understandings of concepts like "the economy" – are always spatially constituted and constituting. Thus, discourses are shaped by the socio-spatial contexts within which they are created and words like "the economy," "freedom," and "democracy" mean different things in different times and places. As the so-called "cultural turn" in economic geography (Crang 1997) has taught us, concepts like "the economy" cannot be invoked uncritically as objects of analysis but must be located within particular historical and spatial contexts and interrogated to understand how they serve to fix meaning and ways of conceptualizing the world at particular times in particular places. Concepts, then, "perform" in myriad social and spatial contexts to shape how we exist in the world – rather than referring to "something solid, coherent and identifiable that lies beyond themselves ('the real economy', 'cultural practices' etc.)" (Castree 2004: 206), discursive constructions like "the economy" (or "globalization") serve to discipline how we think. Thus, conceptualizing "the economy" as something which is constituted within the bounds of nation-states has operated as a "powerful organizing practic[e] that [has served to] create the material effect of the economy as an apparently self-contained structure . . . – material in the sense that the everyday force of the political order of capitalism [has been] structured out of these discursive effects" (Mitchell 1998: 93–4). Put another way, if we think of economies as things that are constituted within nation-states, then we will act as if that were indeed the case until something comes along in a particular time and place ("globalization"?) to disrupt such a way of thinking.

How can we think of the economy as a spatial entity?

So far, we have explored the idea of "the economy" as spatially constituted in two ways: first, the very category of "the economy" is historically and geographically specific; second, beginning in the mid-twentieth century, the pre-Keynesian abstraction of "the market," which had no implied spatial definition, was increasingly replaced by the concept of the "economic system as a whole," a system whose geographical extent was territorially circumscribed by the boundaries of the nation-state. However, accepting that there is, in fact, something that we have come to call "the economy," how might we think of this as having spatial expression?

As intimated above, the economy has frequently been viewed as an essentially aspatial entity, with relationships of production and consumption, supply and demand merely projected onto space to form a kind of spectral economic shadowland. In such a view, space is little more than the stage upon which the economic action takes place, with the economic landscape the reflection thereof – whilst history is seen as dynamic and readily shaped by social forces, geography is viewed as static, merely the platform upon which social, economic, cultural, and political forces and relations play out historically. Even in the case of Keynesian-inspired understandings of "the economy" geography is only implicated in a very limited fashion, in the form of spatially defined boundaries which are coincident with the geographical extent of particular nation-states and which mark the edges of their economies – for Keynes, then, the nation-state serves as a spatial container of the economic action, but that is pretty much it. However, as the past 20 years or so of theorizing within the field of economic geography have shown, space is intimately insinuated in how economies operate, with capitalism's economic organization shaping its geographical organization but, also, with its geographical organization fundamentally shaping the accumulation process. This current way of thinking about "space economies" – that is to say, about economies as spatial entities – is the culmination of theoretical developments stretching back to the nineteenth century.

Briefly stated, we can divide the intellectual history concerning the idea of the "space economy" into four time periods. First, many nineteenth-century economists like Karl Marx tended to privilege time and historical transformation over space and geographical transformation in analyses of how capitalist (and other) economies work. Hence, Marx focused principally upon how capitalists can reduce the time it takes to turn over their capital (without really considering how the spatial organization of production and consumption might facilitate this) and upon historical transitions in the mode of production (as from feudalism to capitalism). Although he occasionally had some quite prescient geographical insights – as when he talked about how transportation improvements in Victorian Britain were leading to the "annihilation of space by time," as the time taken to travel from place to place was reduced – generally Marx's focus was the accumulation of capital (which required capitalists to control workers' labor time) and the historical transition between modes of production.

Second, environmental determinism's influence in the late nineteenth century had a dramatic impact upon how geography was seen to shape economies and patterns of economic development. At its most extreme, environmental determinism argued that a society's geographical location determined its levels and types of economic development (Peet 1985). Hence, it was argued that Africa was "underdeveloped" because the continent's populations were unlucky enough to live in regions where nature was particularly harsh, an explanation which conveniently forgot that Africa's environment had, after all, produced the entire human race and which also provided justification for European conquest – a neat intellectual trick! Thus, writers like British geographer and imperial apologist L. Dudley Stamp (1937) largely explained patterns of economic development as resulting from favorable or unfavorable climatic conditions. In such a formulation, then, geography played the dominant role in explanations of patterns of economic activity.

Third, beginning in the late 1960s, academic geography began to be transformed. Thus the Vietnam War, student protests in France and elsewhere in 1968, the Soviet invasion of Czechoslovakia, the feminist movement, and decolonization radicalized many geographers and led to a concomitant rethinking of how they thought about the world. Within economic geography this was marked by an engagement with political economy. In particular, geographers like Harvey (1973) began to argue that understanding how economic landscapes were made required understanding the inner workings of capitalism as an economic system. However, concerns that placing too much explanatory weight upon geographical location and space would resurrect charges of environmental determinism led to a focus upon space as principally a mirror of the social relations of production. Certainly, such an approach – one which spotlighted the historical and material processes of uneven development spawned by capitalism's "internal laws" – was more dynamic than previous approaches, which had tended to concern themselves with little more than understanding patterns of spatial distribution across the economic landscape (that is to say, they had focused upon the patterns themselves rather than the deeper structural forces of capitalism which caused them). Yet, the conceptual developments of the 1970s still did not really incorporate a view of space as dynamic, something that played an active role in structuring economic relationships.

Fourth, in response both to the theoretical advances, but also limitations, in the approaches developed in the 1970s, economic geographers began to explore what Soja (1980) has called the "socio-spatial" dialectic. Such an approach argued that space and social relations are mutually constitutive – that is to say, how a social system like capitalism is organized economically shapes how its geography is produced but, also, how capitalism is ordered spatially shapes how social relations subsequently evolve (Harvey 1982; Massey 1984). For example, if processes of suburbanization result in urban landscapes designed around the assumption people will primarily access the city by car, then options for moving about the city in other ways (like walking) are often severely limited. This means that to get around an urban environment it increasingly becomes necessary to have access to some form of motorized transport, often a private car. In turn, as more people become dependent upon cars because patterns of development make it very difficult to walk anywhere, there is less incentive or ability to build walkable cities, a phenomenon that further encourages automobile-oriented development. A place's geographical "path dependence," then, shapes how its social relations evolve, and vice versa (Figure 1.2).

One way in which this idea of a socio-spatial dialectic has shaped our understanding of economies as spatial entities is through the concept of what Massey (1984) has called the "spatial division of labor." Whereas the concept of a *social* division of labor

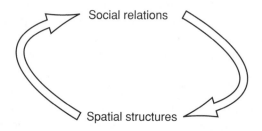

Figure 1.2   **The dialectic between social relations and spatial structures.**

is very familiar, Massey argued that any social division of labor is also implicitly a spatial division, since different types of employment are typically found in different geographic locations. Through an analysis of economic restructuring in Britain she showed how and why various industries located different parts of the production process in different regions – areas like the Southeast became centers of research and development whilst, by the 1970s, the declining industrial heartlands of northern England, Scotland, and South Wales were being reimagined as centers of branch plant light manufacturing. Massey showed how this spatial division of labor developed in response both to firms' internal production and management needs but also in response to the geographic environment within which such decisions were being made, with the existing spatial division of labor having significant impacts upon subsequent patterns of investment across the economic landscape – the boom regions which served as the basis for the nineteenth-century industrial revolution were in decline by the early post-war era and so became attractive locations for branch plants due to their high unemployment (and thus low wages). Through this work she argued that the geography of capitalism could productively be thought of in terms of a geologic-type continuous depositing across the landscape of layers of investment, sometimes deep, sometimes shallow, with such investment interacting with pre-existing spatial arrangements of production and consumption to produce new geographies of uneven development.

Such work to conceptualize the active role space plays in shaping how economies operate was complemented by geographers like Harvey (1982), who argued that economic systems not only play out across space but are, in fact, reliant for their functioning upon how the landscape itself is structured. In particular, Harvey argued that capitalists must create particular geographical configurations – what he called "spatial fixes" – of their investments if they are to accumulate and realize profits. This might necessitate ensuring that transportation networks link factories with sources of raw materials or that finished commodities can be readily sent to markets. For Harvey, then, capitalists' ability to access particular spaces is essential for capitalism to operate, whilst their failure to do so can severely hamper their accumulation activities. For example, prior to the development of high-speed transportation poor spatial connectivity meant that supermarkets in New York City could only source fresh produce that was grown locally (perhaps in New Jersey or eastern Pennsylvania), regardless of cost. Today, in contrast, fresh produce can be sourced quite literally from all over the world, a phenomenon which allows large food corporations and supermarkets to take advantage of producers' low wages in places like central Mexico in ways which previously they could not. Through transforming the spatial relationships between places and creating new "spatial fixes," then, entrepreneurs have been able to "open up" new markets and new source regions and so give themselves greater geographical flexibility with regard to their profit-making activities. Evidently, how capitalism's geography is made is of no mere peripheral concern but is, in fact, central to capitalists' ability to function and reproduce themselves over time.

However, as I have argued elsewhere (Herod 2001), how capitalism's geography is made is of great importance not just to capitalists but also to workers. Thus, ensuring there is work available within the geographical vicinity of where they live is crucial if workers are to sustain themselves on a daily or generational basis. Yet workers are sometimes spatially limited in their ability to access employment opportunities, particularly where there is a "spatial mismatch" (Kain 1968) between where unemployed workers are located – say, central-city areas – and where new job opportunities are opening up – perhaps the suburbs. Only through understanding how workers are spatially trapped in the central city – unable to secure jobs in the suburbs because they lack transportation to get there but unable also to live in the suburbs due to high housing costs – can we understand why some jobs remain unfilled in metropolitan areas with high unemployment. Whereas neoclassical economists may explain such a situation as resulting from workers having the wrong skills or of not being willing to work for the low wages that are perhaps being offered, in fact it is the spatial structure of the urban environment and the limits on workers' ability to cross space which really explain how some are shut out of the local labor market. Consequently, to solve problems of jobs going unfilled whilst many suffer unemployment, workers and their advocates may seek a geographical solution, namely restructuring the spatial relationships between different parts of the urban environment through, say, building a mass transit system linking the urban core with the suburbs.

Similarly, workers may try to mold the spatial relations between places through collective bargaining activities. For example, some workers may decide it is beneficial to negotiate a single wage rate for their industry across an entire region or country, thereby preventing employers from whipsawing plants against one another on the basis of wage rates. Others, though, may prefer a highly spatially variegated wage surface in the belief that their willingness to work for wages lower than the industry average gives their communities a competitive advantage. Through their abilities to structure the geography of wage rates in different ways, then, workers can powerfully shape how the economic landscape unfolds. Whereas uniform wage rates mean capitalists will be unable to play one community against another on the basis of wage differentials, a variegated wage surface allows them more readily to bully workers by threatening to relocate to lower-waged regions, although it may also allow some workers to secure work at below-market rates. Different groups of workers, then, clearly have a vested interest in making economic landscapes in different ways, and their ability to do so is a potent form of social power.

Thinking of economic landscapes as dynamic social products which are constituted by political and economic praxis but also constitutive of such praxis provides us, then, with a more active notion of space than one which conceives of space as simply the stage upon which, or the container within which, social life plays out. Thus, it allows us to think of space as plastic and malleable, as something that can be bent and reshaped through, say, constructing new highways which shorten travel times between places. Like history, space is clearly subject to political, economic, and other forces which shape it in particular ways, ways which may enable or constrain different actors' social practices. However, we must also remember that, much as significant events from the past can continue to have powerful influences on the present and future, even after the principals involved are long dead, so does a place's path dependence mean the economic landscape is always a palimpsest – that is to say, even as contemporary economic and social processes reshape the landscape, elements of prior landscapes are never completely erased and some continue to influence greatly how present and future landscapes are made.[8] Much as a ship's momentum means it may take considerable time to alter direction once the captain has ordered a course change, places' path dependence mean it takes time for the economic landscapes of the past to lose their grip on the making of those which come after them. The landscapes of the past, then, shape the creation of contemporary economic landscapes even whilst such creation erodes these very past landscapes.

Of Globalization, Shrinking Worlds, and the Paradoxes of Geography

So far, we have examined how economies can be thought of as spatial systems. In this final section I want to bring the discussion back to the issue of globalization, for whatever else they may be the contemporary economic and political transformations often taken to be "globalization" are reworking the spatial relationships between different parts of the globe. Arguably, this reworking has been represented most strikingly by the image of the "shrinking globe," in which revolutions in transportation and telecommunications technologies are seen to have brought places ever closer together. Of course, such an image is not meant to suggest that the globe has shrunk geologically. Rather, it denotes that the *relative* distances between places have been reduced as the time taken to get from location to location has dramatically diminished in recent years. Put another way, whilst the *absolute* distances between places (which are measured in some Euclidean metric like miles or kilometers) are always invariant – barring major tectonic activity in the North Atlantic, London and New York City will always be 3 400 or so miles apart – the *relative* distances between places (typically measured in terms of either time or cost) are subject to political and economic forces which can dramatically transform the geographical relationships between different parts of the globe. Often, this phenomenon has been represented visually in the form shown in Figure 1.3, in which different sizes of the world are associated with different forms of travel, such that the world appeared much bigger when information, people, commodities, and capital moved about at the speed of a swift horse than at the speed of a swift stroke on a computer keyboard or a supersonic jet. Thus, whereas it took 11 days to cross the Atlantic by clipper ship in the middle of the nineteenth century, about five days by steamship at the beginning of the twentieth century, and John Alcock and Arthur Whitten Brown took approximately 16 hours to make the first crossing by air in 1919, by the 1980s travelers could have breakfast in London, fly at twice the speed of sound on Concorde to New York, have lunch there, and be back in London in time for an evening meal. The metaphor of the shrinking globe, then, is arguably

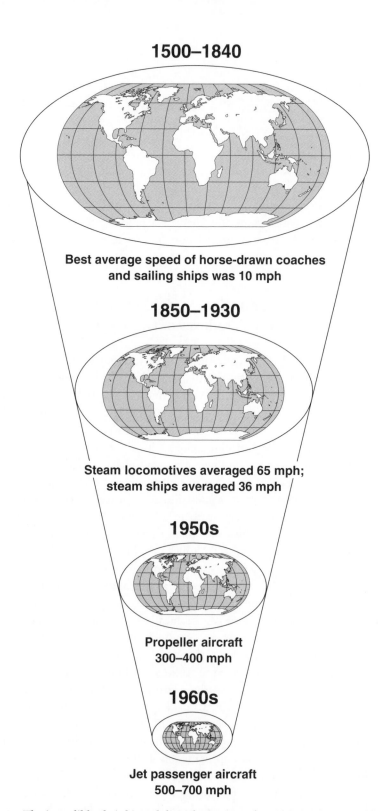

1500–1840

Best average speed of horse-drawn coaches and sailing ships was 10 mph

1850–1930

Steam locomotives averaged 65 mph; steam ships averaged 36 mph

1950s

Propeller aircraft 300–400 mph

1960s

Jet passenger aircraft 500–700 mph

Figure 1.3 The incredible shrinking globe? The impact of revolutions in transportation upon relative distance.

Source: Dicken 2003, Figure 4.3.

the ultimate example of what Marx (1973: 539) referred to as the "annihilation of space by time," in which "space" – in the sense of the physical distances between places – is rendered less significant an impediment to economic interaction as the time taken to cross it is reduced by technological developments and revolutions in the spatial organization of society.

Despite the prevalence of adopting wholesale the picture of the shrinking globe as presented in Figure 1.3 as a way of thinking about globalization – such images, for example, have been widely used in commercial advertisements (Figure 1.4) – there are, however, a number of problems with this imagery. Primarily, of course, although the representation makes a powerful visual impact it is nevertheless a representation which, paradoxically, is both a-geographical and ahistorical – that is to say, it suggests that the annihilation of space by time and the shrinking of the globe is a universal and uniform process impacting all parts of the planet simultaneously, without regard for economic, political, or cultural differences between places or the trajectories of their historical development. This, though, is patently not the case. Rather, the process of annihilating space by time is one that is uneven both historically and geographically – different parts of the globe are impacted at different historical moments in different ways. Thus, whilst much is made of how the internet is making space irrelevant and bringing places together, it is doing so unevenly – in 2004 internet bandwidth between North America and Africa was a mere 1 673 Mbps, less than 0.33 percent of the 504 512 Mbps available between North America and Europe (PriMetrica 2004) (Figure 1.5), a disparity which means that North America and Europe are significantly more connected than are North America and Africa and are apparently becoming more so (the equivalent figure for 2003 was 0.35 percent).[9] Likewise, although the top speed of large commercial jet aircraft is today around 650 miles per hour, not all communities have airports that can service such aircraft and so must be linked to the outside world by slower modes of transportation (small propeller planes, cars, trains, horses, walking), which makes them relatively more isolated than are the major metropolitan regions of the world. The process of increasing inter-connectivity has equally been historically uneven – whereas in the 1970s computer systems communicated via telephone at about 300 bits per second, today they do so at over 100 million.

Additionally, it would be erroneous to assume that contemporary processes are only *shrinking* the globe. Given that relative distances between places are subject to economic and political forces, it is quite conceivable that, on occasion, relative distances between particular places might actually increase, even if the overall trend is towards a decrease. Hence, the US government's limitations on direct flights to Cuba mean that many Americans who wish to go to the island must fly first to places like Cancún, Nassau, or Toronto, and then on to Havana, a process which effectively makes Cuba much farther away than those Caribbean islands that can be accessed directly from the US mainland. Moreover, even when places do become closer together in absolute terms (though not in terms of absolute distances, which never change) – that is to say, if it takes less time to travel between North America, Europe, and Africa today than it did 50 years ago – different parts of the globe may become relatively further

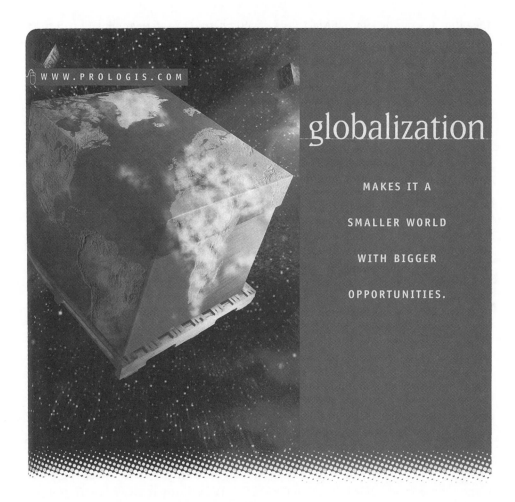

ProLogis **is a leading global provider** of integrated distribution facilities and services, with 204.6 million square feet of distribution facilities owned, operating or under development throughout North America, Europe and Japan. ProLogis has built and continues to expand the industry's first and only global network of distribution facilities with the primary objective to increase shareholder value. The company expects to achieve this objective through the ProLogis Operating System® and its commitment to be 'The Global Distribution Solution' by providing exceptional corporate distribution services and facilities to meet customer expansion and reconfiguration needs globally. To learn more about ProLogis and how we can help your business on a worldwide basis visit **www.prologis.com** or call 800-566-2706.

The Global Distribution Solution

©2002 ProLogis. All rights reserved.

Figure 1.4   Example of the imagery of a shrinking globe used in advertising.

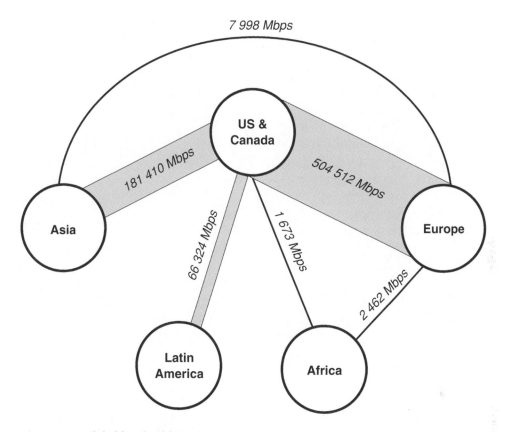

Figure 1.5   Global bandwidth, 2004.

apart if the speed at which distances between them shrink varies. Hence, as much of the "developing world" slips further behind the industrialized world as the technology gap between the two widens, Africa may be "closer" to North America today than it was half a century ago but this distance is not being reduced as quickly as is the distance between North America and, say, Europe. For example, whereas the cost of making a telephone call from the US to much of Africa decreased in the 1990s, the cost of calls to most of Europe and Latin America decreased much more rapidly, bringing them relatively much closer together (compare Figures 1.6 and 1.7). In effect, then, Africa fell further behind and became relatively more isolated globally, even as the world as a whole became more connected.

Finally, it is important to realize also that our understanding of this process of the shrinking of relative distances between places may vary significantly, depending upon how we choose to measure relative distance. Hence, the relative distances between places are likely to be quite different if we measure them in terms of different modes of transportation – how far I can drive in an hour by car will, generally, be a shorter distance than how far I can fly in an hour. Equally, the relative distances between places may be quite different if we measure them in terms of the *time* it takes to travel from one

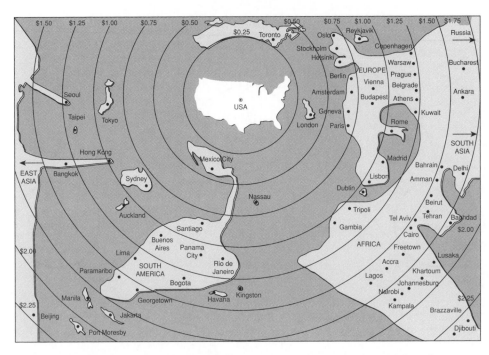

Figure 1.6 Cost per minute of calling from the US to the rest of the world, based on MCI peak rates, June 1994.

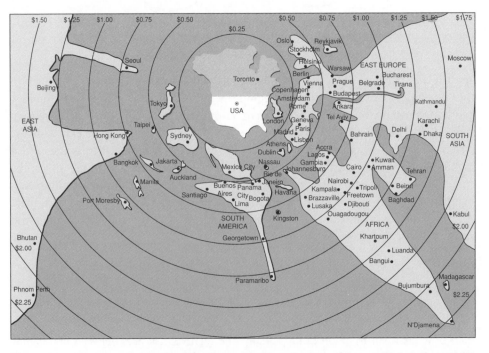

Figure 1.7 Cost per minute of calling from the US to the rest of the world, based on MCI peak rates, July 1998.

**Telephone distance from Syracuse
(based on a 2-minute call)**

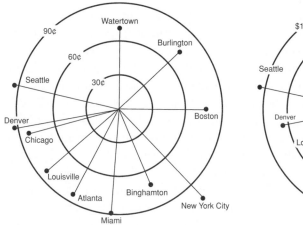

Airline distance from Syracuse

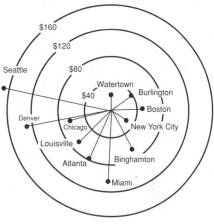

Figure 1.8 Relative distances from Syracuse, NY, as measured by the cost of telephoning and flying.

to another compared with the *cost* of interacting between them – whereas I can mail a letter to anywhere within the US for a standard rate (meaning that all places are equidistant in terms of cost), it will take different amounts of time for the letter to arrive in various locations, since it must be physically transported. Moreover, even when using the same metric – cost – this may vary considerably if different modes of interaction are being measured. Hence, as Monmonier (1977) has shown when comparing the cost of flying from Syracuse, New York, to several other cities across the US and the cost of making a standard-length telephone call between the same cities, some places are closer together when using one measure than they are using the other (Figure 1.8): when measured by the cost of a telephone call, New York City is "farther away" from Syracuse than is Miami, Florida, though in terms of the cost of flying the reverse is true.[10]

In thinking about globalization as a process which is transforming the relative distances between different parts of the globe, then, two important issues emerge which will shape how we understand what is occurring in our contemporary world. First, whilst much of the rhetoric associated with "globalization" (and we shall explore this term more fully in Chapter 3) has portrayed the process and the phenomenon in terms of images like the inexorably "shrinking world," in fact within a generalized process of shrinkage some places may become further apart, either absolutely or relatively, depending upon how the economic and political processes that are reshaping our world's geographical interconnectedness play out in particular places at particular times. Not only is the shrinking of our globe *not* a unidirectional phenomenon which is bringing all places closer together – some places are actually becoming more distant from each other – but it is quite clear that there is a complex and particular historical geography to how the world is being rewired spatially: the process proceeds at different paces in different historical time periods and between different places, a reality which challenges narratives that see globalization as a unidirectional and universal process (again, we shall return to this issue in Chapter 3).

Second, as the "annihilation of space by time" proceeds, and as the time it takes to get from place to place appears to get ever shorter (notwithstanding what I have just said about how this is a complex and not necessarily unidirectional process), one might be tempted to argue that geography and geographic location are increasingly irrelevant since, in a world of virtually instantaneous communication and ease of travel, corporations can now locate their operations pretty much anywhere in the globe and service pretty much anywhere else just as easily as if they were situated right down the street. Certainly, a not-insignificant number of commentators have suggested as much, publishing books with evocative titles such as Frances Cairncross's (2001) *The Death of Distance* and Thomas Friedman's (2005) *The World is Flat* or claiming that "in the age of the global economy, physical location is much less important" than previously and that it "no longer matters where a company is based" (Ohmae 2005: 13, 94). However, such views, I think, fundamentally misread how the reorganization of the spatial relationships between different parts of the planet is impacting our world. *Contra* the pundits who suggest globalization is rendering geography increasingly irrelevant, I would argue that, in a shrinking globe, geographic location actually becomes *more* – not less – important because firms with the capacity to locate their operations almost anywhere will undoubtedly be more discriminating in choosing between different places when making their investment decisions. Thus, a hundred years ago it mattered less how much someone in a foreign country was paid or what the regulatory conditions were under which they labored there because they were located so far away that capitalists usually could not take ready advantage of such conditions – many investment opportunities in overseas markets would have come and gone before foreign capital could be deployed to take advantage of them.

Today, by way of contrast, the ability to send billions of dollars around the globe in the twinkling of an eye or to transport rapidly highly perishable commodities (like cut flowers grown in East Africa for consumption in Europe) mean that circulating capital now has many more locational choices around the globe and can more readily exploit relatively minor differences in wage rates or regulatory regimes between places than when these spaces were beyond its reach – or, as Harvey (1989: 294) has put it: "[a]s spatial barriers diminish so we become much more sensitized to what the world's spaces contain." Paradoxically, then, whereas in a "larger" and "slower" world spatial restrictions could limit how economies developed, so that many goods and services had to be produced locally or regionally, regardless of what this meant for capitalists, now the ability to overcome distance – to annihilate space by time – has made spatial variations between places in conditions of capital accumulation more significant than ever before in the locational calculus of investment decision making. The more the world is opened up spatially to capital investment, apparently, the more significant do geographical considerations become for it.

Questions for Reflection

- How does the way in which we conceptualize economies shape how we understand the process of "globalization"?
- How does the way in which we conceptualize space shape how we understand the process of "globalization"?
- How might our geographical location shape how we think about "globalization"?

Notes

1. Complicating matters, although General Mills manufactures Yoplait in the US using milk from local dairy producers, it does so under license from French company Sodiaal. Moreover, whilst many boycotters chose to buy, instead, Dannon yogurt in the belief this was a US brand, it is in fact owned by Groupe Danone.

2. The NUMMI (New United Motor Manufacturing, Inc.) plant, owned in equal parts by Toyota and General Motors, produces two-thirds of all Corollas sold in the USA (the remainder are Canadian-produced). It uses the same production platform to build the Chevrolet Prizm, and also manufactures the Toyota Tacoma, the Pontiac Vibe, and the Toyota Voltz (which is exported to Japan!).

3. On the West Japan Railway, for instance, a train is considered late if it is more than one minute behind schedule, whereas for London's Thameslink it is five minutes and for the New York Metro-North Railroad it is six (*New York Times*, April 27, 2005).

4. For its part, the idea of "the market" had been solidified in the early modern period with the development of concepts and practices related to accounting. As Poovey (1996: 2) argues with regard to the invention of double-entry bookkeeping: "[i]n the domain of commerce . . . the development of a system for representing commercial transactions permitted early modern English merchants to conceptualize experiences that were heterogeneous by nature as comparable in kind, and then to generalize from these transactions a 'market' that appeared to be a separate and law-abiding domain; this conceptual abstraction was eventually institutionalized, in banks and instruments of credit, so that it actually became a domain separate from politics and theology." Hence, diverse activities – selling cows, transporting bread, buying bricks – increasingly came to be seen merely as variations on a theme: engagement in something conceptualized as "the market."

5. Post-colonial scholarship argues that concepts and ideas about the world (like what is considered normal, natural, or best) have been shaped by colonialism and assume implicitly that "the West" represents the standard against which progress and normalcy should be measured. Consequently, such scholars argue, these ideas of what constitutes progress and normalcy need to be deconstructed as a strategy of liberation of formerly colonized peoples.

6. This conflation has raised methodological issues concerning how to measure levels of foreign direct investment (FDI). Whereas some have suggested that investments in colonies be excluded from FDI measures, since these were not "foreign" in the same way as were investments in non-colonies, Jones (1996: 39) argues that the administrative and

legal contexts within which imperial metropole firms had to operate in the colonies were sufficiently different from investing domestically that their investments should be considered FDI. Furthermore, whereas British colonies like Canada and Australia eventually evolved into independent states, it is difficult to determine at what point they became sufficiently "foreign" that British investments in them should be considered FDI, with the result that early investment might not be considered FDI but later investment would be so considered.

7. The overseas *départements* of Réunion, Guadeloupe, Martinique, and Guyane remain part of France (and of the European Union). Three "Overseas territories" ("*territoires d'outre-mer*") – French Polynesia, Wallis and Futuna, and the French Southern and Antarctic Territories – and three special-status territories ("*collectivités territorialles*") – Saint Pierre and Miquelon, Mayotte, and New Caledonia – are part of France but not of the EU. This colonial legacy can be observed in everyday life in France – advertisements for telephone calling rates on the Paris *Métro*, for example, typically indicate that these refer only to "Metropolitan France" (to make the distinction between "mainland" France and non-European *départements*).

8. In the Middle Ages it was common to reuse parchment by scratching out what was written on it and writing a new text over the old one. Typically, the earlier writing was incompletely erased and so was still partially visible. Such manuscripts are "palimpsests." One can think of landscapes in similar ways.

9. If we were to examine this phenomenon at different geographical scales, we would find that the levels of connectivity between North America and different countries within Europe and Africa, or between different cities in North America and in Europe and Africa, would also vary considerably.

10. This is the result of several economic and political considerations. When flying, a person must be physically transported, whereas when making a telephone call nothing actually moves between two points. The nature of competition and government regulation also plays a role. Thus, the manner in which the telephone industry was regulated at the time of the study meant that it generally cost more per minute to call locally within a region like New England than it did to call long-distance between regions, whilst some of the smaller towns within New England were only served by a single airline flying from Syracuse, such that the lack of competition meant airlines could charge monopoly prices on those routes. As this example illustrates, the cost of transporting weightless goods (cf. Quah 1999) will usually be less reflective of the absolute distance between places than will be the cost of transporting things which have mass.

Further Reading

Barnes, T. (1996) Place, space, and theories of economic value: Context and essentialism in economic geography. In: T. Barnes, *Logics of Dislocation: Models, Metaphors, and Meanings of Economic Space.* New York: Guilford, pp. 53–79.

Cameron, J. and Gibson-Graham, J. K. (2003) Feminising the economy: Metaphors, strategies, politics. *Gender, Place and Culture* 10.2: 145–57.

Curry, M. R. (1996) On space and spatial practice in contemporary geography. In: C. Earle, K. Mathewson, and M. S. Kenzer (eds.), *Concepts in Human Geography.* Lanham, MD: Rowman & Littlefield, pp. 3–32.

Harvey, D. (1985) The geopolitics of capitalism. In: D. Gregory, and J. Urry (eds.), *Social Relations and Spatial Structures*. New York: St. Martin's Press, pp. 128–63.

Herod, A., Rainnie, A., and McGrath-Champ, S. (2007) Working space: Why incorporating the geographical is central to theorizing work and employment practices. *Work, Employment and Society* 21.2: 247–64.

Soja, E. (1985) The spatiality of social life: Towards a transformative retheorization. In: D. Gregory and J. Urry (eds.), *Social Relations and Spatial Structures*. New York: St. Martin's Press, pp. 90–127.

Electronic resources

Globalization101.org – A project of the Carnegie Endowment (www.globalization101.org)

KOF Index of Globalization (http://globalization.kof.ethz.ch)

The Globalization Website (www.sociology.emory.edu/globalization)

Chapter 2

Envisioning Global Visions

Chapter summary: This chapter surveys how images of globality have developed historically, particularly in the West. It shows that efforts to develop global frameworks of time and space (like time zones) have a long history, though they picked up speed in the nineteenth century as economic and political processes increasingly linked different parts of the globe together. The nineteenth- and twentieth-century fascination with globality is seen in various cultural forms, like art and literature.

- Universalism, Cosmopolitanism, and Globalism
- Cartographic Convention and Power
- Globalizing Time
- Images of Global Integration in Popular Culture

"I maintain," said Stuart, "that the chances are in favour of the thief, who must be a shrewd fellow." "Well, but where can he fly to?" asked Ralph. "No country is safe for him . . ." "Oh, I don't know that. The world is big enough." "It was once," said Phileas Fogg . . . "What do you mean by 'once'? Has the world grown smaller?" "Certainly," returned Ralph. "I agree with Mr. Fogg. The world has grown smaller, since a man can now go round it ten times more quickly than a hundred years ago. And that is why the search for this thief will be more likely to succeed." "And also why the thief can get away more easily." . . . But the incredulous Stuart was not convinced . . . "You have a strange way, Ralph, of proving that the world has grown smaller. So, because you can round it in three months —" "In eighty days," interrupted Phileas Fogg. "That is true, gentlemen," added John Sullivan. "Only eighty days, now that the section between Rothal and Allahabad, on the Great Indian

Peninsula Railway, has been opened." "I'd like to see you do it in eighty days ...
I would wager four thousand pounds that such a journey, made under these
conditions, is impossible." "Quite possible, on the contrary," returned Mr. Fogg.
(Jules Verne, 1872, *Around the World in Eighty Days*)

A hallmark of much contemporary discourse about globalization is the representation of the globe as a seamlessly integrated whole. Whether in the titles of books like Ohmae's (1990) *The Borderless World* or in advertisements lauding the transformation of our Earth into Planet Hollywood, Disney World, and Planet Reebok, today images of globality surround us. Although conceptions of the world as an integrated whole have been with us for hundreds, if not thousands, of years (particularly in theological discourses), such a *Weltanschauung* (literally, a "world view") perhaps reached its culmination with – and received fresh impetus from – the famous 1968 photograph taken by the Apollo 8 astronauts showing Earth rising above the Moon's horizon (Figure 2.1). Arguably, only the ability to view our planet in its entirety from space truly seared into the popular imagination representations of the world as a distinct, unified global entity whose constituent parts are fitted together into a single whole.[1]

Given that images of the globe as an integrated whole are central to many rhetorics associated with globalization, in this chapter I explore several noteworthy historical

Figure 2.1 View of rising Earth about five degrees above the lunar horizon.

events and intellectual developments that have played important roles in shaping contemporary views of "the global." However, in suggesting that events like photographing the Earth hovering iconically in space have had significant implications for how we think about globality, I am *not* implying that contemporary discourses of globalization necessarily represent the teleological end-point of some intellectual process that began millennia ago. Rather, there have been conflicts and divergences over how we think about the planet and its globality, and about what such globality means for comprehending the connections between the Earth's various parts. Thus, whereas the dominant rhetorics and iconographies of globalization are typically seen to be a creation of the Western imagination, such that "globalization" is viewed by many as little more than the "Westernization" – or, to be more geographically specific, the "Americanization" – of the planet, other cultures have also viewed the world in global or unitary terms – the Maya worldview, for instance, connected the Earth to the heavens and the underworld through the idea of the *yaxché*, the "world tree."[2] Nevertheless, given that the idea of "the global" upon which much contemporary discourse rests has been tied to historical developments within, particularly, North America and Western Europe – developments like the nineteenth-century creation of time zones establishing Greenwich, England, as the point of reference for planetary space and time – here I will focus primarily upon how "the global" has been imagined within the discursive terrains of "the West." Whilst I recognize this is necessarily an incomplete view, I would also argue that "the West's" economic, political, and cultural power means that its view is having most bearing upon how contemporary images of global integration – and thus understandings of globalization – are playing out across the planet.

The chapter proceeds as follows. First, I explore ideas of universalism and cosmopolitanism stretching back at least to the ancient Greeks. I then outline how the planet's physical geography has been imagined in the Western mind and how Western cartographic conventions have reinforced particular representations of the globe. Third, I chart how new practices of disciplining time and space emerged in the nineteenth century and their effect. Finally, I survey images of globality in popular culture. The chapter's goal, then, is to trace trends in how globality has been envisioned as a basis for contemporary discourses of globalization.

Universalism, Cosmopolitanism, and Globalism

Frequently, globalization is presented as an unstoppable economic and political juggernaut. Indeed, it has become fashionable in some circles to talk about the emergence of a "TINA" ("There Is No Alternative") discourse concerning globalization – in their book *Market Unbound: Unleashing Global Capitalism*, for instance, Bryan and Farrell (1996: 160, emphasis added) have suggested that, "If governments recognize that the events taking place are largely *inevitable*, they can accept the changes." Much of this rhetoric of globalization as an inevitable and unstoppable force – a rhetoric central to neoliberal political and economic projects – relies for its power upon images of the planet as a coherent, integrated, and seamless whole, images which have a long history stretching back at least to the early Greeks. In fact, 2 500 years ago Thales of Miletus is often credited as having been the first to envisage the Earth as spherical, a position supported by Pythagoras and Aristotle. Certainly, by the time of Eratosthenes in the third century BC, reasonably accurate estimates of the Earth's size were being made. Indeed, Eratosthenes' figure was more accurate than that of later scholars like Posidonius, who significantly underestimated the Earth's circumference, a miscalculation which would have important consequences for Renaissance explorers and cartographers – Posidonius's figure was repeated by Claudius Ptolemy, who miscalculated the westward distance from Europe to Asia in his second-century AD treatise *Guide to Geography*, a treatise that influenced Columbus's navigational calculations and led him to believe he had landed in Asia ("the Indies") when, in fact, he was in the Caribbean.[3] Although today we clearly have more accurate measures of the Earth's size and its landmasses than did the Greeks, such early descriptions of the world continue to shape contemporary depictions of globality in two important ways.

First, the idea of a global unity had great significance for Greek and Roman Stoic philosophers and, later, for Christian theology (Cosgrove 2003), particularly concerning notions that local attachments to family, community, and ethnic background are less important than one's membership in a universal humanity – such a scorn for localism is evident in Socrates' famous statement that "I am not an Athenian or a Greek but a citizen of the world" and in Paul's Letter to the Galatians (3: 26–9) declaring that "We are all God's children."[4] In this regard it is significant that medieval Christian iconography frequently used circles (as with halos) and spheres – which have neither beginning nor end – to emulate divine perfection (a favored rendering was Christ either holding or standing upon a globe), whilst in the nineteenth century images of the globe were favored emblems of various missionary societies (Cosgrove 1994). Ideals of a singular humanity and an associated global cosmopolitanism were also invoked in Enlightenment concepts of universal reason and human rights, and can be seen in projects like the 1948 Universal Declaration of Human Rights and the invention of Esperanto. Thus eighteenth-century German philosopher Immanuel Kant drew parallels between the Earth's physical shape and notions of universal rights, arguing that the world's "spherical form forces [humans] to support one another, since they [are] not able to disperse themselves to infinity" (quoted in Cosgrove 2001: 176) – given that its curved surface means there are no literal "ends of the Earth" to which people can go to avoid one another, all humans' lives are intertwined to some degree. As various nations expanded their colonial enterprises, the globe was again commonly used as an emblem of unity and universality. Today, globes remain a popular icon with organizations aspiring to universalist claims, and cosmopolitan universalism is often invoked in contemporary discourses of globalization, whether in neoliberal arguments that global capitalism's spread will lead us to view ourselves in terms of new sets of post-national identities (wherein "old fashioned" and "parochial" identities based upon religious affiliation, citizenship, or ethnicity will be replaced by images of ourselves as "globalized consumers" of Planet Reebok and McWorld) or in the ecological arguments associated with the Gaia Hypothesis, amongst others.[5]

Of course, there have been challenges to such images of universalism and seamless integration, particularly through philosophical engagement with post-modernism (with its anti-universalism) and post-colonial scholarship, which argues that uncritical acceptance of the idea that "the West" serves as the privileged measure of world values is not tenable. Equally, as Cosgrove (1994) argues, whilst concepts like "one-worldism" and "whole-Earthism" have both been invoked to describe a sense of unified globalism, these terms can be read in starkly contrasting ways – "one-worldism" carries within it a greater sense of singularity than does "whole-Earthism," which tends more towards the notion that the Earth is made up of many diverse, if connected, parts. Nonetheless, images of global unity and seamlessness are central to rhetorics of globalization, particularly those of proponents of the TINA view that the spread of "free market capitalism" and "Western civilization" are both inevitable and beneficial to all humans.

The second way in which images of the globe going back to the ancient world are important for considering contemporary discourses of globalization relates to the development of modern cartography. In particular, although clearly incorrect in its suggestion that the Earth stood at the center of the universe, the Ptolemaic view continues to have significance in at least one way for contemporary representations of global space (and so for representations of globalization): as far as we know, Ptolemy's *Guide to Geography* was the first to project the spherical world onto a plane, thereby depicting the globe as a whole as an object which could be imagined from the God's-eye point of view of some detached observer hovering above the planet (Figure 2.2). Although such a perspective was downplayed in much medieval and early modern mapping and art (in which the point of view rendered was as if the map

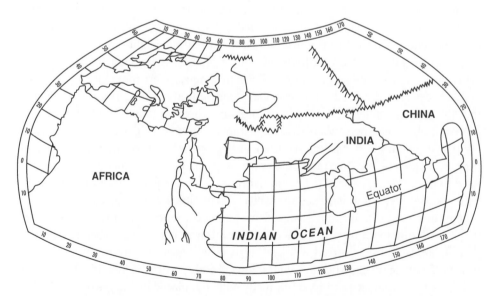

Figure 2.2 A simplified version of Ptolemy's map of the world shows that Europeans' knowledge of India and China was fairly well developed 2000 years ago.

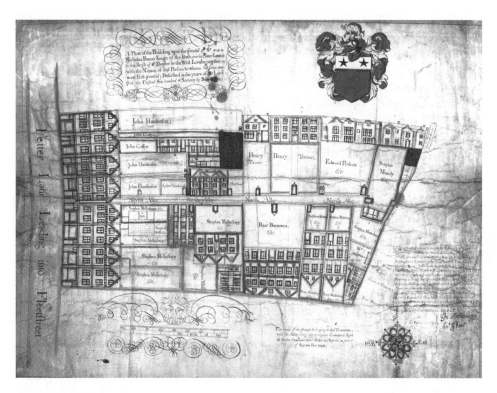

Figure 2.3 A 1670 map of part of London presents houses from the point of view of an observer standing in front of them.

reader were located within the landscape (Figure 2.3) and/or in which the relative size of buildings and people was determined not by their location relative to each other and the observer but by their importance), the rediscovery of this God's-eye perspective in the early Renaissance period (Edgerton 1975) and its promulgation in art and cartography was an important aspect in the development of mapping during the sixteenth and seventeenth centuries, for it led to the establishment of cartographic practices which still play central roles in our contemporary views of spatial representations.

Cartographic Convention and Power

Issues of cartography and spatial representation are crucial to discourses of globalization, for they inculcate within us mental images of where places are located in relation to each other and, consequently, of what appears to be a "normal" (and hence, also, an "abnormal") sense of global space. This is especially important, given that the contemporary transformations in our everyday lives which we frequently take to be instances of globalization (like the shrinking of relative distances between places) are reworking the spatial organization of our world and connecting its different parts together in new ways. Whereas maps are often assumed to be objective scientific representations of the Earth which, depending upon a cartographer's skill and the state of the technology s/he uses, are more or less accurate, in fact they are socio-cultural products which both reflect and reproduce particular social relationships (Harley 1988, 1989). Put another way, maps are more than simple mimetic accountings of the Earth's surface but are a form of discourse and an expression of power, a way of "articulat[ing] symbolic values as part of a visual language by which specific interests, doctrines, and even world views [may be] communicated" (Harley 1983: 22) and through which we "gain a sense of our place in the world [and] orient ourselves" spatially (King 1996: 40). Thus, many European maps of the Middle Ages incorporated not the scientific beliefs of the Greeks and Romans, who had produced maps for navigation and property demarcation purposes, but, instead, theological concerns. Accordingly, various "T–O" maps, in which the world's known landmasses were reduced to a T-shaped form contained within an O-shaped boundary of the Earth, became quite popular (Figure 2.4). Such maps were not accurate in a modern cartographic sense but were reflections of what their makers (often monks) believed were important aspects of their own worldview – hence, they were usually centered upon Jerusalem whilst the portrayal of the three continents took on the symbolic representation of the Holy Trinity and/or the repopulating of the Earth after the flood (each of Noah's three sons being apportioned to one of the continents). For its part, the T-shape at their heart could be taken to represent the St. Anthony's cross (*crux commissa*) which the Romans commonly used for crucifiction.[6] Many such maps

Figure 2.4   A typical T and O map, attributed to Augustine of Hippo (AD 354–430). Tanais is the ancient name for the River Don in Russia, regarded at the time as the divider between Europe and Asia; Gog and Magog are biblical references to, depending on the interpretation, people, supernatural beings, or foreign lands.

were also oriented to associate Heaven with the East and Hell with the West (the direction of the setting sun), associations which earlier civilizations like the Egyptians had also made.

With the coming of the Renaissance, however, T–O maps were replaced by the world maps with which we are more familiar today. This transition in mapping practices resulted both from a transformation in the locus of social power *within* Europe, as theologically based knowledge was replaced by knowledge derived from science, and of the growing political and economic power *of* Europe. Hence, the need to map distant lands accurately, both in terms of how to journey to them (and back!) and how to lay claim to them – mapping colonies was considered crucial to appropriating them – greatly stimulated European cartography, whilst the development of the printing press meant that world maps like those based upon Ptolemy's began to circulate widely. Despite suffering the problem that it erroneously portrayed the Mediterranean as stretching across some 75° of longitude, almost twice its actual longitudinal spread, by the 1480s Ptolemy's map – and those of others – had begun to implant new worldviews in Europeans' minds, as well as in those of indigenous populations in Africa, Asia, and the Americas who were increasingly subjected to European ways of knowing. Although Ptolemy's longitudinal error was finally corrected by the Flemish geographer Gerardus Mercator in the sixteenth century, Mercator's own projection – designed to facilitate navigation because a straight line drawn on it is always a line of constant compass bearing, even if it is not necessarily the shortest distance between two points – had important implications for subsequent understandings of the globe's geography. For instance, because Mercator's projection represents landmasses at high latitudes as being much larger than in reality, it still exaggerated Europe's size and centrality (and so subliminal understandings of its importance) on world maps – hence the Danish possession Greenland appears significantly larger than Africa, even though it is actually about one-fourteenth its size (Figure 2.5). This is significant, for Mercator's representation of the relative locations and sizes of the Earth's landmasses quickly became the new standard amongst European cartographers.

These developments are critical because they established several conventions for representing the Earth's surface which still have great influence upon our mental images of places' relative locations and, so, of how we think about how economic processes are transforming the geographical relationships between places. Hence, whereas T–O maps often put East at the top (due to associations between the rising of the sun, Christ's rebirth, and Heaven being "above" us), maps after Mercator have tended to put North at the top, thereby visually placing Europe atop the other landmasses, with all the associated symbolism of domination this brings with it. Certainly, not all maps adopted Mercator's conventions – more recent projections like the Peters "equal area" map have become popular and do not overplay Europe's relative size. Equally, it has been quite common for different cultures to assume they sit at the center of things – thus the Chinese call their country *zhōngguó* (literally, "center country" or "Middle Kingdom"). However, with the rise of European colonialism it was Mercator's map, with Europe at its center, which became the dominant global

Figure 2.5 Mercator projection showing greater distortion of landmasses closer to the Poles. Europe is centrally located.

representation and was the world image invariably presented to colonized peoples in the textbooks Europeans would later ship out to their empires. Consequently, Mercator's projection remains the taken-for-granted mental image of the distribution and relative sizes of the planet's landmasses internalized by billions of people. Furthermore, it facilitated the adoption of Eurocentric conceptions of the relative locations of other parts of the world, including terms like the "Near East," the "Middle East," and the "Far East," terms adopted by many around the world but which only make sense from the perspective of Europe.[7] Indeed, the ingrained nature of such representations is revealed by how normal the Mercator projection appears to most of us and how abnormal seem to be those projections and cartographic conventions which challenge it (see Figures 2.6 and 2.7).

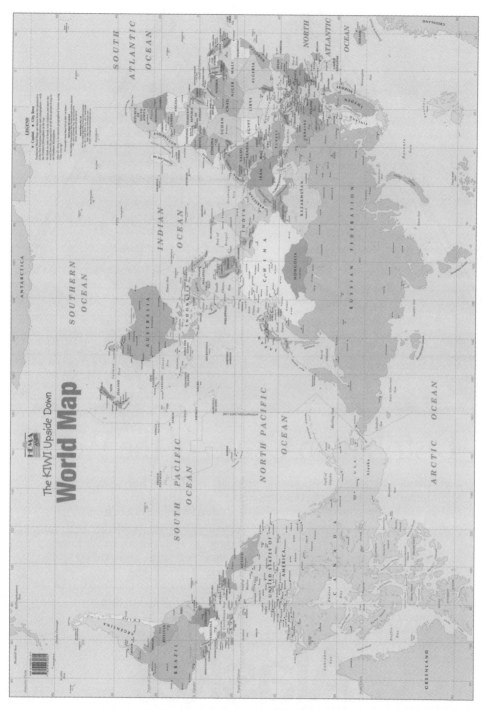

Figure 2.6 An "upside down" map of the world reveals the degree to which placing North at the top of the map has been internalized by so many of us as "normal" – placing South at the top appears "abnormal."

Distribution Center of the World

Alaska is perfectly positioned as a global distribution hub — especially
for manufacturers doing business between the TRIAD economies.
Japan, Europe, and the Eastern United States are just hours from
Alaska.

Alaska has the resources to accommodate any size operation. We have
airport land, ample warehousing, a foreign trade zone, and over 20
international and domestic cargo carriers.

Put some distance between you and the competition by bringing your
markets closer together. Put yourself in Alaska — the distribution
center of the world.

Alaska!
Alaska International Airport System

Figure 2.7 An unusual view, used in an advertisement, places Alaska at the map's center,
showing how that state gives easy access to North America, Europe, and Asia.

Whilst such cartographic conventions were being developed in Europe, various "voyages of discovery" were generating geographical knowledge with which to fill in the globe's "blank spaces," geographical knowledge which reinforced a European worldview that was then exported to the rest of the planet. The arrival of Europeans in Australia in particular had a profound impact upon their view of global unity, especially as the antipodes were imagined to represent the obverse of Europe, a place where everything was quite literally stood on its head. Significantly, the existence of a southern landmass of great size, one supposed necessary to balance out the northern hemisphere landmasses, had been postulated long before its existence was confirmed – Ptolemy believed the Indian Ocean was bounded by a southern continent, whilst early Christian scholars argued that a great equatorial ocean divided the landmasses of the northern and southern hemispheres, ideas only finally disproved in 1770 by Captain Cook's arrival in Australia. Nevertheless, although Cook's journeys debunked certain European ideas, the European "discovery" of Australia, arguably more so than the "discovery" of the Americas, continued to have significant effects upon European notions of globality. Indeed, confirmation of such a large landmass (notwithstanding the fact it was not the southern landmass of Ptolemy or of Christian theology) seemed to offer proof of the harmony in God's world, proof that the Earth was indeed in equilibrium – such hemispherically balanced landmasses attested to the essential geographical unity of the planet.

However, the antipodes also played another important role with regard to European notions of a unified world, for belief in a *terra australis* (southern land) was not simply on account of the need for an aesthetic or pseudo-scientific desire for geographical symmetry (Ryan 1996). Rather, the antipodes served to reinforce Europeans' views of themselves and their own place at the center of things. Even before Cook's journey the antipodes (particularly Australia) had became a rhetorical place wherein European fantasies about backwardness, disorder, difference, unnaturalness, and aberration could play out, a place in which people were imagined to have feet where their heads should be and in which bizarre and outlandish animals, plants, and people lived – a "Land of the Dawning" wherein could be found "the strange scribblings of Nature learning how to write," as Australian author Marcus Clarke put it in 1874 (quoted in Smith 1975: 135). Such images were only reinforced by Europeans' actual exposure to Australia's plants, animals, and physical landscapes. Hence, animals like the duck-billed platypus – an egg-laying mammal that suckles its young, with the male of the species being poisonous – did not fit into European systems of classification then being developed to make sense of the world's flora and fauna, whilst the indigenous human populations were invariably seen as "backward" because they did not conform to European ideas of what it meant to be civilized. Paradoxically, then, notions of global wholeness incorporated within themselves tensions between reason and anti-reason, normalcy and abnormalcy, rationality and irrationality, and these were played out spatially: Europe served as the place of reason, normalcy, and rationality, whilst Australia was its opposite and Other, a place requiring disciplining through the introduction of European ways of thinking and living (e.g., European-style agriculture to correct the "wasteful" use of land by the Aboriginals). Australia's distance from European norms seemed to reflect its physical remoteness, a place at the edge of the world where "Nature plays with greater freedom secretly . . . than she does openly and nearer us [Europe] in the middle" (Ranulf Higden, quoted in Ryan 1996: 107).

Importantly, such long-standing notions of Australia (and New Zealand) as places at the edge of the world continue to be reinforced in many contemporary maps (in which these countries appear near the map's border), as well as in popular culture, wherein they are sometimes lauded and sometimes lamented as being so far from European "culture" – places that are wild, exotic, and primitive yet simultaneously free from orthodoxy and conformity.[8] Equally, the idea that parts of the globe need to be disciplined, that order needs to be brought to them such that reason, normalcy, and rationality may prevail, is a central part of much contemporary discourse about globalization, particularly that articulated by neoliberals who argue that the geographical spread of "free market capitalism" will bring such things to those parts of the world presently lacking them. Hence, discourses of rationality/irrationality, normalcy/abnormalcy, order/disorder have been widely used to describe the economic and political changes in Eastern Europe after the collapse of communism and, more recently, the Bush administration's Middle East policy, in which these regions are viewed as chaotic and unruly places on the edge of the sane world, regions in which the establishment of capitalist market societies and "Western values" will bring democracy, prudence, and enlightenment (cf. Said 1978).

If the discovery of Australia filled in significantly the Western view of what lay at the "globe's edge," efforts to reach the North and South Poles in the first decades of the twentieth century were seen as truly completing the human encirclement of the planet – arguably, only Mount Everest's scaling in 1953 and the descent of a submersible 35 813 feet into the Marianas Trench in 1960 even came close to the sense of finally bringing the Earth's extremities under humans' sway that did reaching the Poles. Portrayed as the last two cardinal points which needed to be reached to complete first-hand human experience of the primary coordinates of the global web of parallels and meridians that allow the mapping of the absolute spaces of the Earth's surface, the race to the Poles also reflected imperial rivalries – whoever reached them first could lay claim to the honor of closing out the world map. Indeed, in the imagination of American Robert Peary, who in April 1909 claimed title as discoverer of the North Pole, the achievement of being the first man to "stand with 360 degrees of longitude beneath his motionless feet and for whom East and West . . . vanished" (quoted in N. Smith 2003: 482) was a feat only equaled by Columbus's "discovery" of the Americas.[9] When Norwegian Roald Amundsen's party reached the South Pole in December 1912 this seemed quite literally to round out the Earth, though it is noteworthy that for some this closure of the Earth's last two geographic frontiers was viewed as marking the end not of geography but of history – thus, American Geographical Society President Isaiah Bowman would argue in 1924 that whereas "[i]t has become the fashion to say that major exploration is at an end because the North Pole and the South Pole have been attained and the general design of the mountains, deserts, and drainage systems of the earth has become known . . . [t]he map is still crowded with *scientific* mysteries [even] though its great *historic* mysteries have been swept away" (quoted in N. Smith 2003: 54, emphasis in original). Intriguingly, such connections between geographical expansion and the passing of historical epochs would be echoed six decades later in conservative commentator Frances Fukuyama's pronouncement that the end of the Cold War and the geographical spread of the capitalist market economy into lands formerly officially off-limits to it behind the Iron Curtain represented the "end of history," the endpoint of humankind's ideological evolution.[10]

Globalizing Time

If cartographic conventions have affected how we conceive of the Earth, providing a disciplining system for the spatial narrative by which we comprehend our planet's geography and thus globalization, then the nineteenth-century erasure of local times by globally coordinated time in the form of time zones based upon Greenwich, England, has also played an important role in fostering a global outlook. Whereas for millennia humanity had myriad local times based upon the sun's height, growing transnational economic and political integration brought about a situation in which some form of standardization was deemed necessary. Whilst lack of standardization had been recognized as a problem in the eighteenth century, such that teamsters on long-distance coaches were often provided timepieces designed to gain or lose time, depending upon whether they were traveling eastward or westward, the railroads' expansion in the 1840s exacerbated such problems, for the multiplicity of local times not only brought confusion for passengers transferring between rail lines but could also prove dangerous, as trains operating on different times often used the same tracks. The lack of standardization also meant that different official times frequently operated simultaneously in the same location, a phenomenon which undoubtedly had interesting psychological aspects – passengers from Brest traveling to Paris, for instance, experienced three times: local Brest time outside the train station, Paris time in the waiting room, and "train time" (set 27 minutes ahead of Brest local time) in the track area (Galison 2003: 99). The lack of a standard also had important implications in the geographically expanding financial world. Thus, on what date and in what year did someone who had purchased life insurance in Bristol but who died in London at 12:01 a.m. London time on January 1, 1850 really die when, according to Bristol time, it was still 11:51 p.m. on December 31, 1849? As the spread of the railroads encouraged the annihilation of space by time, and as financial entanglements increasingly criss-crossed space to link distant places together, the need for a system of standardized time to annihilate local time became greater.

One of the earliest efforts to standardize time was the 1792 London adoption of "local mean time" to replace the city's myriad times based upon the sun's movement across the sky (by which London's western suburbs were a minute behind Greenwich Royal Observatory time). These efforts further developed in the nineteenth century, though largely on a voluntary basis. Hence, whilst in 1848 Britain became the first country to standardize time across its entire territory by designating Greenwich Mean Time as the nation's time (an act which helped reinforce notions of national unity), many towns continued to observe local time (for a map showing the situation in February 1852, see Howse 1980: 110). At about the same time US railroads began forming associations to run their trains according to a standardized time, though the fact that they were voluntary undermined their goal of reducing the potential for accidents – in 1849 most (but not all!) New England railroads adopted Boston time, though two accidents in 1853 quickly encouraged the laggards to follow suit (Shaw 1978). The electric telegraph's development provided further impetus by allowing towns physically distant from places like Boston and London to know exactly what time it was in these cities, courtesy of a signal sent down the wires at certain times of the day (like noon), although the need for a physical infrastructure of wiring meant that the erasure of local time could proceed only as quickly as the telegraph poles which carried the electronic signal were erected – by necessity, the geography of its erasure mirrored the unfolding geography of the network's own construction. Transnationally, the first international cable to be a commercial success was laid between Dover, England, and Calais, France, in 1850–1 (White 1942), whilst the 1858 laying of the first transatlantic telegraph cable (which John Bigelow, US Minister to France, described as the "umbilical cord with which the old world is reunited to its transatlantic offspring" [cited in Weber 1992: 105]) dramatically increased the speed with which news could be sent between Europe and North America and stimulated interest in establishing standardized time on a world – rather than simply national – scale.

Desires for a global time standard, however, raised the question of which meridian should serve as its basis. Whereas efforts to establish a standardized meridian by which to measure longitude stretch back at least to the ancient Greeks – Hipparchus proposed a meridian passing through Rhodes, whilst Ptolemy argued for one located at the Canary Islands, viewed as the westernmost edge of the known world – the growth of nationalism in the nineteenth century made instituting an internationally recognized prime meridian contentious. Hence, during the War of 1812 anti-British US residents called for the American National Observatory in Washington, DC, to serve as the basis of any global prime meridian (O'Malley 1990), whereas many Britons, arguing that Britain was the most powerful seafaring nation and that determination of a prime meridian was most essential for sailors' ability to calculate how far east or west of land they were (the miscalculation of which could have fatal results), maintained that the Greenwich Observatory should function as such. Others variously suggested Paris, Cadiz, Pulkovo near St. Petersburg, Naples, New York, Yokohama, and Rio de Janeiro, whilst some with a less nationalistic approach proposed the prime meridian pass through the Great Pyramid at Giza, seen to be worthy of such an honor because its construction was considered humanity's greatest achievement (Galison 2003).

This struggle to secure one's own (usually capital) city as the location of the global prime meridian was inextricably tied up with broader rivalries to leave a distinctive national mark upon the emerging global sense of time and space. Increasingly, science and nationalism became intertwined as British, US, and French engineers competed to lay the telegraph cables that would create what French naval officer Octave de Bernardières called "an immense geodesic network that would encompass the entire globe, fixing precisely its form and dimensions" (quoted in Galison 2003: 141). Thus, with the building of cable stations in towns and villages across the planet, more accurate measures of time could be broadcast throughout the globe via the electronic pulses coursing through the telegraph lines which, in turn, allowed a more precise mapping of territory.[11] The conquest of territory and the control of time, then, went hand in hand, and empires of space were also empires of time. Keen to impress a Gallic mark upon this emerging global grid, the French Bureau of Longitude quickly moved to link the far-flung reaches of France's empire, dispatching scientists to Dakar, Senegal, to determine the port's exact latitude and longitude as a crucial first step to extending the French cable system along the West African coast all the way to the Cape of Good Hope. Nevertheless, it was the British, seeking to link their geographically dispersed colonies, who led the way and by 1880 some 90 000 (mostly British) miles of cable – what sculptor Hendrik Christian Andersen (1913: xiii) would dub telegraphy's "wireless girdles around the world" – criss-crossed the globe, binding together every continent. In the struggle for colonial mastery of distant lands, then, "[l]ongitude, train tracks, telegraphy, and time-synchronization reinforced each other . . . [for e]ach showed a different facet of a new global grid" (Galison 2003: 175).

The prime meridian's determination, however, was inextricably linked to the determination of an "anti-meridian" International Date Line, the prime meridian's Other and partner. Since the days of the first circumnavigation of the globe, European sailors traveling westwards across the Atlantic and then the Pacific and around Africa to get back to Europe had noted that they had somehow lost a day – the date according to their ship's log was one day behind the local date in their home port. Equally, those circumnavigating the globe by traveling eastwards gained a day, a phenomenon which was the basis of Phileas Fogg winning his wager in Jules Verne's *Around the World in Eighty Days*.[12] The sense of confusion this caused was felt particularly in places like the Philippines, for sailors who had traveled westwards from Europe found their reckoning of the date to be one day behind those who had come eastward via the Cape of Good Hope, as English navigator William Dampier noted in 1686: "among the *Indian Mahometans* here, their *Friday*, the day of their Sultans going to their Mosques, was *Thursday* with us; though it were *Friday* also with those who came Eastward from *Europe*" (Dampier 1697: 377). Such temporal differences raised the issue of which date islands in the Pacific should actually keep, for different colonial powers had come to the Pacific via different routes – the Portuguese, the Dutch, the French, and the British primarily came by way of the Cape of Good Hope, whereas the Spanish came via the Americas, with the result that although they both are located at approximately 121° E, until 1844 Manila in the (Spanish) Philippines and Malili in (Dutch) Indonesia kept different dates. The arrival of American trappers in Alaska in the mid-nineteenth century likewise complicated matters, for they determined the day of the week to be one day behind what the Russians believed it to be (an American Sunday was a Russian Monday), a matter made worse by the fact that the Russians continued to use the old Julian calendar whilst the Americans and much of the rest of Europe used the Gregorian calendar (Howse 1980).[13] As the globe was increasingly integrated by transoceanic travels, then, the lack of a date line caused progressively more confusion, although some were able to turn this to their advantage: in 1892 the King of Samoa, seeking to curry favor with US business interests, defied the British Governor of Fiji and changed the date in his kingdom from the British-determined Antipodean Time to the date as calculated by the US, in the process ensuring that July 4 was celebrated twice that year (Leigh-Browne 1942).

With international trade growing in the nineteenth century, though, problems of determining time, dates, and longitude encouraged the Great Powers to seek to impose greater temporal and spatial order on the globe. Some, like the Canadian railroad promoter Sandford Fleming, wanted to establish a singular world "Cosmopolitan Time," suggesting in 1879 that every clock across the planet display the same time simultaneously (Blaise 2000). Ever the dutiful colonial servant, Fleming argued that Greenwich should serve as the basis for any prime meridian around which world time would be structured, a location which had the added advantage that the 180° "anti-meridian" at which Fleming's universal day would begin would pass, for the most part, through the landless parts of the Pacific. Others argued that a world time made little sense – "what does a man living in Ireland or Turkey care about Cosmopolitan Time [?]" argued one opponent (Galison 2003: 120) – and preferred either to continue with the multitude of local times determined by the sun's passage overhead (arguing that it was these which made innate sense to people), or to establish uniform "national times" (astronomers at the US Naval Observatory argued for a single time zone for the entire US, based upon Washington, DC, time). With much riding on such questions, in 1884 41 delegates from 25 countries as diverse as Chile, Italy, Hawaii (then an independent kingdom), Austria-Hungary, Liberia, and El Salvador, together with the major colonial powers, met in Washington, DC, to come to agreement regarding standardizing world time. Ultimately, several conventions were determined, including that Greenwich would serve as the prime meridian and that there would be a universal day of 24 hours beginning at Greenwich midnight, though the Conference specified that this should not interfere with the use of local or standard time. By default, such decisions also rendered the 180° meridian the International Date Line, though this was not actually defined by any treaty and continues to be delineated according to British and US naval conventions.

The 1884 conference, then, imposed a new sense of global order to time and, hence, to space as, for the first time, there would be an internationally agreed-upon standard by which both would be regulated. Certainly, some locales did not follow the schema devised in Washington. Hence, Detroit continued keeping local time until 1905. Equally, some countries did not adopt time zones based upon the Greenwich meridian until several decades into the twentieth century, whilst others only partially base their time on Greenwich – China uses a single time zone based upon Beijing local time, even though its land area spans some 60° of longitude. Nevertheless, despite some local variations based upon political or other considerations, the agreements reached meant that most of the world had become part of a single temporal and spatial grid, with populations' sense of time and space being determined by their geographic location relative to Greenwich.[14] Such senses of global time were reinforced in 1912 when France, having lost the prime meridian to its great rival, hosted an International Conference on Time which adopted a uniform way of transmitting planetarily accurate time signals.

Delegates agreed that the observatory at Paris would take astronomical readings to deter-mine precise time and transmit them to the Eiffel Tower, from which they would then be sent via telegraph to eight stations located around the globe. The first such signal was broadcast worldwide on July 1, 1913, and Paris remains the center of world time coordination as the seat of the International Bureau of Weights and Measures.

Efforts to develop globally coordinated systems of time matured in the twentieth century as the need for ever-more accurate temporal measures developed, particu-larly in response to advances in computing and navigation. Thus, in fields requiring greater precision than that offered by Greenwich Mean Time, "Coordinated Universal Time" (UTC) – based upon atomic measurements rather than the Earth's rotation – has, since 1986, become the global standard for telling time. Though still based on the location of the prime meridian, UTC is accurate to within a nanosecond per day and is used by airlines and many internet protocols, reinforcing not only a global sense of time but also ensuring that London remains centrally located in people's global consciousness, even if unconsciously. Indeed, such a historical legacy of basing time upon Greenwich meant that there was some debate among scientists, theologians, and others as to whether the third millennium in the Christian era could only truly be said to have begun once midnight December 31 in London had been passed (as opposed to January 1 dawning in the middle of the Pacific, at the International Date Line). How much different might our worldview and perceptions of global issues be if the prime meridian – and hence our systems of coordinating time and space – had been located in New York, Paris, Berlin or Tokyo?

Images of Global Integration in Popular Culture

Mary Louise Pratt (1992) has argued that a "planetary consciousness" began emer-ging in the European psyche in the eighteenth century as the slave trade increasingly linked West Africa with the Americas. Others, like Kirby (2002), have posited that the blossoming of astronomy in the sixteenth century – as improvements in glass making allowed for progress in telescope making – encouraged people to view the Earth and its relation to the broader cosmos in novel ways. Still others, like Cosgrove (2001), have suggested that a global consciousness has deeper historical roots, going back to the pre-Christian era. Whatever the origins of any such planetary conscious-ness, though, it is unquestionable that nineteenth-century developments like the telegraph dramatically transformed people's views of the world and their place in it by allowing users to influence events thousands of miles away virtually instantaneously, a development which encouraged a sense of "panoptical time" wherein global history was now consumed, God-like, "at a glance – in a single spectacle from a point of privileged invisibility" (McClintock 1995: 37). In particular, the establishment of new temporal and spatial orders had significant impacts upon images of global coming-together within literature, art, and film. Whilst some saw benefits in moves towards globalism – Hendrik Christian Andersen (1913: xiii–xiv), for instance, wanted to estab-lish an "International World Capital [that would] unite all leaders of national labour

unions as well as representatives of capital . . . [so as to avoid] those misunderstand-
ings . . . which cause social dislocation, affect the peace of nations and interfere with
industry, commerce and transportation" – others saw danger.

The sense of both possibility and contradiction in the new world being forged in
the late nineteenth century is particularly manifest in the writings of people like Jules
Verne, who wrote several novels touching upon the new spatial and temporal horizons then being opened: *The English at the North Pole* (1864), *Journey to the Center of
the Earth* (1864), *From the Earth to the Moon* (1865), *Twenty Thousand Leagues under
the Sea* (1869–70), *Around the World in Eighty Days* (1872)), and *Master of the World*
(1904). Such novels mixed utopian rationalism with a sense of foreboding and
paradox. Thus, in *Around the World in Eighty Days* Phileas Fogg and his friends contemplate how the shrinking of the globe caused by transportation improvements would
make it easier for the authorities to catch a thief yet also facilitate the thief's escape
from his pursuers. Ironically, it seemed, the smaller the world became, the wider were
its possibilities, whilst mastery of the new time and space order could have material
benefits. Hence, Fogg believes he has lost his bet to traverse the world in 80 days until
his servant Passepartout realizes that, in traveling east, they had over-counted by
a day: not only did this conqueror of time and space win money as a result of his
servant's discovery – a narrative which slotted seamlessly into Victorian mores about
class structure – but he also gained a wife in his exploits, a parting touch by which
Verne confirms the heteronormality of the master navigator of the new global order.[15]

Similar images of globality and human subjugation of time and space can be dis-
cerned in the works of other popular writers. H. G. Wells, for instance, explored themes
both of time and industrial capitalism in his 1895 *The Time Machine*, and gave
expression to fears of global domination in his 1898 *War of the Worlds*. Paradoxically,
perhaps, the growing integration of the world augured by empire and technologies of
global spatial and temporal control like the telegraph and Standard Time appeared to
bring forth in Verne, Wells, and others a fascination with the global and the relationship
of the Earth to the "extra-global." Equally, the smugness of powerbrokers in London,
Paris, and New York transmitting orders (whether military or financial) globally
seemed to occasion circumspection by some concerning humans' insignificance
within the broader cosmos – or, as Wells puts it in the opening paragraph of *War of
the Worlds*: "With infinite complacency men went to and fro over this globe about
their little affairs, serene in their assurance of their empire over matter." Furthermore,
whilst fears of Martian conquest had their corollary in British dread of invasion from
Continental Europe, for Wells it was neither the supremacy of human technology and
ingenuity nor a global alliance of nations which would save Earth from alien attack
but, rather, terrestrial microbes to which the Martians had no natural resistance. For
all of humanity's ability to master time and space globally through technology, it seemed
that a remarkably low-tech solution to planetary annihilation would save the species.

Fears of global control can be seen in other writing genres. Hence, in *The Octopus*,
Frank Norris's 1901 novel about California agriculture and industrialization, linkages
between wheat growing in California, the corporate power of the railroads (whose
abilities to annihilate space by time had facilitated San Joaquin Valley wheat farmers'

subjugation to commodity traders in distant cities like Chicago and New York), and the vagaries of the broader world are made plain to see.[16] Thus, because ranches were

> connected by wire with San Francisco, and through that city with Minneapolis, Duluth, Chicago, New York, and at last, and most important of all, with Liverpool [, f]luctuations in the price of the world's crop during and after the harvest thrilled straight to the [ranch's] office . . . At such moments [wheat farmers] no longer felt their individuality. The ranch became merely the part of an enormous whole, a unit in the vast agglomeration of wheat land the whole world round, feeling the effects of causes thousands of miles distant – a drought on the prairies of Dakota, a rain on the plains of India, a frost on the Russian steppes, a hot wind on the llanos of the Argentine. (Norris 1967 [1901]: 51)

For Norris, then, the whirring of the commodity ticker epitomized the transformation in the familiar spatial and temporal parameters of the world at the end of the nineteenth century, allowing farmers to know instantly the price of crops in far-off markets. Like many commentators of his time, though, Norris was ultimately ambivalent about the annihilation of space by time which the railroads and "the market" were auguring. Thus, whilst he was a critic of the forces of capitalism transforming the landscape and allowing railroad barons to expropriate farmers' lands, the railroads' ability to transport California grain over ever-greater distances nevertheless meant that "in a far distant corner of the world a thousand lives are saved [, for] all things, surely, inevitably, resistlessly work together for good" (1967: 652). (As an aside, it is perhaps no coincidence that the term "*los pulpos*" ("the octopi") is widely used in Latin America today to refer, negatively, to transnational corporations.)

The sense that a new time–space discipline had been unleashed upon the public by the 1884 Meridian Conference and that not all favored what they saw as the centralizing of control over time and space is also explored forcefully in Joseph Conrad's 1907 novel *The Secret Agent*, in which anarchists determine to challenge the new global order through a bombing campaign. After considering several options, they pick a target whose destruction they think will wreak havoc upon the capitalist world order: astronomy. Concluding that the Greenwich Observatory would serve as the perfect target, Conrad's anarchists determine to explode a bomb there in the belief that "blowing up . . . the first meridian is bound to raise a howl of execration" whilst also providing "the greatest possible regard for humanity" by avoiding civilian casualties (Conrad 1947 [1907]: 34–5). What is significant about Conrad's novel, however, is that its central event is based upon actual occurrences, for an alleged anarchist did (unsuccessfully) attempt such a feat in February 1894. Destruction of the Observatory, it seems, was seen as a way of bringing the new system of ordering time and space – one which facilitated capitalist commerce and made workers slaves to the factory clock and the new time and space disciplines it embodied – crashing down in an act of temporal and spatial rebellion. Truth, it seems, is indeed sometimes stranger than fiction!

Similar fears of global capitalist domination were explored in film. Thus, the 1922 movie *Nosferatu*, the first big-screen adaptation of Bram Stoker's novel *Dracula*, contains several allusions to global capitalist expansion, especially foreign direct investment in the local economy (although from Carpathia, the vampire Count Orlok invests in the German property market, for instance). In particular, in a scene cut from director F. W. Murnau's final version, the vampire is attacked in the street and stabbed through the heart, though instead of blood it is gold coins which gush from his chest. Significantly, Murnau's linking of global capitalist expansion and vampirism have deep historical origins. Thus, for Marx (1976 [1867]), capitalists are "vampire-like" (1976: 342), economic parasites whose "vampire thirst for the living blood of labour" (p. 367) can never be quenched "while there remains a single muscle, sinew or drop of blood to be exploited" (p. 416), whilst even earlier associations between the expansionary nature of capital and the parasitical predations of various blood-sucking satanic creatures can be seen in early Protestant texts – sixteenth-century religious reformer Martin Luther, for instance, pronounced that, "Usury is a great huge monster, like a were-wolf, who lays waste all" (quoted in Marx 1976: 740).

More recently, such dread of supernatural tendencies towards globality have been manifested in contemporary fundamentalist Christian beliefs concerning propensities towards one-world government. Originating in the Book of Revelation, in which Satan is said to seek to dominate humanity through establishing a single, planetary-wide rule, they have been articulated most recently in the popular *Left Behind* series of novels by Jerry Jenkins and Tim LaHaye, in which the Antichrist (Nicolae Carpathia) becomes, first, President of Romania and then United Nations General Secretary.[17] Under Carpathia's Secretaryship, the UN establishes a world government wherein all nation-states merge into a single political unit. Ultimately, Carpathia is declared "Potentate of the Global Community," and it is left to a group of Christians from a Chicago church (the "Tribulation Force") to defeat him. Although the series has been criticized by theologians (e.g., Jackson 2000) for its incorrect biblical interpretations, it has achieved great commercial success – selling over 60 million copies as of 2004 and with an estimated one in eight Americans having read books in the series (Gates 2004) – whilst some pundits see in it an eerie parallel with George Bush's 2003 Iraq invasion, given that in the novels Baghdad is viewed as Satan's headquarters.

If many representations in popular culture have been suspicious of tendencies towards globalization, other ideas more favorably inclined towards globality have also shaped the popular consciousness. Hence, the Canadian media critic Marshall McLuhan propagated the phrase "global village" to describe a world in which electronic media have so compressed time and space that the world and its peoples are so interconnected that "'time' has ceased, 'space' has vanished" (McLuhan & Fiore 1967: 63). Given its bucolic imagery, it is perhaps no surprise that the term "global village" has been commandeered by many in the corporate world – the US telecommunications firm Zoom Technologies, for instance, produces and markets modems, digital web cameras, and other communications products under its Global Village® brand name. Likewise, whilst many religious conservatives have expressed concerns about globalization and apparent moves towards centralized global political control, many economic conservatives have sought to inculcate amongst the population ideas concerning globalization's benefits. Thus, Hillis et al. (2002) have shown how US schoolchildren are being taught to think of neoliberal globalization as a necessary, inevitable, and unquestioned process through a well-financed and slick educational cum media campaign sponsored by the Virtual Trade Mission (VTM). Established in 1996 by private business and the President's Export Council (an advisory board on matters of international trade), the VTM provides videos, literature, websites, and simulation games to schools in an effort to make students think of themselves as "global entrepreneurs" who must identify with US corporate interests so that the US can maintain its hegemony in the new globalizing marketplace. In a neoliberal twist on the cosmopolitanism of the Greek Stoics, then, the VTM promotes a shift in student self-identity away from place-based concepts of citizenship and towards a globalized sense of entrepreneurship.

Images of globalism have also worked their way into the contemporary popular consciousness through anxieties about various types of planetary calamity, whether natural or humanly spawned. Hence, a slew of science fiction movies at the end of the twentieth century explored Earth's vulnerability in the face of superior alien technology (*Independence Day, Mars Attacks!, Men in Black*) in much the same way as did H. G. Wells at the end of the nineteenth (Kirby 2002). Fears of global catastrophe, though, are manifested not just in science fiction but also in science proper. Thus, the Severe Acute Respiratory Syndrome (SARS) outbreak which began in China in 2003 and quickly spread to Canada rapidly caught the public imagination as merely the latest in a series of global disease pandemics which appeared to know no geographical boundaries – SARS was referred to as a "global killer" by the Discovery Channel TV network, for example.[18] Likewise, some of the discourses around planetary climate change highlight how globalism has come to pervade everyday perceptions – whilst increased greenhouse gases threaten to make some parts of the planet (like Western Europe) colder, at least in the immediate near future, for millions contemporary planetary climate change has come to be known simply as "*global* warming." Such is the power of the discourse of globality, then, that the complexities of "*global* warming's" impacts in particular geographic locations – warming here, cooling there, drought here, increased rainfall there – have been submerged for many within a universalist language of planetary heating.

The importance of linguistic nuances concerning matters of globality have also been explored somewhat whimsically in the popular press, with conservative political columnist William Safire (2004) ruminating on the terms used to describe the structure of the contemporary economy:

> This month, a group calling itself the Coalition for Economic Growth and American Jobs ... decided to oust *out* from *outsourcing*, proposing instead *worldwide sourcing*. Within Cegaj, as the coalition has not yet become widely known, *worldwide* was chosen over *global* because the adjective *global* had become too warm – that is, the noun formed from the adjective's verb, *globalization*, had acquired a pejorative connotation, in turn casting a pall over the root *global* itself. The use of *world* as an attributive noun, however, is still acceptable; that use as a modifier has been long established in World Series, World Cup, World Bank, World Economic Forum, world class, etc. This is despite the fact that the word, as a regular noun, is now eschewed by concerned liberals, who much prefer *planet*. Forget *international*. This soporific modifier has been rejected by naming committees not on ideological grounds but because it is too long a word to fit in a one-column

headline. It remains in old and revered institutions, like the International Monetary
Fund and the International House of Pancakes, but is not being used in the newest
nomenclature.

For Safire, then, how people think about contemporary changes in our world – and
whether they use terms like "global," "international," or "worldwide" – may be the
result of something as mundane as newspaper editors trying to save space. Certainly,
the International Bank for Reconstruction and Development is more commonly
referred to as the "World Bank," whereas the latest incarnation of the *International
Trade Organization* (first established in 1944) goes by the name *World* Trade
Organization (these organizations are addressed in Chapter 7). Moreover, some organizations have modified their names in an effort, perhaps, to signal a changed
outlook on the world – in 2002, for instance, the International Trade Secretariats,
representing workers in industries from steelmaking to journalism, renamed themselves Global Union Federations (see Chapter 8).

However, if the term "international" is, in fact, being passed over by some
contemporary naming committees, this is probably suggestive of something more than
simple journalistic convenience, for the concerns about headline size Safire expresses
do not appear to have affected naming committees in the past – even accepting that
there may have been more newspapers using a broadsheet format a century ago, between
1900 and 1914 at least 231 new transnational organizations were chartered, most
of which used "International" in their names, with "World" being somewhat less
common (Lyons 1963). Moreover, both "International" and "World" remain popular
designations. Hence, the index of the 2003 edition of the *Encyclopedia of Associations
International Organizations* (a reference encyclopedia listing some 20 000 multinational
and national membership organizations from across the globe in diverse fields)
reveals there to be over 3 300 institutions using the word "International" in their
name, compared to 600 or so using "World" or "Worldwide," and a mere 46 using
"Global." Interestingly, though, whereas the *Encyclopedia*'s 1989 edition reveals almost
parallel numbers of "International" and "World"/"Worldwide" organizations, it
records a mere 12 "Global" organizations. Contrary to Safire's (admittedly tongue-
in-cheek) commentary, then, it is not so much that "International" is becoming less
used but that "Global" is becoming a much more common moniker. Indeed,
although "global" was available to nineteenth-century wordsmiths as a taxonomic option
– the *Oxford English Dictionary* dates its first English use to 1676, whilst an 1848
edition of the *London Magazine* referred to the Earth as "being global" and references
to the "global" are quite common in late nineteenth-/early twentieth-century writings
– they did not seem to favor organizational names incorporating "Global," which is
a relatively recent naming practice. Likewise, usage of adjectival terms related to
things "global" appears to be fairly recent. Thus, the word "globalized" did not appear
in print in English until 1959 in an article in *The Economist*, whereas "globalization"
was first listed in Merriam–Webster's *New International Dictionary* in 1961 and
appears to have first shown up in the popular press on October 5, 1962, when the
British magazine *The Spectator* stated that "Globalisation is, indeed, a staggering

concept." Such changes in nomenclature and the growing use of "global" and "globalization" towards the end of the twentieth century, then, are suggestive of a significant change in the popular *Weltanschauung*, changes which we shall examine in the following chapters.

Questions for Reflection

- How does our internalized image of the globe affect how we understand the process of "globalization"?
- How have ideas about globality been struggled over historically and in different places (for instance, how did nineteenth-century imperialism subject colonial subjects to new concepts of time and space and how did they resist these concepts)?
- In what ways did the physical linking together of the planet through technologies like telegraphy and the transnationalization of economic and political processes shape images of globality and with what effect?

Notes

1. The first photographs of the Earth from space were shot over New Mexico on October 24, 1946 by a camera attached to a German V-2 rocket (Goddard 1969: 381–2).
2. The *yaxché* forms the axis around which the world turns. With its roots symbolically in the underworld, the tree links the underworld with the Earth and the heavens.
3. Ptolemy indicated that Asia extended farther east than it does. Given that Columbus knew of Asia's existence from Marco Polo, many argue that by traveling some 3 000 miles west from Spain he believed he would land in Japan.
4. Arguments about a universal humanity have not always been universally accepted. Thus, the conquest of the Americas led to intense debate amongst European theologians concerning whether its indigenous peoples were part of the "Family of Man" (Todorov 1984).
5. Formulated by James Lovelock, the Gaia Hypothesis argues that the Earth functions as a single organism (Lovelock 1979).
6. Modern iconography usually pictures Christ being crucified upon the Latin cross (†). In fact, the Romans used several crosses, including those in the shape of a Y and an X. Matthew 27:37 states that the inscription (the *titulus*) giving Christ's crime was placed over his head, which has led many to assume a Latin cross must have been used. However, the *titulus* was frequently nailed to the crossbar (*patibulum*) of the T-shaped *crux commissa* (Davis 1965).
7. Such terms have a historical geography. Prior to the Ottoman Empire's collapse, "Near East" was often used in English to refer to the Balkans and what are now Turkey, Syria, Lebanon, Israel, Jordan, and Iraq, whereas "Middle East" referred to countries like Persia (Iran), Afghanistan, and occasionally those of the Central Asian Republics of the former

Soviet Union. The "Far East" included countries like Japan. Today, Turkey, Syria, Lebanon, Israel, Jordan, and Iraq are more commonly referred to in the English-speaking world as part of the "Middle East," although they are still often considered the "Near East" in Germany.

8. Such ideas are used today by advertisers to give these places a sense of allure. In a 1998 address to the New Zealand Tourism Industry Association, Saatchi and Saatchi CEO Kevin Roberts argued that efforts to market New Zealand tourism should leverage the image that "*It's the edge of the world* . . . [a] country on the fringe of mainstream global culture. A place where you meet an extraordinary nation of people who could only have arisen on the fringe. This idea is . . . about freedom from orthodoxy and conformity. It's about exciting new ideas, interesting people unconstrained by the need to conform to the mores of New York or Berlin. A holiday in a place like this is about escape. Adventure. Liberation. Discovery . . . We are not isolated. We are the edge."

9. As it turns out, Peary may have falsified his claim (N. Smith 2003: 97–107).

10. Fukuyama (1992: xii) states what he means by the term "end of history": "[W]hat I suggested had come to an end was not the occurrence of events, even large and grave events, but History: that is, history understood as a single, coherent, evolutionary process, when taking into account the experience of all peoples in all times. This understanding of History was most closely associated with the great German philosopher G. W. F. Hegel. It was made part of our daily intellectual atmosphere by Karl Marx, who borrowed this concept of History from Hegel, and is implicit in our use of words like 'primitive' or 'advanced,' 'traditional' or 'modern,' when referring to different types of human societies. For both of these thinkers, there was a coherent development of human societies from simple tribal ones based on slavery and subsistence agriculture, through various theocracies, monarchies, and feudal aristocracies, up through modern liberal democracy and technologically driven capitalism. This evolutionary process was neither random nor unintelligible, even if it did not proceed in a straight line, and even if it was possible to question whether man was happier or better off as a result of historical 'progress.' Both Hegel and Marx believed that the evolution of human societies was not open-ended, but would end when mankind had achieved a form of society that satisfied its deepest and most fundamental longings. Both thinkers thus posited an 'end of history': for Hegel this was the liberal state, while for Marx it was a communist society. This did not mean that the natural cycle of birth, life, and death would end, that important events would no longer happen, or that newspapers reporting them would cease to be published. It meant, rather, that there would be no further progress in the development of underlying principles and institutions, because all of the really big questions had been settled."

11. Accurate determination of time difference between a location and a prime meridian would enable the precise longitude of all points on the Earth's surface to be delineated. The inability to measure time differences precisely – and hence to determine longitude – had greatly stimulated the eighteenth-century English clock-making industry.

12. Upon leaving London at 8:45 p.m. on October 2, 1872, Fogg traveled eastwards, arriving back on what he thought was December 21 at 8:50 p.m. (i.e., 5 minutes late), a time which meant he would have lost his wager. However, he had really arrived on December 20 at 8:50 p.m. after 79 days and so won his bet, the explanation being that, without realizing it, he had gained one full day: for each degree east he traveled, he gained four minutes' time ($360°$ of circumference $\times 4$ minutes $= 1440$ minutes $= 24$ hours). Or, as Verne puts it: "While Phileas Fogg, going eastward, saw the sun pass the meridian eighty times, his friends in London only saw it pass the meridian seventy-nine times."

13. The Gregorian calendar was adopted at different times. Catholic countries adopted it immediately upon Pope Gregory XIII's issuance of his 1582 papal bull. Protestant countries followed later – the calendar was only adopted in Britain and its colonies in 1752, with Wednesday, September 2 followed immediately by Thursday, September 14, a change which caused people to demand the government "give us back our eleven days" on the basis that landlords had charged them for a full month's rent yet they had only got to use their lodgings for nineteen days. Russia did not adopt the new calendar until 1918, whilst Turkey waited until 1927.

14. Nevertheless, contemporary systems of global time coordination may still cause confusion and lead to the development of different conceptions of significant events. Hence, history records the Japanese attack on Pearl Harbor as occurring on December 7, 1941 (local time) whilst the Japanese landings in Thailand and Malaysia are dated as occurring on December 8, 1941 (local time), even though the latter actually occurred several hours previously.

15. Heteronormality refers to the idea that heterosexuality is a normal state of affairs between men and women and that homosexuality is an abnormality.

16. Henderson (1999) provides an insightful analysis of Norris's novels.

17. In choosing Romania as their anti-hero's home, Jenkins and LaHaye draw upon powerful allusions to vampirism well known in the West (both Bram Stoker's Dracula and Murnau's Orlok are from Carpathia). However, given that vampirism has often been associated with the predations of capitalism, it is significant that Jenkins and LaHaye are avowedly right-wing and pro-capitalist in their personal politics.

18. See www.discoverychannel.co.uk/reality/sars (accessed March 2, 2008).

Further Reading

Harley, J. B. (1989) Deconstructing the map. *Cartographica* 26.2: 1–20.

Harvey, D. (1990) Between space and time: Reflections on the geographical imagination. *Annals of the Association of American Geographers* 80.3: 418–34.

Howse, D. (1997) *Greenwich Time and the Longitude*. London: Philip Wilson.

Kern, S. (1983) *The Culture of Time and Space, 1880–1918*. Cambridge, MA: Harvard University Press.

Pickles, J. (2003) *A History of Spaces: Cartographic Reason, Mapping, and the Geo-Coded World*. New York: Routledge.

Thrift, N. (1996) *Vivos voco*: Ringing the changes in the historical geography of time consciousness. In: N. Thrift, *Spatial Formations*. London: Sage, pp. 169–212.

Electronic resources

Greenwich Mean Time (wwp.greenwichmeantime.com)

Mapping History (www.bl.uk/learning/artimages/maphist/mappinghistory.html)

Modern Medieval Map Myths: The Flat World, Ancient Sea-Kings, and Dragons (www.strangehorizons.com/2002/20020610/medieval_maps.shtml)

Chapter 3

Interpreting Globalization

Chapter summary: This chapter first compares and contrasts concepts of "internationalization" and "globalization." It then explores empirical instances of both processes. Rather than suggesting an "either/or" approach to the two concepts, the chapter argues for a "both/and" approach, one that shows how processes of internationalization and globalization unfold in a complex historical geography. The chapter does so by examining how new and how global are processes often taken as emblematic of globalization.

[T]he world is increasingly one economic area.
　　　　　　　　　　　　(C. B. Fawcett, "The question of colonies," p. 309)

Twenty years ago, globalization was a term, a theoretical concept. Now it is a reality.　　　　　　　　*(Kenichi Ohmae, The Next Global Stage, p. xxiii)*

Understanding the ambiguity of the concept of globalization is crucial to understanding the emergence of globalization as a fact.
　　　　　　　　(Jens Bartelson, "Three concepts of globalization," p. 181)

As I have argued in the preceding two chapters, much discourse concerning "globalization" has relied for its strength upon two sets of ideas – that the world has become increasingly small and that the world economy is increasingly developing into a seamless whole, "one huge arena for economic activity, no longer compartmentalized by barriers" (Ohmae 2005: 5). Following from the discussion at the end of Chapter 2, in this chapter I examine two issues that relate to how "globalization" as a process is understood. First, I explore what advantages, if any, there are to distinguishing analytically between two terms often used interchangeably: "internationalization" and "globalization." Hence, is "globalization" simply a newer term for describing long-standing processes formerly called processes of "internationalization," or does it refer to something else and, if so, what? Second, as a way to evaluate whether present-day events constitute a fundamentally new way of organizing economic, political, and cultural life, I examine several historical and contemporary processes and patterns concerning the geography of the world economy and how it has been wired together spatially during the past two centuries.

Internationalization and Globalization

For many, there is no distinction between "internationalization" and "globalization." Rather, these terms refer to the same sets of processes, with "globalization" simply a novel way of describing what was previously described as "internationalization." For such observers, then, the question "What is the difference between globalization and internationalization?" is meaningless. Others, however, have suggested there is indeed validity in considering these two terms as referring to different states of affairs. In adopting the latter position, two sets of questions immediately arise: to what might these terms variously refer, and at what historical moment (if at all) is one form of economic, political, and cultural organization ("globalization") replacing the other ("internationalization") and how is this playing out in different parts of the globe? The key issue in this regard concerns the status of the nation-state and whether it is being transformed by contemporary processes.

Internationalization

By its very definition, the word "*inter*national" makes implicit reference to activities and processes that reach beyond national boundaries. As such, it suggests that the various nation-states into which the Earth is divided are the primary structurers of social life. Indeed, the term allows us to make sense of one realm of social life ("the international") only through its intrinsic contrast with another ("the national"). The nation-state's importance for debates concerning globalization, then, relates to the fact that nation-states have typically been conceptualized and represented as containers for social life, discrete and distinct spatial units whose boundaries are seen as fairly impermeable geographical circumscribers of particular practices, processes, and identities (Herod 2007a).

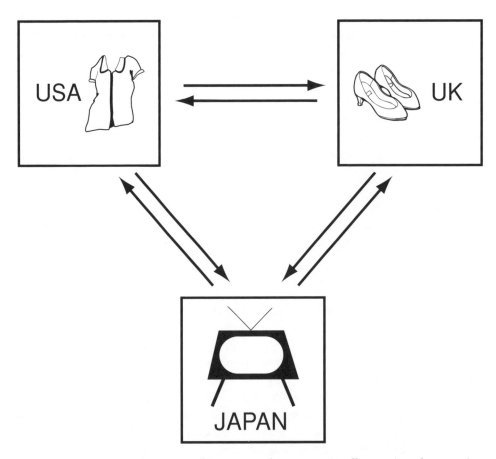

Figure 3.1 Internationalization: exchange occurs between nationally constituted economies.

Considering the world as an entity that is divided into a series of fundamentally discrete absolute spaces called nation-states has significant implications. With regard to the economic realm, it means that the nation-state is seen as the primary unit within which economic relationships are organized – thus manufacturing takes place wholly within the boundaries of particular nation-states, with finished goods consumed either within that nation-state or exported overseas. To refer to the world economy as an entity undergoing "economic internationalization," then, is to refer to a situation in which we may witness increased degrees of economic interaction between nation-states – as when shoes manufactured in the UK are traded for televisions produced in Japan – but this is a system in which economic activity like manufacturing is still primarily organized *within* national boundaries (Figure 3.1).

In the political realm, to refer to processes of internationalization is to see the world as made up of social actors who view their political identities as primarily manifested through their relationships with the nation-state (Dicken et al. 1997). Thus, even if different nation-states come together in organizations like the United Nations, the political identity of their populations and governmental actors remains such that they see their primary allegiance as being to their own nation-state. Although there are several ways in which political identity may be constructed with reference to the nation-state, probably one of the most common is through the notion of citizenship, which generally binds individuals to particular nation-states and imbues them with certain rights (e.g., voting) yet requires of them certain responsibilities (e.g., compulsory military service). For sure, the connection between citizenship and an individual's membership within a community defined by the nation-state's territorial boundaries has varied historically. Thus, during much of the nineteenth century US citizenship was *de jure* limited to certain classes of individuals (e.g., whites), whilst between 1776 and 1926 some 40 states and federal territories adopted a form of political representation ("alien suffrage") allowing non-citizens to exercise what has generally come to be seen as the *primus inter pares* of political rights (voting) in local, state, and even federal elections (Hayduk 2006). Equally, there have always been individuals who have viewed their identities not in national terms but either in broader "citizen of the world" terms or in terms of being ethnic or religious minorities within nation-states, such that their loyalties have either been constructed along non-territorial lines or have been focused upon territorial units smaller than the nation-states within which they live (e.g., Basques, Kurds). Furthermore, whereas in some instances nationalism – the practices which serve to patrol one's allegiances to particular territorial units – preceded the creation of modern nation-states, in others it followed it.[1] Nevertheless, as nationalism grew during the nineteenth century, binding individuals to an imagined political community which had as its spatial expression the bounds of the nation-state, there was a growing convergence in many parts of the world between nation-states' physical territory, their citizenry, and various political rights.[2]

In the cultural realm, as in the economic and the political, in a world which is seen to be undergoing processes of internationalization the nation-states' boundaries are viewed as the primary spatial delineators of cultural practices and identities, such that nation-states are seen to contain cultures – the French nation-state is the container of French culture, for instance. Certainly, this representation of culture relies upon a particular fiction, namely that cultures have ever been distinct and organized within nation-states. Thus, in the case of Britain, a place invaded by Celts, Romans, Vikings, Saxons, and Normans and ruled by a series of imported German and Dutch monarchs, a destination over the centuries for refugees from many parts of the globe, and an absorber of "foreign" cultural practices, ideas, and linguistics through its conquest of places like India, viewing "British culture" as the product of the landmass located off the northwest coast of Europe relies upon such a representational invention. Nevertheless, the idea of the nation-state has served as a powerful discipliner to encourage a view of cultures as things which are constructed within national boundaries and which are thus distinguishable from cultures developed by other nation-states' populations. The result is that the cultural differences within nation-states like France, for all their regional idiosyncrasies (Normans, Basques, Bretons, Burgundians, Gascons, etc.), are viewed as smaller than the differences between them and other nation-states' inhabitants, such that, even within an increasingly interconnected world, it is still possible to distinguish different national cultures from each other.

In sum, "internationalization" describes a process whereby economic, political, and cultural exchanges or movements across national boundaries increase but do not fundamentally change the nature of the nation-state.

Globalization

By way of contrast, to argue that our planet is undergoing "globalization" is to describe a fundamentally different set of relationships being established between, on the one hand, nation-states and, on the other, economic, political, and cultural practices. Specifically, it is to argue that, rather than the world being stratified into a series of fairly discrete units (nation-states) whose interior spaces are clearly distinguishable from the exterior spaces beyond them, we have entered a time in which the globe is "intelligible as a *single place*" (Bartelson 2000: 187). In the economic realm, this suggests that a fundamental reorganization has occurred, particularly that there has been a significant degree of functional integration among what may be geographically quite dispersed activities (Dicken et al. 1997). Although there are several areas in which this might be seen, arguably it is the stretching of production chains across national boundaries where it is most significant. Thus, when a company like General Motors assembles cars in the United States but uses components sourced from Mexico and Asia, this represents a vitally different way of organizing production than when it manufactures cars using parts sourced from within the US and then sells those cars domestically or overseas (Figure 3.2).

In the political realm, processes of globalization may be taken to be those which do not simply imply that various nation-states are becoming more involved in matters beyond their national boundaries but that the nation-state's privileged role as a political actor and focus for political action is itself being undermined. This is a complex process, for there are at least three ways in which the relationship between the nation-state and the extra-national may be thought of: (1) the actors who shape nation-states' activities are domestically focused and largely uninterested in, and disconnected from, events which transpire beyond their borders (they are isolationists); (2) the actors shaping nation-states' behavior may affiliate with organizations whose job it is to manage "international space" (e.g., the UN) but do so largely to pursue national interests, with domestic political concerns being foremost; and (3) the allegiances of nation-states' citizenries are secured by supranational entities and the domestic political sphere is increasingly constrained *de jure* by such supranational entities. Whereas the second of these options may, according to our definitions here, be thought of as an example of the nation-state's "internationalization," the third indicates a fundamentally different state of affairs, what I am here calling "globalization."

Certainly, recent years have seen examples of the undermining of the correspondence between nation-states' physical territory, their citizenry, and various political rights, at least when compared to the nineteenth and twentieth centuries, which is often seen as the nation-state's "golden age." For instance, in New Zealand the century-long link between citizenship and voting was broken in 1975 when all permanent residents were granted the right to vote in national elections, regardless of whether they were

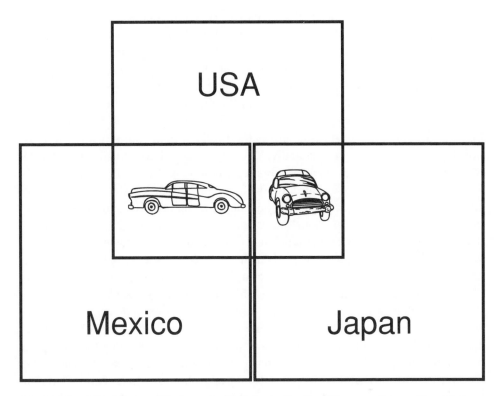

Figure 3.2 Globalization: The boundaries between "national" economies are blurred as production chains cross them.

citizens, and today over 40 countries permit non-citizen voting (Hayduk 2006: 5). Moreover, in several countries where one could already vote as a non-citizen in certain elections, immigration has encouraged efforts to further decouple voting from citizenship by allowing non-citizens to vote in a wider variety of elections. Growing immigration also raises the question of migrants developing transnational identities, where they come to view their identities not as bound within particular nation-states but as transcending them, thereby creating a host of new transnational identities – what Hall (1988) calls "new ethnicities."

Part of the problem with analyzing challenges to nation-states' status as the primary agents of political change and sites of struggle, of course, is that even during its so-called "golden age" there were many who did not align their allegiances to the nation-state within which they lived. Nevertheless, it is possible to point to some developments that might be read to indicate that the power of individual nation-states is being challenged by entities which may themselves be beginning to serve as focal points for new supranational identities. Early examples include the brief union of Egypt and Syria to form the United Arab Republic, or perhaps the union of Zanzibar and Tanganyika to form Tanzania, both of which represent examples of two separate states merging into a single one to create new identities. However, it is the European Union (EU) which exhibits the most advanced tendencies in this regard, adopting many of the practices typically associated with individual nation-states' efforts to develop allegiance from their citizenry – a flag, a "national" anthem, an (ill-fated) constitution, a single (for the most part) currency, a central bank, EU passports.

Unquestionably, the EU's creation can be interpreted in different ways: for some it is the embodiment of a "post-national" dream for Europe, whereas for others it simply represents different nation-states coming together to pursue their own national interests more effectively through the mechanism of the EU than they could individually. However, what distinguishes the EU from many other types of international organizations is that once individual nation-states have joined, the organization's institutions can exert considerable legal power over them in certain regards. Furthermore, the political architecture of the individual nation-states – parliaments and legal systems – are being, if not supplanted, then at least challenged by the creation of a permanent supranational infrastructure of governance. The point, then, is that if growing numbers of EU inhabitants increasingly come to see themselves as "European," rather than as German, French, or Irish, such a realignment of loyalty away from individual nation-states and towards this supranational entity would mark an important reterritorialization of citizenship and political identity.

Finally, with regard to the cultural realm the nation-state's undermining as a container of national culture can be seen in a multifaceted reworking, cross-fertilization, and interpenetration of cultural forms and practices, one marked by the emergence of a "globalized" popular culture in which national distinctiveness is increasingly eroded. As is the case with transformations in the political realm, for such a representation to be maintained it is necessary to view cultural practices as things which, at some point in the past, were actually contained within nation-states, such that the "globalization" of culture represents a fundamental reorganization of the relationship between the nation-state and the cultural practices of populations resident therein. Whether this representation is sustainable depends heavily on the view adopted concerning what constitutes "national" culture. Nevertheless, the growing coalescence of consumption patterns measured, at least amongst the world's elites, through the preponderance of people across the planet consuming similar products (eating at McDonald's and drinking Coca-Cola whilst listening to their iPods or discussing the latest global blockbuster movie or events covered by CNN), together with the greater spatial reach of what we might call the forces of cultural projection (e.g., the worldwide syndication rights for television programs like *Baywatch*, *Star Trek* and its various spin-offs, *Desperate Housewives*, and/or the development of worldwide television franchises like *Pop Idol/American Idol*, *Who Wants to Be a Millionaire?*, or *Weakest Link*), seems to suggest that, at least as far as "national" cultures are imagined by millions of people across the globe, some degree of cultural convergence may be occurring, one which challenges the notion of "French" or "Brazilian" or "Japanese" culture as discrete and separate entities. The result is that many people, like former Gillette company CEO Alfred Zeien (quoted in Barber 2001: 23), appear to "not find foreign countries foreign" in the way in which they used to.

As this discussion has perhaps illustrated, then, trying to distinguish between "internationalization" and "globalization" is tricky, because the distinction rests on determining the position of the nation-state within economic, political, and cultural realms, something which is easier to do in certain regards than in others – it is arguably easier to establish when production chains have been stretched across national boundaries than it is to ascertain whether the nation-state remains the focal point of political identity, although even here there are complexities (are the importation of unprocessed wood from Henry Ford's Brazilian plantation Fordlândia [Galey 1979] in the 1940s for use on cars manufactured within the US and the importation of finished electrical components from Mexico for use on US-assembled automobiles the same thing?). Nonetheless, attempting to make a distinction between "internationalization" and "globalization" is crucial, I think, for two related reasons.

First, given that much discourse has argued that the nation-state's power is being undermined by contemporary processes, it is essential to consider what roles the nation-state has been seen to play in the past. In other words, for the claim that we are witnessing unfold something new called "globalization" to work rhetorically, it is necessary to see how the nation-state served materially, and was represented discursively, as a container of past economic, political, and cultural processes and practices (I shall particularly return to this in Chapter 7). Second, drawing a distinction between "internationalization" and "globalization" gives us a much sharper analytical knife with which to dissect contemporary events, for if it *is* occurring, globalization – at least as just described – represents not simply a deepening of processes of internationalization but a sharp and fundamental break with them. Thus, as Dicken et al. (1997: 162) have put it:

> It is not simply the case that in a globalizing economy there is more international trade (though there may be), that transnational corporations have extended their global reach and significance (though they may have), that local and national economies are increasingly exposed to global competition (though they may be), or that nation-state powers and capacities have been eroded (though they may have) . . . Rather, globalization is associated with a *qualitative reorganization* of the structural capacities and strategic emphases of the nation-state [such that] the impact of the process of globalization is measured not in the crude terms of whether there is "more" or "less" of the nation-state, but in its changing structure and orientation.

Given, then, that for some a "globalized" world may be emerging out of the shell of an "internationalized" one, it is important to understand the historical geography of this birth, especially as the naissance of new orders can be painful – as Gramsci (1971: 276) cautioned us, in the period between when an old order is dying and the new cannot yet be born, "a great variety of morbid symptoms appear."

Toward a Globalized World? Full Steam Ahead!

In his latest exploration of the contemporary scene, globalization guru Kenichi Ohmae (2005: 28) dates the global economy's birth precisely: 1985.[3] He points to four events in that year which, whilst not constituting a "Big Bang" of globalization, sowed the seeds "of a variety of plant not previously grown, belonging to a totally novel and unknown genus and species" (2005: 28): (1) Mikhail Gorbachev became General Secretary of the Soviet Communist Party and ushered in reforms which would eventually end the Cold War; (2) the Plaza Accords were signed, presaging greater flexibility in global currency markets; (3) in the United States the Gramm–Rudman Act to reduce the federal budget deficit and restore "fiscal discipline" was passed; and (4) Microsoft launched the first version of Windows. What, though, might a "globalized" world look like? Although different commentators have different perspectives on this question, Korten (2001: 133) has argued that "the architects of globalization" generally envision a perfectly globalized world as containing six core elements:

1 The world's money, technology, and markets are controlled and managed by gigantic global corporations.
2 A common consumer culture unifies all people in a shared quest for material gratification, such that people see themselves as consumers rather than as workers, citizens, Christians, atheists, Africans, Australians, etc.
3 There is perfect global competition among workers and localities to offer their services to investors at the most advantageous terms.
4 Corporations are free to act solely on the basis of profitability, without regard to national or local consequences.
5 Relationships, both individual and corporate, are defined entirely by the market, such that there is no sense of social obligation to anyone except shareholders.
6 There are no loyalties to place and community.

Undeniably, there do seem to be trends towards the emergence of such a world. For example, an analysis of the world's 100 largest economies shows that in 2005 47 were corporations, of which 30 were headquartered in the United States, 15 in Europe, one in Japan, and one in China (Table 3.1). With a combined market value of $6.63 trillion, the planet's 50 largest corporations control wealth equivalent to 15 percent of world GDP (gross domestic product), whilst the two largest corporations (General Electric and Microsoft) have a combined market value greater than the GDP of sub-Saharan Africa. Moreover, in recent years corporations' values have been growing at a faster rate than have the economies of many nation-states – between 1983 and 1999 the combined sales of the world's 200 largest corporations grew from 25 percent of world GDP to 27.5 percent (Anderson & Cavanagh 2000: 3) and by 2005 had reached 30.1 percent.[4] Equally, whilst expanding their operations geographically many corporations have attempted to inculcate an image for themselves as entities who are bringing the world's populations together into a single, common unity defined by their consumption of particular products. Arguably, one of the earliest and most widespread efforts to do this was undertaken in the 1970s by the Coca-Cola Company, which viewed its global "I'd Like to Teach the World to Sing/I'd Like to Buy the World a Coke" advertising slogan as a "United Chorus of the World," a "World's National Anthem" to sell "not a soft drink to please a palate with a unique taste accompanied by the

Table 3.1 The world's largest 100 economies, 2005

Ranking	Country/company	Location of headquarters (if company)	Primary Economic Sector	Economic worth ($bil)[1]
1	United States			**12 455.07**
2	Japan			**4 505.91**
3	Germany			**2 781.90**
4	P.R. China			**2 228.86**
5	United Kingdom			**2 192.55**
6	France[2]			**2 110.19**
7	Italy			**1 723.04**
8	Spain			**1 123.69**
9	Canada			**1 115.19**
10	Brazil			**794.10**
11	South Korea			**787.62**
12	India			**785.47**
13	Mexico			**768.44**
14	Russian Federation			**763.72**
15	Australia			**700.67**
16	Netherlands			**594.76**
17	Switzerland			**365.94**
18	Belgium			**364.74**
19	Turkey			**363.30**
20	Sweden			**354.12**
21	General Electric	US	Conglomerates	**328.54**
22	Saudi Arabia			**309.78**
23	Austria			**304.53**
24	Poland			**299.15**
25	Indonesia			**287.22**
26	Microsoft	US	Software & services	**287.02**
27	Pfizer	US	Drugs & biotechnology	**285.27**
28	Norway			**283.92**
29	ExxonMobil	US	Oil & gas operations	**277.02**
30	Citigroup	US	Banking	**255.30**
31	Denmark			**254.40**
32	Wal-Mart Stores	US	Retailing	**243.74**
33	South Africa			**240.15**
34	Greece			**213.70**
35	Intel	US	Semiconductors	**196.87**
36	Ireland			**196.39**
37	Iran			**196.34**
38	American Intl Group	US	Insurance	**194.87**
39	Finland			**193.18**
40	Argentina			**183.31**
41	HSBC Group	UK	Banking	**177.96**

Table 3.1   (*Continued*)

Ranking	Country/company	Location of headquarters (if company)	Primary Economic Sector	Economic worth ($bil)[1]
42	Hong Kong (China)			177.72
43	Thailand			176.60
44	Vodafone	UK	Telecommunications services	174.61
45	BP	UK	Oil & gas operations	173.54
46	Portugal			173.09
47	IBM	US	Technology hardware & equipment	171.54
48	Cisco Systems	US	Technology hardware & equipment	166.09
49	Royal Dutch Shell Group	Netherlands/ UK	Oil & gas operations	163.45
50	Johnson & Johnson	US	Drugs & biotechnology	160.96
51	Berkshire Hathaway	US	Insurance	141.14
52	Venezuela			138.86
53	Procter & Gamble	US	Household & personal products	131.89
54	Malaysia			130.14
55	Coca-Cola	US	Food, drink & tobacco	125.37
56	GlaxoSmithKline	UK	Drugs & biotechnology	124.79
57	Israel			123.43
58	Czech Republic			122.35
59	Colombia			122.31
60	Bank of America	US	Banking	117.55
61	Singapore			116.76
62	Total	France	Oil & gas operations	116.64
63	Novartis Group	Switzerland	Drugs & biotechnology	116.43
64	Toyota Motor	Japan	Consumer durables	115.40
65	Chile			115.25
66	Altria Group	US	Food, drink & tobacco	111.02
67	Pakistan			110.73
68	Hungary			109.15
69	New Zealand			109.04
70	Merck & Co	US	Drugs & biotechnology	108.76
71	Nestlé	Switzerland	Food, drink & tobacco	106.55
72	Nokia	Finland	Technology hardware & equipment	104.30
73	United Arab Emirates			104.20
74	Verizon Commun	US	Telecommunications services	103.97
75	Algeria			102.26

Table 3.1 (*Continued*)

Ranking	Country/company	Location of headquarters (if company)	Primary Economic Sector	Economic worth ($bil)[1]
76	Nigeria			**98.95**
77	Romania			**98.56**
78	Philippines			**98.31**
79	Wells Fargo	US	Banking	**97.53**
80	Roche Group	Switzerland	Drugs & biotechnology	**95.38**
81	ChevronTexaco	US	Oil & gas operations	**92.49**
82	PetroChina	China	Oil & gas operations	**90.49**
83	Royal Bank of Scotland	UK	Banking	**90.21**
84	Egypt			**89.34**
85	Dell	US	Technology hardware & equipment	**88.46**
86	PepsiCo	US	Food, drink & tobacco	**86.73**
87	Telefsnica	Spain	Telecommunications services	**86.39**
88	UBS	Switzerland	Diversified financials	**85.07**
89	Deutsche Telekom	Germany	Telecommunications services	**84.18**
90	AstraZeneca	UK	Drugs & biotechnology	**83.03**
91	Amgen	US	Drugs & biotechnology	**83.02**
92	SBC Communications	US	Telecommunications services	**82.93**
93	Eli Lilly and Co	US	Drugs & biotechnology	**82.53**
94	Home Depot	US	Retailing	**82.29**
95	JP Morgan Chase	US	Banking	**81.94**
96	Ukraine			**81.66**
97	United Parcel Service	US	Transportation	**79.62**
98	Peru			**78.43**
99	Time Warner	US	Media	**77.95**
100	Fannie Mae	US	Diversified financials	**76.84**

Notes:
1. Measured as "Market Value" for companies, as GDP for countries.
2. Data include French overseas *départments* of: French Guiana, Guadeloupe, Martinique, and Réunion
Sources: Forbes (2006); World Bank (2006); collated by author.

pleasant lifting of spirits that caffeine supplies" but to use a soft drink to encourage and express "a tiny bit of commonality between all peoples" (Backer 1993: 6–7, 205). Having in mind a consumer who was "of no particular race, color, creed, sex, or country of origin [but who] was everyone and named 'The World' " (1993: 31–2), the executives who devised the advert saw Coca-Cola bottles "as symbols of harmony and understanding between peoples" (p. 203). To emphasize the globality of their product and consumer base, these executives sought to project a certain placelessness through the visual images they used, seeking a shooting location for the advert that was "both out of this world, yet in it . . . [a] . . . nowhere land" (p. 160). Their ultimate goal was to show diverse peoples "united by a common wish – that everyone in this world could have a secure home full of love, good food, worldwide harmony, and, of course, the daily pleasure of keeping each other company over their favorite refreshment, Coca-Cola. In that sense it was a chorus of a hundred 'I's wishing the same wish" (pp. 270–1). Such sentiments, then, reflected the belief articulated by not a few neoliberal theorists that peoples in different parts of the world "are *basically* the same," that "political bound-aries do not circumscribe psychological or emotional attitudes" (Fatt 1967: 61). Although the Coca-Cola campaign was one of the earliest global advertising campaigns to articulate a vision that, as consumers, the world's people share a certain common-ality that might overcome whatever other differences they have, other companies have followed suit – in the 1980s Ford sought to develop a "world car" (the Escort) that would appeal to consumers across the planet and, a decade later, announced it would develop a "global car," the suitably named "Mondeo" (Anon. 1994).

Likewise, a pointed social experiment by consumer advocate and erstwhile US Presidential candidate Ralph Nader seems to support the notion that many corpora-tions are seeking to shed their national colors by declaring they hold loyalty only to the marketplace – that they are "homeless" (Ohmae 2005: 246). Thus, in response to Nader's 1996 writing to the 100 largest US corporations to request they open their shareholders' meetings with a pledge of allegiance to the US flag – a pledge schoolchildren make every morning – August Busch, the CEO of beer-maker Anheuser-Busch, stated he would not do so because although the "company headquarters remains in St. Louis, we are a global company," whilst Ford Motor Company declined by declaring that it did "not believe that the concept of 'corporate allegiance' is possible" (Buchanan 1998). For its part, the manufacturer Motorola asserted that pledging allegiance to the flag "would be introducing political and nationalistic over-tones which have nothing to do [with] the true purpose of a stockholders' meeting" (Domenich 1998: 5). Indeed, only a single company – Federated Department Stores – believed such a show of nationalism a good idea, leading Nader to suggest this was "because they can't relocate overseas" (Nader 1999). Evidently, Ohmae's (2005: 247) argument that for companies to be successful in the contemporary world economy there "must no longer be any sentimental attachment with an old nation-state in which the company has its headquarters" appears to have been internalized by a large number.

The growth of transnational sourcing also seems to fit in with such a view of what globalization means, such that today one-third of all world trade is simply transactions among various units of the same corporation (Anderson & Cavanagh 2000). Whereas in the mid-1990s some 40 000 corporations worldwide had an estimated 250 000 foreign affiliates constituting $2.6 trillion worth of foreign direct investment (FDI) and sold over $5 trillion worth of goods and services overseas (UNCTAD 1995: xx), by 2004 these figures had increased to 70 000 corporations with 690 000 overseas affiliates constituting $9 trillion of FDI stock and having sales of almost $19 trillion (UNC-TAD 2005: xix). In the case of the world's largest economy (the United States), in 2005 US-based firms had overseas FDI valued at $2.1 trillion and "foreign" companies had invested $1.6 trillion of FDI in the US (Koncz & Yorgason 2006). Correspondingly, many companies have developed standardized manufacturing operations so as to more easily move from place to place and/or more easily supply diverse markets – a phenomenon that led Levitt (1983: 101) to argue that it made sense for manufacturers to treat the Earth "as flat" a full two decades before Thomas Friedman (2005) articulated such visions. Hence, when Hoover Ltd., then the UK's leading manufacturer of automatic washing machines, set out in 1965 to develop a new line of machines, it designed a product that was sufficiently standardized to be sold across Western Europe (Buzzell 1968). For its part, by the 1990s the Gillette company had established global manufacturing specifications for its plants and standardized management and training policies to allow easy personnel transfer between its different units. Gillette also consolidated its advertising operation in Boston, so that its television commercials would be identical across the globe (though dubbed into local languages) (Kanter 1995). McDonald's has adopted a similar strategy with its recent "I'm lovin' it" campaign, which uses the same slogan (in local languages) across the planet, and maintains standardized specifications for its equipment technology, product offerings, customer service, cleanliness, and operational systems (Samiee & Roth 1992).

Similarly, the growth of global capital markets in which billions of dollars can swoop in and out of nations in minutes suggests that truly global markets may be emerging in the fashion highlighted by Korten. Hence, it is argued, now that "[i]nformation defies barriers, whether physical or political . . . [and m]oney can flow unrestricted to the areas of highest return" (Ohmae 2005: xxiv–xxv), national financial markets are "losing their separate identities as they merge into a single, overpowering marketplace" (Bryan & Farrell 1996: 4), a marketplace marked by the "operation of an international law of one price" (McKinsey & Company 2005: 7). Such claims seem to be supported by some contemporary developments. Cross-border flows of capital (purchases of foreign equity and debt securities, FDI, and transnational bank lending) now total $5 trillion annually, more than triple what they were in 1995, whilst in 2005 overseas investors held 14 percent of US equities, 27 percent of US corporate bonds, and 52 percent of Treasury securities (up from 4 percent, 1 percent, and 20 percent in 1975) (McKinsey & Company 2005: 19; 2006a: 13). Moreover, with the development of an increasingly integrated system of trading across international financial markets, investors can now match the spinning of the Earth, jumping out from the Tokyo market and into Singapore, and from there perhaps to Dubai and on to London and New York, using sophisticated computer software to trade simultaneously on dozens of stock markets across the globe.

In the case of the foreign exchange market (the "Forex"), an increasingly unitary planetary market has emerged allowing global currency trading from the start of business in New Zealand on a Monday morning until the close of business in North America on the following Friday afternoon (Saturday morning in New Zealand), with the overlap in real time between market openings and closings in London and Tokyo and London and New York enabling traders to buy in one region of the globe and sell in another. Growing from a trading volume of $70 billion daily in the early 1980s to a current daily volume averaging about $1.9 trillion (though reaching $3 trillion on some days), the Forex is the largest and most liquid market in the world, dwarfing the world-wide daily market for equities (roughly $204 billion). In fact, such is the market's significance that the volume of currency traded on any given day is some 55 times the daily value of international trade in merchandise and services.[5] Likewise, the international bond markets – particularly on the secondary market – have become increasingly intertwined. For its part, the global equities market has become so integrated that some 12 percent of the opening price variation on New York's S&P 500, 14 percent on Tokyo's Nikkei 225, and 30 percent on London's FTSE 100 can be explained by what happened overnight in each of the other two markets, with the largest influences on the New York and Tokyo markets coming from those markets which trade immediately prior to them (London and New York, respectively), though in the case of London it is New York rather than Tokyo which has the greatest influence (Milunovich & Thorp 2006).[6]

For its advocates, then, such developments show that globalization is "a process of global optimization," the creation of a global market "based on a world in which borderlessness is no longer a dream . . . but a reality" (Ohmae 2005: 122, 18) and which is unconstrained by "distortions" like national boundaries, allegiances to place, or ethnic and kinship ties. Indeed, their goal is to create a "superconductive" (Bryan & Farrell 1996) and "friction-free" (Gates et al. 1995) capitalism in which capital can flow across the globe's surface in the "twinkling of an eye" (Marx 1973 [1858]: 548–9). Certainly, there is some diversity of opinion as to what this may all mean. Hence, whereas Coca-Cola's executives envisioned a world in which diverse peoples are "united by a common wish" (Backer 1993: 270), Bryan and Farrell (1996) argue globalization will herald a simultaneous segmentation and integration of the global economy as, on the one hand, there will continue to be variations in patterns of consumer demand globally but, on the other, producers of goods and services will find a growing convergence of opportunity across nations. For them, then, globalization is about changing one's viewpoint and creating a new planetary gestalt, such that whilst any given national market may become "increasingly segmented, when your perspective is the world, all those individual segments will look like an integrated global market opportunity" (1996: 167). Likewise, for Bill Gates the future will be one in which "[a]ll goods for sale in the world will be available for you to examine, compare, and, often, customize" but in which the "information highway will be able to sort consumers according to much finer individual distinctions" so as to bring to them tailored products (Gates et al. 1995: 158, 171). For his part, Ohmae (2005: 81) sees the emergence of the global economy in triumphalist terms, as "one grand continent of opportunity whose exact contours still remain vague in places but that will amply reward brave exploration."

Despite their slightly different visions, however, two things are common to the views of advocates of neoliberal globalization. First, globalization is seen as proceeding apace. Hence, for Ohmae (2005: 20), although the world "is not yet completely border free, as nation-states still have reasons to maintain controls on the movement of people and goods in the interest of security and public safety," in terms of what he calls the "four Cs" (communication, capital movement, corporate activity, and consumer behavior), "the world has already attained the position of being effectively without borders." Furthermore, not only is globalization argued to be already with us, but the growing mobility of capital means the "process of globalization will move much faster" in the future (Bryan & Farrell 1996: 153). Second, globalization is seen as unstoppable and self-sustaining. Thus, Ohmae (2005: 18) has argued that "[t]he global economy has its own dynamic and its own logic . . . [and] is going to grow stronger rather than weaker . . . It is irresistible . . . There is no use complaining about it . . . People will have to learn to live with it," whereas Bryan and Farrell (1996: 10) maintain we are now at a stage from which "[w]e cannot go backward[, for w]hat has been done cannot be undone without truly destructive consequences." Significantly, this belief is pro-mulgated not just by academics and commentators but also by policy makers – Roger Ferguson (2005), Vice-Chairman of the Board of Governors of the US Federal Reserve System, for instance, has argued that, "I think that globalization is here to stay and that further globalization is inevitable." In such a TINA view, then, we are leaping into a brave new world of instantaneous and global response to market fluctuations, one in which the future will always represent a more integrated and open global economy (Figure 3.3).

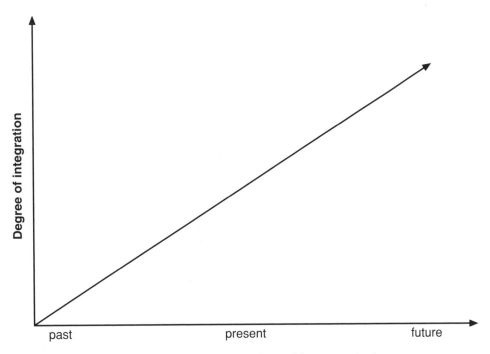

Figure 3.3 In the TINA view of globalization, the world economy is always more integrated in the future than in the past or present.

Toward a Globalized World? Not So Fast!

For Ohmae (2005: 59) and others, "[t]he global economy is something that has developed only in very recent terms" and its appearance represents a fundamental break with the past. As we saw in Chapter 1, central to such arguments has been the notion that the world has gone "from being small to [being] tiny" (Thomas Friedman, quoted in Pink 2005) and that people, commodities, money, and data can travel across large distances more quickly and cheaply than ever before.[7] Certainly, the disruption of the old temporal and spatial certainties that billions of people across the planet had previously internalized when the world was, by implication, "larger and slower" seems to have occasioned a significant feeling of awe as many individuals' senses of distance, time, and place are being transformed by the thought of engaging in instantaneous interaction with others thousands of miles away or of boarding a plane in one continent and a few hours later getting off it in another or of consuming strange and exotic commodities from far-off lands.[8] Indeed, such a sense of spatial and temporal disorientation is pivotal to claims that we are witnessing the emergence of something new – a rapidly globalizing world. However, when assessing proclamations that contemporary developments represent new phenomena, it is perhaps important to bear in mind that, modifying Ecclesiastes 1: 9, "What has been will [often] be again, what has been done will [often] be done again; there is [often] nothing new under the Sun."

Is "globalization" so new?

Without doubt, the contemporary period is one in which the shrinking of relative distances has augured a perception among many that things have changed in unprecedented ways. However, in casting an eye towards history, it becomes apparent that the senses of speed and distance shrinking taken to be representative of contemporary globalization may not be as novel as they appear. Thus, the 1858 completion of the first transatlantic telegraph cable led the opinion-shapers at the *Times* of London to argue in strangely familiar terms that "[d]istance as a ground of uncertainty will [now] be eliminated from the calculation of the statesman and the merchant," for as the "[d]istance between Canada and England is annihilated [and] the Atlantic is dried up . . . we become in reality as well as in wish one country. . . . To the ties of a common blood, language, and religion, to the intimate association in business and a complete sympathy on so many subjects, is now added the faculty of instantaneous communication, which must give to all these tendencies to unity an intensity which they never before could possess." With the completion eight years later of a more reliable cable, the sense of global simultaneity was even greater, as the Atlantic was "found to be so little impediment to conversation" that "now, for the first time, it [is] possible . . . [to hold a] conversation at the same time with Europe and America . . . [and for] all the civilized world to communicate instantaneously." The sense that the world was shrinking was palpable: "England and the United States, the Old and the New World, [which] were but lately separated by a ten days' voyage on a stormy sea," proclaimed the *Times*, "are now, for some of the most important purposes of intelligent intercourse, as closely connected as England and France."[9]

The cable's completion, then, elicited a sense of optimism – even utopianism – amongst many commentators closely mirroring that of many twenty-first-century advocates of globalization's benevolence. What Standage (1998) has called the "Victorian internet" – the Atlantic cable and others, like that laid in 1864 linking India with Britain – was seen as a medium for encouraging greater understanding across diverse cultures and populations. As the *Times* put it, whereas "[f]or a long time the nations of the civilized world seemed to grow up apart and independently, . . . obtaining false or exaggerated ideas of each other," the interconnectivity afforded by the rapidly unfolding global telegraph network would facilitate greater planetary *amité*, for "[w]hen we know people intimately, and hear from day to day all about their doings, we cannot but feel a lively interest in them . . . We cannot have such constant knowledge of each other without being always in each other's thoughts . . . We not only hear from each other; we actually touch." Greater connectivity would also encourage cultural homogeneity, *Times* editors believed, as the cable meant that "America cannot fail to live more in Europe and Europe in America [such that f]or the purposes of mutual intercourse the whole world is fast becoming one vast city." Indeed, the "three old Continents" were now considered "a single mass."[10]

Furthermore, the sense of life speeding up – dramatically illustrated with the 1872 completion of a telegraph connection cutting from 45 days to 24 hours the time to send messages between Britain and Australia – led to a growing conviction that rapid communication would lead to "a quickened and more energetic life[, for t]he more men are brought together, the keener does their life become, the more vigorous, rapid, and energetic are their thoughts" (*Times* July 30, 1866: 8), sentiments which resonate with contemporary claims that, in the future, the "process of globalization will move much faster" (Bryan & Farrell 1996: 153). Equally, there is a discernible corollary between the current appropriation of the language of physics to suggest that contemporary technologies are allowing capital to become so mobile that a "superconductive" (Bryan & Farrell 1996: 154) capitalism is being created and nineteenth-century suggestions that governments give "the electric fluid" of the telegraph "the freest possible course through their countries" to carry information about markets and other "intelligence of universal interest" (*Times* September 21, 1866: 6). Not only would the telegraph bring people together, then, but allowing this flow of information would be of such immense benefit to mercantile interests that many could "well conceive the capitalists of England and America being anxious to make ventures for the success of the enterprise, for that success will be worth millions to them" (*Times* July 27, 1866: 9) – a fact highlighted when, in the 1870s, US newspapers began regularly reporting on the previous day's European stock market activities.

Other technological advances had similar impacts upon people's sense of space and time. Hence, the railroads not only revolutionized how markets operated by allowing capitalists to turn over their capital more readily and consumers to enjoy goods from farther afield but they also dramatically affected people's understandings of how they were connected to places near and far. As trains criss-crossed the nineteenth-century landscape at ever greater speeds, there emerged a growing feeling that society was plunging full-throttle into the future (Schivelbusch 1986). Myriad developments, from the bicycle to the telephone to the phonograph (which allowed the dead to "speak" to the living), foreshortened people's sense of the past and contributed to their feeling they were rushing into the future, a feeling reflected in popular culture, literature, and art (Kern 1983). As is the case today, commercial advertisers played upon such sensations – an 1858 advertisement used the occasion of the first transatlantic cable's completion to suggest that Queen Victoria had asked President Buchanan, "Whose are the best SEWING-MACHINES in the United States?," to which Buchanan had replied "WEED'S PATENT" (*New York Daily Tribune* August 14, 1858: 7).

Furthermore, given that proponents of the view that we are now living in globalized times have made much of how new technologies have driven contemporary processes – Ohmae (2005: xxvi), for instance, argues that "the global economy is powered by technology" – it should not be forgotten that many technologies taken today as emblematic of the high-paced world of twenty-first-century global capitalism have their origins in much earlier times, even if they have been perfected and improved upon significantly since. For example, the first patent for an early version of what has arguably become a central icon of the modern world – the fax machine – was issued in May 1843, a full three decades before any patent for the telephone.[11] Whilst this first fax system was too costly to gain widespread popularity – though a commercial fax service (the *pantélégraphe*) operated between Paris and Lyon and Paris and Marseille in the 1860s (transmitting some 5 000 faxes in a year), as did an experimental one between London and Liverpool – later developments revived interest in it, such that by the early twentieth century newspapers in Berlin, London, and Paris were using radio waves to exchange photographs via fax. For sure, such systems were not as fast as today – in the 1920s it took six minutes to send a single page – but this suggests that what we have witnessed subsequently is simply a speeding up and perfecting of the technology, rather than a fundamental leap in it. Equally, if one of the hallmarks of contemporary globalization is taken to be an ever increasing rapidity of economic and social life and shrinking of distances, then it is important to recognize that earlier time periods have seen more rapid changes than have contemporary eras. Hence, whereas in the 1920s it took 10 days to fly from Britain to Australia, by the early post-World War II period flying times had shrunk to 48 hours and by the 1960s to some 27 hours (Thomas & Smith 2003: 10, 31) – in other words, there was a much greater relative reduction in flying times between the 1920s and the early post-war period than there was in the subsequent decades.

Several other developments taken to herald the naissance of a fundamentally new order – the existence of enormous corporations, the evolution of singular, global markets, the emergence of transnational consumer cultures and identities, decreased loyalty to places and their communities – are, likewise, little different from those seen in earlier times. For example, whereas the emergence of huge corporations is taken as indicative of globalization, an examination of the early twentieth century reveals this is not unprecedented. Hence, in 1907 United States Steel reported having assets of $1.76 billion, equivalent to almost 6 percent of the nation's GNP (gross national product), whereas relative to its share of US GDP Standard Oil's assets of some $600 million would be valued today at approximately $246 billion (Moody's 1909: 2859; US Census Bureau 1975). Similarly, by 1920 US Steel's gross sales and earnings ($1.76 billion) were greater than the GDP of Norway ($1.24 billion) and Denmark ($1.17 billion), and almost as great as that of the Netherlands ($2.16 billion), Sweden ($2.52 billion), and Mexico ($2.44 billion).[12] Although smaller, by 1905 the largest British industrial company (Imperial Tobacco) was worth £17.5 million (equivalent to some £11 billion today) (Yonekawa 1974: 241). By the same token, a harmonization of economic booms and crises had already become evident in Europe by the 1840s and labor markets became more transnationally integrated as British employers increasingly turned to foreign strike-breakers during periods of industrial unrest (Hobsbawm 1975). Equally, facilitated by the telegraph, in the 1870s financial markets in different parts of the world began to synchronize, such that in the emerging transatlantic capital market crises of over-accumulation in Britain were "solved" by exporting surplus capital to the US, and vice versa (Thomas 1973).

Patterns of FDI in the late nineteenth/early twentieth century also indicate a significant degree of transnational interest on the part of corporations. For instance, prior to the outbreak of World War I Ford Motor Company had established branches in Argentina, Australia, Britain, Canada, and France, and by 1925 had facilities in Belgium, Brazil, Chile, Cuba, Denmark, Germany, Ireland, Italy, Japan, Mexico, the Netherlands, South Africa, Spain, Sweden, and Uruguay (Wilkins & Hill 1964: 434–5). Hall (1963: 201) indicates that in 1914 37 percent of the companies registered in England and Scotland and listed in the *Stock Exchange Official Intelligence* were operating either primarily or wholly overseas, a proportion little changed since the 1880s. For his part, German industrialist Ludwig Knoop had helped equip 122 spinning mills in Russia between 1839 and 1894, owning shares in many of them. He also opened an office in Mumbai in 1864 and had links with the Egyptian cotton market (Thompstone 1984). British textile manufacturers like Rylands ran branches in Paris and Rio de Janeiro and had mills in India, whilst other British companies owned distilleries, tea plantations, salt works, mines, and newspapers in southern and eastern Africa (Chapman 1992). Already in 1870 there was $1.4 billion of long-term foreign investment in the United States [2005 = $2.2 trillion], of which $243 million [$389 billion] was invested in railroads and $10 million [$16 billion] in mining activities, whilst by the late 1880s over $200 million [$185 billion] net worth of foreign investment was flowing into the US annually. British overseas investments totaled £1.3 billion by 1881 (roughly equivalent to the UK's GDP), whilst by 1913–14 British, French, German, and Dutch foreign investments constituted approximately 25 percent, 15 percent, 10 percent, and 18 percent respectively of these countries' national wealth (Wilkins 1989: 91, 151, 158).[13] By 1914, US manufacturers had $200 million of FDI in Europe, $221 million in Canada, $10 million in Mexico, $20 million in the Caribbean, $7 million in South America, $10 million in Asia, and $10 million in Oceania (Wilkins 1974: 31).[14] Indeed, the level of capital flows between countries was sufficient to lead Russian revolutionary Leon Trotsky (1970 [1930/1906]: 107) to argue in 1905 that "capitalism has converted the whole world into a single economic and political organism."

In similar vein, Hirst and Thompson (1999: 28) have shown that, relative to their GDPs, international capital flows amongst the world's wealthiest nations were actually higher in the 1880s and 1910s than they were in the early 1930s and the 1960s, 1970s, and 1980s. Equally, whereas the volume of trade (which can be taken as a measure of how open is the world economy) amongst the world's leading industrial economies expanded 4.3 percent per annum between 1853 and 1872, it expanded only 3.1 percent between 1872 and 1899, 3.9 percent between 1899 and 1911, 0.5 percent between 1913 and 1950, 9.4 percent between 1950 and 1973, and 3.6 percent between 1973 and 1984 (Hirst & Thompson 1996: 22). More recently, the volume of world trade increased by an average of 7 percent per annum from 1988 to 1997 but only 6.4 percent between 1998 and 2005 (with a yearly high during this latter timeframe of 12.1 percent [2000] and a low of 0.3 percent [2001]) (International Monetary Fund 2006: 205). These figures illustrate, then, that growing economic openness – at least measured by the degree of trade – has proceeded in quite identifiable fits and starts. Moreover, as Madsen (2001: 848, 851) has shown, there have been periods in recent history where the world has actually become more, rather than less, closed – imports and exports between the industrialized nations declined about 30 percent between 1929 and 1932 and nominal trade for the world as a whole declined about 50 percent. More contemporaneously, although the value of international capital flows has increased over the course of the 1990s, from $1.17 trillion to $5.57 trillion, by 2003 it had fallen to $4.23 trillion, largely the result of a significant (57 percent) cutback in global FDI (McKinsey & Company 2006b). Concomitantly, whilst the business cycles of the major national economies enjoyed growing synchronization in the mid-1970s and early 1980s – which Hirst and Thompson (1999: 41) take as an indication of increased integration and openness – this process had reversed itself by the late 1980s.

Putting all of this together suggests several significant things. Certainly, it casts doubt on the claim that many contemporary developments in the world economy are novel. Furthermore, the fact that rates of world trade expansion have increased and decreased at varying speeds means that rather than there having been a relatively smooth and unidirectional process whereby the world economy becomes indomitably more integrated, a more accurate representation would be one in which processes of integration are seen to play out in historically uneven fashion – whereas sometimes economic integration seems to rush at us with the speed of a greyhound, at others the pace is more snail-like (Figure 3.4). Moreover, it suggests that prior historical periods may sometimes experience greater states of integration than do those that come after, as protectionist laws (like the 1930 Smoot–Hawley Tariff Act), wars, and economic depressions reduce openness.

Is "globalization" so global?

Whereas, then, there seems to be a certain degree of similarity between some of what is taken to characterize contemporary globalization and past situations and processes, a second set of issues concerning claims that we are either in, or are heading inexorably towards, a globalized world relates to the question of just how global such

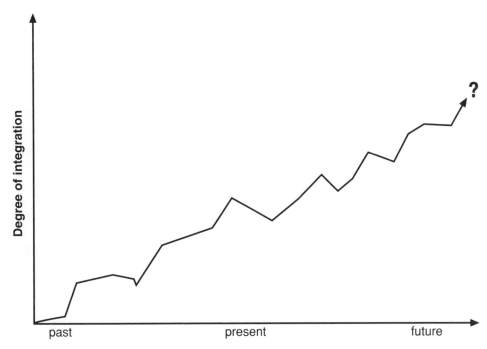

Figure 3.4 Global integration occurs in a historically uneven manner, being faster in some times and slower in others – or even moving in reverse.

trends towards globalization in fact are. For example, despite the rhetoric that "national borders are far less constrictive than they once were" (Ohmae 2005: 20) and that firms are no longer confined to national horizons but are now free to explore global opportunities, by the early twenty-first century – nearly two decades after Ohmae's declaration that globalization had arrived on the world stage – fewer than 5 percent of the US's approximately 5 700 000 firms were involved in trade beyond the United States and fewer than 3 000 (0.05 percent) had overseas affiliates.[15] Furthermore, of the firms engaged in exporting, 58 percent traded with only a single foreign country and only 0.3 percent shipped to 50 or more countries, although this latter group did account for 49 percent of the known export value (US Census Bureau 2006: 5). More US companies (83 659) exported to Canada than to any other country, followed by Mexico (42 521), the UK (40 078), Germany (28 174), and Japan (27 626), with these countries having remained the same five largest recipients of US exports for several years. Moreover, these five accounted for 52 percent of the total value of US company exports in 2004 (US Census Bureau 2006: 21). Meanwhile, there were only 24 600 affiliates of foreign firms operating in the US in 2002, compared to almost 225 000 in the People's Republic of China (including Hong Kong and Macao) and 199 000 in the European Union (UNCTAD 2005: Table A.I.8). Indeed, the US economy – the world's largest – ranks fairly low on the UN's "Transnationality Index," below that for China, many European countries, and even the average for "developed nations" (UNCTAD 2005:

16).[16] Put another way, much of the economy of the nation which is frequently seen as a key driver of neoliberal globalization and which has the world's largest economy remains decidedly nationally oriented, and that which is looking beyond its borders is not looking particularly far or wide.

A broader examination of current capital flows likewise puts into question claims that a singular, integrated global economy – "one grand continent of opportunity," in Ohmae's (2005: 81) words – is emerging, even as the volumes of such flows may be increasing. For instance, although net inflows of FDI to what the International Monetary Fund calls "developing countries" increased 76 percent annually on average between 1995 and 2001 (from $122.4 billion to $215.4 billion) – thereby suggesting a possible spatial broadening of corporate investment strategies – inflows to industrialized nations increased by 150 percent (from $205.5 billion to $513.8 billion). Of the $5.9 trillion FDI invested globally between 1990 and 2001, only 27 percent went to "developing countries," with 48 percent of that ($764.0 billion) going to Asia and $349.8 billion going specifically to China (excluding Hong Kong). In the case of the Western hemisphere, of the $510.9 billion FDI invested in what were considered to be "developing countries" 69 percent went to just three nations – Argentina ($79.9 billion), Brazil ($158.5 billion), and Mexico ($116.4 billion). Africa, on the other hand, was largely redlined, receiving only 1.2 percent of global FDI flows during this period (International Monetary Fund 2003: 9–10). As a result, in 2004 the G-3 nations (the US, the EU-15, and Japan) still accounted for 74 percent of FDI outflows and 46 percent of inflows, whereas the world's 50 "least developed" countries accounted for 0.02 percent and 1.7 percent respectively (UNCTAD 2005: Table B1). Meanwhile, in 2001 about half the global stock of FDI came from just four countries – the US, the UK, France, and Germany. Only 12 percent (some $800 billion) of the world stock of outward FDI represented investment from developing countries (International Monetary Fund 2003: 12). Measuring FDI flows in terms of numbers of facilities rather than value reveals comparable patterns, with the "developed" nations receiving 43 percent of all new and expansion FDI projects in the early twenty-first century and the "transition" economies of Eastern Europe and the former Soviet Union receiving 10 percent. Although Asia and Oceania received 35 percent of the world total, nearly half of this went to China. Africa, for its part, received 3 percent (Table 3.2).

Similarly, an examination of global financial stock shows that whereas its value increased from $53 trillion in 1993 to $69 trillion in 1996, to $96 trillion in 1999, and to $118 trillion in 2003, in terms of such capital's geographical spread there has been a growing concentration in the US and Europe: in 1993 the US and Europe accounted for 36 percent and 27 percent respectively of global totals, whilst by 2003 these figures were 37 percent and 31 percent. For their part, Japan saw its share fall from 23 percent to 15 percent and China increased its proportion from 3 percent to 4 percent. The rest of the world, meanwhile, only increased its portion of the planetary total from 11 percent to 12 percent (McKinsey & Company 2005: 17). Despite Ohmae's (2005: 111) claim, then, that "[a]rbitrage and leverage make sure that any residual national feeling in the money markets is dissolved," the fact that between them the G-3 controlled 83 percent of global totals indicates that there is a distinct topography to

Table 3.2 Number of new and expansion FDI projects (announced and realized) 2002–2004

	2002	*2003*	*2004*	*Total*
World	5 656	9 303	9 796	24 755
Developed countries	2 721	3 843	4 070	10 634
EU	1 770	2 565	2 851	7 186
UK	326	414	482	1 222
Hungary	210	231	211	652
Spain	153	215	241	609
Germany	130	264	247	641
France	126	155	201	482
Poland	91	154	230	475
US	414	588	578	1 580
Canada	218	241	223	683
Japan	106	132	152	390
Australia	137	180	139	456
Developing countries & territories	2 355	4 446	4 758	11 559
Africa	169	306	262	737
L. America & Caribbean	562	794	794	2 150
Brazil	175	287	258	720
Mexico	129	168	154	451
Argentina	44	64	73	181
Asia & Oceania	1 624	3 346	3 702	8 672
China	581	1 299	1 529	3 409
Hong Kong	57	90	122	269
Macao	2	3	6	11
India	250	457	685	1 392
E. Europe and CIS	580	1 014	968	2 562
Russia	202	432	377	1 011

Source: UNCTAD (2005: Table A.I.2).

the international financial market, as capital flows follow particular contours of the world economy.

In terms of the planetary distribution of overall wealth generation, just five nations – the US, the UK, Germany, France, and Japan – accounted for 37 percent of global GDP in 2005, despite having only 10 percent of the world's population. For their part, the 146 "emerging market and developing countries" contained 85 percent of the planet's population yet controlled only 48 percent of its GDP (64 percent and 32 percent without China), whilst the 29 most indebted "poor countries" contained 7 percent of world population but controlled only 1 percent of global GDP (International Monetary Fund 2006: 170). The vast majority (171) of the world's 200 largest companies (ranked by sales) are headquartered in these same five countries (of the remaining 29, all but five originated in Europe). Moreover, there has actually been a concentration of ownership relative to the early 1980s, when these five nations accounted for a mere (!) 169 of the largest companies (Anderson & Cavanagh 2000: Table 1). Meanwhile, an examination of the 100 largest TNCs ranked by foreign assets shows that only four originated in developing countries. Furthermore, though Bryan and Farrell (1996: 153) have suggested that the "process of globalization will move much faster" as we move into the future, the pace at which these largest 100 corporations are expanding transnationally seems actually to have slowed in recent years (UNCTAD 2005: xix–xx). Finally, although many neoliberal commentators suggest that such concentrations will last only in the short term as globalization matures, such proclamations merely beg the question of how short the short term is.

If such an examination of corporate activities and global capital flows appears to throw doubt on claims that "[i]ndividual national financial markets are losing their separate identities as they merge into a single, overpowering marketplace" (Bryan & Farrell 1996: 4) and that "the global economy does not respect national borders" (Ohmae 2005: 89), several other contemporary developments raise questions about just how globalized the world is at the beginning of the twenty-first century. For instance, despite the fact that there were some 191 million people living outside the countries of their birth at the end of 2005, this number represented only about 2.9 percent of world population – that is to say, despite the image of globalization as characterized by large-scale flows of labor from locations of oversupply to locations of labor demand, in fact most people still stayed home. What is more, three-quarters of these migrants lived in just 28 nations, and one in five lived in the US (United Nations 2006: 12). Put another way, just as the migration of capital across the surface of the globe appears to be highly focused on specific geographical locations, differentially connecting a relatively few regions of the world economy, so is that of labor. Rather than seeing the emergence of a highly integrated global capital and labor market à la neoliberal commentators' claims, then, in many regards the world economy seems to be becoming more spatially compartmentalized.

Moving beyond the realm of capital and labor flows, another neoliberal claim is that regarding firms' supposed lack of loyalty to particular places – in the new globalizing business environment, it is argued, firms will shed themselves of pre-global loyalties to place. However, whereas corporations often argue they must exhibit primary loyalty to shareholders and so should be free to seek out the most profitable locations without regard to "irrational" sensibilities like people's attachment to particular locales, in reality they must often wrap themselves in a fiction of loyalty to specific places. They must, in other words, walk a tightrope between global investment opportunity, on the one hand, and people's continuing sense of – and allegiance to – place, on the other. Hence, even as increasing numbers of North American or European corporations outsource customer inquiries to call centers in, say, India, the workers responding to these inquiries are encouraged to conjure up identities which give the impression that calls are being answered more locally. Thus, it is not unusual for a US customer of General Electric, American Express, or Amazon.com to believe they are speaking to someone like "Betty Coulter," a "typical 21-year-old college graduate from Illinois [who] wears bell-bottom jeans and is a faithful fan of 'Friends' and 'Buffy the Vampire Slayer'" when, in fact, they are speaking with Savitha Balasubramanyam, a call-center worker in Bangalore who has created the persona of Betty to put her US customers at greater ease. Indeed, such is the importance of maintaining these illusions that Indian call-center workers receive "months of speech training in American or British accents, depending on the client they represent . . . bon[ing] up on sports terms and slang and a good dose of 'Baywatch' and 'Friends' to bridge the cultural divide between Bombay and Boston" (*Island Packet*, July 1, 2001: p. 8A).

Equally, whereas a company like McDonald's operates in nearly half of the countries of the world and is often seen to epitomize global standardization – it is no accident that Barber (2001) has referred to rising localist cultural tendencies in terms of "Jihad versus McWorld," nor that Friedman (2000: 248), in what he calls his "Golden Arches theory of conflict prevention," has argued that no two countries with McDonald's have fought wars against each other since McDonald's arrived, nor that *The Economist* (1986) has derived a "Big Mac Index" to measure the relative costs of living across the planet – its menus actually vary significantly across the globe.[17] Hence, the company sells chicken and vegetarian burgers in India, beer in Europe, a McOz burger (containing beetroot) in Australia, a McKofta meatball sandwich in Pakistan, gazpacho in Spain, and operates kosher and non-kosher restaurants in Israel. Moreover, even within particular countries there are significant variations in the company's menu – pineapple in Hawai'i, crabcakes in coastal Maryland, lobster rolls in New England, and a cajun chicken sandwich in Louisiana. Such variety, then, suggests that, contra the image presented by many hypers of globalization, even in a world in which "the Sun never sets on the Golden Arches" – and revising Gillette CEO Zeien's dictum – it is still possible to find foreign countries feeling foreign, even whilst sitting in a McDonald's. Furthermore, the fact that McDonald's has recently closed its stores in several Latin American and Caribbean nations (including Bolivia, Jamaica, and Trinidad and Tobago) would seem to challenge, to some degree, the argument that contemporary processes of "globalization" are unidirectional.

Between Internationalization and Globalization

Clearly, there is significant debate as to the degree to which the world has become globalized. For many, we have entered a period of unstoppable globalization, a new era in the development of humanity from which there is no retreat. Some of these commentators have happily welcomed contemporary events, suggesting that not only have we crossed over a historical event horizon and are now firmly entrapped in globalization's gravity, but that even if it were possible to undo history, to turn back the economic clock, we should not do so. Others approach such events with a sense of resigned inevitability, not happy about them but convinced that there is little option but to accept the new rules and realities of the global economy. For both such groups, most – if not all – contemporary developments are generally attributable to "globalization," which is seen as simply the next natural stage in the social evolution of civilization. In such a view, although there may be differences across the planet in how things are, the world is only really intelligible as a single, integrated unit wherein the global is "a wholly autonomous domain of thought and action" (Bartelson 2000: 191). On the other hand, for many others globalization "is little more than a mirage" (Dicken et al. 1997: 159). Hence, for writers like Hirst and Thompson (1999), contemporary events are little more than a replay of late nineteenth-century ones; there seems to be no major tendency towards the growth of truly global companies; capital mobility is not producing a significant equalization of global economic development; and what cultural homogeneity is emerging tends to be seen only in the consumption patterns of a small, privileged minority of the global population.

Several important matters emerge out of this discussion. The first relates to questions of what, materially, is happening around us. Thus, some writers have sought to tally various economic, demographic, technological, political, cultural, and other measures to form myriad "Globalization Indices" so as to "prove" either that we have or have not entered a period of globalization. This can certainly be an interesting exercise, but it misses a crucial element: it rests entirely on the definitions of globalization with which we begin our empirical analysis – the broader our definition of "globalization," the more things we will take to be evidence of our subjection to it. For instance, is having a planetarily integrated market the same thing as experiencing globalization? Is growth in the use of the internet or purchase of goods from abroad equivalent to globalization? We cannot, in other words, escape the power of categorization.

The second issue concerns how we understand globalization as a temporal and spatial process. Thus, even accepting that we can come up with a definition of what it means to be globalized and can use that to measure empirically how far along in the process we are, it is important to not evaluate claims of globalization in either/or terms – to claim that *either* we live in a world which is globalizing/globalized *or* we live in one which is not. Rather, unless we see no distinction between the terms "globalization" and "internationalization," it is important to recognize that aspects of globalization, internationalization, and even "deglobalization" and "deinternationalization" can be occurring simultaneously. This is not to suggest a middle-of-the-road position in which our world is more than internationalized but not yet globalized. Rather, it is to recognize that social processes have historical geographies to them. Hence, unless we believe that globalization can appear instantaneously on the world scene fully formed, then we have to accept that processes of economic integration and political and cultural transformation play out unevenly in space and time – even as some economies appear to be becoming ever more globally integrated (the G-3), others are largely bypassed (Africa). There is, in other words, a complicated historical geography of global integration and disintegration, one not readily captured by "either/or" assessments.

The final issue takes us back to questions of discourse, though in a different way from the first. Specifically, even if we can agree on what constitutes "globalization" and are able to determine empirically to what degree, if any, there has been a qualitative change in the world economy recently, how is the *idea* of globalization used to shape how we think about contemporary processes and what does this mean for how we make sense of them? Regardless of the empirical reality of the world, whose interests are served by arguing either that we have entered an era of globalization or that we have not? These are crucial points, for if we believe that we live in one kind of world rather than another (a globalizing rather than a non-globalizing one), then that will affect our behavior, thereby perhaps encouraging a degree of self-fulfilling prophecy. It is to questions of how globalization is represented discursively, then, that I now turn.

Questions for Reflection

- How is the nation-state implicated in the concepts of internationalization and globalization?
- How does viewing the growing geographical connectivity between different parts of the globe as instances either of internationalization or of globalization shape how we interpret this growing connectivity?
- How does the way in which we define either "internationalization" or "globalization" shape how we interpret empirical changes in the world economy's operation?

Notes

1. Hobsbawm (1990) argues the French nation-state had already been constituted through suppression of regional dialects and languages like Provençal and their replacement with a standardized "national" French language before a sense of something called "the French people" emerged, whereas in countries like Germany it was the recognition of the existence of something called "the German people" which facilitated the German nation-state's creation.
2. Significantly, in some nations this convergence was achieved quite late. Thus, Varsanyi (2005: 124) argues that it was only by about 1970 that there emerged a "historically unique convergence between the physical territory of the US, its citizenry, and (*de jure*) its enfranchised population . . . In other words, unlike prior periods . . . in which, for instance, not all persons resident within the territory of the US were permitted to be citizens (for example, Native Americans, African Americans, Chinese), not all citizens were able to vote (for example, women), or those who were not citizens were able to vote (for example, declarant alien suffrage), by 1970, non-citizens could no longer cast votes at the state scale, the number of non-citizens in the population was at a historical low, the vast majority of persons living within the territory of the US were citizens and those persons were guaranteed the right to vote, at least by law."
3. Another globalization guru, *New York Times* columnist Thomas Friedman, argues that fundamental changes in the world economy's structure can be dated to around 2000, although he does point to the 1989 collapse of the Berlin Wall and the 1995 public

tendering of Netscape as also being crucial events (Pink 2005). A famous 1998 news-paper advertisement by investment firm Merrill Lynch, which ran in several US news-papers, argued that, "The world is 10 years old. It was born when the Wall fell in 1989. It is no surprise that the world's youngest economy – the global economy – is still finding its bearings."

4.   2005 figure calculated by the author, using data from Forbes (2006) and World Bank (2006) databases.

5.   "55 times" figures is based on the following: Bank for International Settlements (2005: 1) for average daily currency traded; WTO (2006: 11–12) for trade figures (according to the WTO, $10.12 trillion worth of merchandise and $2.42 trillion of commercial services were traded internationally in 2005, giving a total of $12.54 trillion in trade and a daily average of $34.36 billion). The World Federation of Exchanges (2006: 70) lists the value of equities traded in 2005 at $51 052 238 400 000; assuming 250 trading days per year gives a daily volume of $204.2 billion.

6.   Milunovich and Thorp (2006) suggest this asymmetry indicates that cultural connections to the US are more important in shaping the early performance of the London market than is overnight news from Tokyo.

7.   Travel's cheapening is illustrated by the fact that in 1945 it cost the average worker 130 weeks' worth of wages to fly from Sydney to London, whereas in 2003 it cost two (Thomas & Smith 2003: 181).

8.   Such senses have perhaps been exacerbated by US parents' ability to hire on-line tutors in India to help their children with school work (*New York Times*, September 7, 2005) or of computer gamers' option to outsource to Chinese workers the job of laboring through early stages of various games so that they may themselves begin several levels into them (*New York Times*, December 9, 2005).

9.   *Times* of London, August 6, 1858 (p. 8), September 21, 1866 (p. 6), and July 30, 1866 (p. 8).

10.   All quotations are from the *Times*, July 30, 1866 (p. 8), except for those of the last sentence, which are from the *Times*, July 27, 1866 (p. 9).

11.   Invented by Scottish watch maker Alexander Bain, the machine consisted of a pendulum swinging over metallic type to close an electric circuit. When wired to a second machine elsewhere, closing the circuit on the first caused an electrical current to flow to the second, which left a brown image on a piece of chemically treated paper positioned beneath the pendulum tip (Burns 1993). Although Bain never himself sent a fax, the English physicist Frederick Bakewell demonstrated a working machine at the 1851 Great Exhibition at London's Crystal Palace.

12.   Figures calculated by the author. Data on national incomes come from Mitchell (2003a, 2003b, 2003c). Information on exchange rates comes from Carter et al. (2006). All figures are for GDP and 1920, except: the Netherlands (Net National Product); and Mexico (1925). In 1920 the US's GDP was $91.5 billion; in 2005 it was $12.46 trillion.

13.   Wilkins (1989: 158) provides a figure of $6.3 billion for "British" (England, Wales, Scotland, and Ireland) overseas investment, which I have converted back to sterling using the rate she provides of £1 = $4.86. The figure for the UK's 1881 GDP comes from Mitchell (1988: 832–3). In 2005 the UK's GDP was £1.2 trillion (Office of National Statistics 2006: Table A1) – that is, the amount of overseas investment in 1881 would be equivalent to £1.2 trillion as a proportion of 2005 GDP. It is important, however, to recognize that there is methodological dispute concerning how levels of early FDI are measured. For

instance, should investment in overseas imperial possessions be characterized as FDI or as something else, and how should portfolio investments made by overseas expatriates be counted (Winder 2006)? My purpose here is not to evaluate which method is better. Rather, it is to indicate that, even using the narrow definition favored by Wilkins (1988) herself, amounts of FDI at this time were substantial.

14. As a proportion of the US's 2005 GDP, these amounts would be worth $68.3 billion (Europe), $75.5 billion (Canada), $3.4 billion (Mexico), $6.8 billion (Caribbean), $2.4 billion (South America), $3.4 billion (Asia), and $3.4 billion (Oceania).

15. Figures calculated as follows: the US Census Bureau (2003) records 5 767 127 firms with paid employees operating in the US in 2003 and 5 607 743 in 1999. The US Census Bureau (2006: Table 1a) lists 231 736 firms engaged in some form of export transaction in 2004, although it suggests this may understate slightly the number. Even allowing for a 20 percent undercount and recognizing that the data are for two separate years, this means that only about 4.96 percent were involved in exporting. The US Bureau of Economic Analysis (2004: Table 3) gives a total of 2 606 US firms with foreign affiliates in 1999.

16. On a scale of 1 to 100, and using a weighted average to account for different ways of measuring "transnationality," the US scored only an 8, whereas the UK scored 17, New Zealand 27, Slovakia 28, Estonia 39, and Ireland 69. China scored 11, whilst the averages for the "developed" and "developing" countries as a whole were 11 and 13.

17. A review of history illustrates the faultiness of Friedman's claim – Israel and Lebanon, and India and Pakistan, have engaged in conflict since McDonald's franchises opened. Friedman (2000: 252), however, defends his theory by arguing that it should not be taken literally, that it is "simply a metaphor for a larger point about the impact of globalization on geopolitics."

Further Reading

Barber, B. R. (1992) Jihad vs. McWorld. *The Atlantic* March, pp. 53–5.

Dicken, P., Peck, J., and Tickell, A. (1997) Unpacking the global. In: R. Lee and J. Wills (eds.), *Geographies of Economies*. London: Edward Arnold, pp. 158–66.

Huntington, S. (1996) The West unique, not universal. *Foreign Affairs* 75.6: 28–46.

Leyshon, A. (1997) True stories? Global dreams, global nightmares, and writing globalization. In: R. Lee and J. Wills (eds.), *Geographies of Economies*. London: Edward Arnold, pp. 133–46.

Electronic resources

Organisation for Economic Co-operation and Development Statistics Portal (www.oecd.org/statsportal/0,3352,en_2825_293564_1_1_1_1_1,00.html)

United Nations Conference on Trade and Development World Investment Reports (www.unctad.org/Templates/Page.asp?intItemID=1485&lang=1)

Chapter 4

Talking Globalization

Chapter summary: This chapter examines discourses of geographical scale, especially how globalization is frequently presented as eviscerating other scales. I explore how scale is conceived of – do scales really exist or are they merely convenient mental devices for dividing up the world? – and how "the global" is generally portrayed as more powerful than "the local." Finally, I ponder how different representations of the relationship between the local and the global (and other scales) shape how we conceptualize globalization.

- Representing Global Capitalism
- The Scales of Globalization
  - Conceptions of geographical scale
  - Rhetorics of scale
- Placing Scale

In the "global village" . . . there is a temptation to see everything in global terms.
(F. Mayor, Comments, p. 437)

The global village . . . is the fantasy of the colonizer, not the colonized.
(L. Spigel, "The suburban home companion," p. 217)

So far, I have explored how the globe as an object has been represented discursively and how numerous technological and social developments have transformed the geographical relationships between different parts of the planet at different times, together with how these developments have been conceptualized variously as internationalization or as globalization. As I have argued, much of the power of contemporary neoliberal discourses lies in representing the globe as an increasingly integrated, seamless whole. Central to this representation is the portrayal of globalization as a process whereby other spatial scales are eviscerated – globalization, in other words, is the "delocalization" (Virilio 1997; Gray 1998) and/or "denationalization" (Sassen 2003) of economic and political life.[1] Such issues of discourse and representation are important because how we think about contemporary processes shapes how we act in the world.

Given this, here I investigate two aspects of discourses surrounding how globalization has been represented by many. First, I survey how the neoliberal advocates of what has come to be called globalization have characterized the planetary spread of capitalist social relations. For them, the "marketization" of economies across the globe – particularly in Eastern Europe and the global South – is equivalent to globalization, whilst globalization is seen as being about spreading capitalism and markets planetarily. There is, in other words, a significant conflation of these terms, a conflation which has considerable implications for understanding what is happening. Second, I probe how the term "the global" is used discursively to shape perceptions of contemporary economic, political, and cultural processes, particularly when contrasted with what is often seen as its Other: "the local." Considering how we think about relationships between different geographical scales like "the global" and "the local" (and others, like "the national" and "the regional") is important because it goes to the heart of what it is to talk about how globalization is supposedly "re-scaling" the geographical resolutions at which our lives are lived. Thus, if globalization is portrayed as a process whereby social, economic, and political relationships are "delocalized," what does that actually mean? Equally, how does imagining the relationship between "the global" and other scales in different ways carry implications for understanding what "globalization" is about?

Representing Global Capitalism

One of the most important ways in which the process of globalization has been presented by its neoliberal advocates as normal and natural – and, therefore, unchallengeable – has been through the representation of capitalism as normal and natural. Rather than being seen as a form of economic organization that has evolved out of a particular time and place – often argued to be late medieval Europe – capitalism is frequently presented as a geographically universal and historically eternal "thing."[2] This representation has been furthered by the common conflation of two quite different entities: capitalism and markets. Hence, in most neoliberal discourse, anywhere markets operate is seen to be capitalist, whereas wherever there is capitalism it is assumed that "free" markets are in operation. Clearly, this is an absurd conflation.[3] Markets allowing people to buy and sell things have existed in non-capitalist societies – both under feudalism (i.e., "before" capitalism) and within the centrally planned economies of the Soviet Union, Cuba, the People's Republic of China, North Korea, etc. – without such societies being viewed as "capitalist" economies. Equally, the existence of capitalist social relations does not necessarily imply the operation of functioning markets. Indeed, it is actually the case that capitalist firms generally prefer not competitive "free" markets but, instead, monopolies, to such a degree that government regulations (e.g., the 1890 US Sherman Antitrust Act) have frequently been necessary to ensure that firms do not develop monopolies and that markets operate "as they should."[4] Nevertheless, such a conflation is a powerful one and, as we shall see, allows a number of narratives to be constructed about globalization. In order to explore how this conflation has been used, however, it is first necessary to give a definition of what "capitalist" social relations – and hence "capitalism" – are. Many definitions exist. However, I choose here to focus upon what I consider the core element of a capitalist mode of production, namely that workers sell their capacity to labor to others and that, in the process, they do not receive remuneration equal to the value they create through this labor (Wolff & Resnick 1987).[5] Such a means of one class extracting surplus labor from another, then, requires that workers do not own the means of their own subsistence and must sell their labor for a wage. Significantly, this is quite different from other ways of organizing the extraction of surplus labor, such as under European feudalism where peasants did not toil for a wage but nevertheless were expected to provide labor to their lords.[6]

With such a definition it is now possible to mark the emergence of capitalist social relations as having a particular historical geography, one related to the historical development and spread of wage labor. Rather than being eternal and universal, then, capitalism has particular origins and has spread over time and across space. Equally, such a definition allows for a more decentered and nuanced understanding of contemporary capitalist economies and how they operate. Hence, "capitalist" economies exhibit, in fact, a number of different types of economic relationships, some of which are capitalist and some of which are not. For instance, when people are engaged in producing goods and own the products of their labor in proportion to the amount of labor they have expended (as in a co-operative) or are self-employed and hire no other workers from whom they extract surplus labor, then they are not engaged *directly* in capitalist relations of production, although they may nevertheless still be embroiled in capitalist social relations (perhaps through using raw materials which have been produced by others under such relations). Likewise, putatively "non-capitalist" economies may exhibit capitalist practices, such as when workers in countries like China generate more value than they are paid for, even if the product is owned by the state (and, therefore, purportedly by "the people"). However, the tendency to conflate markets and capitalism, such that any buying or selling of goods or services is seen as "capitalist," has the effect of making capitalism appear virtually omnipresent, regardless of the manner in which the goods and services were actually produced. In such a conflation, even barter economies are seen as "proto-capitalist."

Within this conflation of capitalism and markets, combined with representations of "capitalist economies" as only being made up of "capitalist" forms of economic relationships, capitalists and capitalism appear hegemonic. Thus, Gibson-Graham (1996: 4) argues that neoliberal discourses have generally presented capital and capitalism as "large, powerful, persistent, active, expansive, progressive [i.e., bringing "modernization"], dynamic, transformative; embracing, penetrating, disciplining, colonizing [of pre-capitalist spaces], constraining; systemic, self-reproducing, rational, lawful, self-rectifying; organized and organizing, centered and centering; originating, creative, protean; victorious and ascendant; self-identical, self-expressive, full, definite, real, positive, and capable of conferring identity and meaning." Such active adjectives present a picture of capital as almost a living and breathing entity, one that is basically omnipotent, omniscient, and omnipresent. Through the conflation of capitalism with markets, this representation has been extended to "the market." Thus, many commentators have suggested that markets themselves (rather than the individuals which constitute them) have agency – Bryan and Farrell (1996: 21), for instance, maintain that "markets freed themselves" from the Bretton Woods system which regulated international financial markets after World War II (see Chapter 7), whereas global capital markets are seen by some as "an independent, motive force" (McKinsey & Company 2005: 7). Significantly, Gibson-Graham argues, these representations have frequently been adopted and promulgated by those on both the political right and the left – those on the right have done so to make capital appear hegemonic, whereas those on the left have internalized such a view, either to make their challenges to globally organized capital appear more heroic or because they truly feel overwhelmed by the thought of articulating such challenges. Delving more deeply into such representations, Gibson-Graham explores how such narratives have been constructed and with what effect. Five elements are particularly pertinent.

First, Gibson-Graham (1996: 7–9) argues that capital is typically represented as the "hero" of the industrial development narrative, the driver of history, the bringer of the future, and the liberator of humanity from the struggle with nature. Central to these claims is the image of an anthropomorphized capital, with capital and capitalism represented as almost human, if not divine. Thus, rather than people making decisions, it is "global capital" and/or "global capitalism" which acts. Arguably, this representation is clearest in the metaphor which is most frequently associated by neoliberal commentators with the operation of capitalist markets, namely Adam Smith's (1961 [1776]) famous "invisible hand" of the market. Through its actions, then, capital and capitalism are seen as the agents who will take us into a bright new future, one in which scarcity and the eternal struggle with nature are overcome – capitalism's geographical extension will universalize "modern" ways of living wherein humans will not be oppressed by the misfortunes of geography or climate but will prevail over them.

Second, and relatedly, capitalism as an economic system is represented as the acme of social evolution, such that attaining this evolutionary zenith will bring with it the end of scarcity, of "backwardness," of ignorance and superstition, of anti-democratic and "primitive" political forms. Ultimately, this is a Hegelian view, one in which the emergence of global capitalism heralds the materialization of humanity's final and most favorable means of arranging economic, political, and social life.[7] Arguably, it is a view most notably expressed in Fukuyama's 1992 book *The End of History and the Last Man*, in which he suggests the Cold War's end has marked capitalism's final victory over alternative ways of structuring economies and that from now on there will be no serious challengers to this model of economic organization. Such a view lies at the heart of discourses about "modernization" – as embodied in "modernization theory" – and arguments that societies in the "developing world" must adopt Western economic and political practices if they are to become developed, civilized, and modern.[8] Indeed, Slater (2002) has argued that current efforts to promote neoliberal globalization and encourage/impose a Western model of development in/on the rest of the globe are merely the latest embodiment of the modernization theory popular in the 1950s and 1960s and are equally universalist – a universalization of the Western experience much in evidence in books like *The Triumph of the West* by John Roberts, who has argued (1985: 431) that "the story of western civilisation is now the story of mankind, its influence so diffused that old oppositions and antitheses are now meaningless."

Such a discourse incorporates two significant spatial imaginaries. First, if the mark of modernity is the adoption of Western-style economic, political, and cultural forms, then modernity diffuses from the "advanced" core regions of the globe to the "less developed" peripheral ones in a one-way direction. Second, the universalization of Western economic, political, and cultural forms through the conflation of "Western civilization" with "the story of mankind" suggests a geographical process whereby non-Western ways of living are eventually erased as the planet moves towards the kind of one-worldism discussed in Chapter 2 – at some point everywhere will become "Western" and "modern." This latter is, paradoxically, a quite aspatial vision, for it suggests that the LDCs/non-capitalist parts of the globe are simply "behind" the West in terms of their historical development and that someday they will "catch up" to become "just like us." Neo-Hegelian modernization theory, then, purports globalization will erase spatial difference by making everywhere Western and modern (i.e., capitalist), such that traveling from the West to, say, sub-Saharan Africa will no longer be viewed as allowing one to journey, as it were, back in time to view how "pre-Modern" societies function.[9] In so doing, though, it also ironically erases history – the pre-Modern parts of the globe are seen not as having fundamentally different experiences from the West but, instead, as simply slower to drag themselves up the greasy pole of economic and political development to reach the pinnacle of social life: "Modernity."

Within this representation of capitalism's arrival as heralding modernity and the erasure of "primitive" political forms is the frequent combining of two distinct things – capitalism and democracy. Indeed, Frank (2000) argues that neoliberal globalization has relied for its rhetorical power upon inculcating the idea that capitalism is the economic expression of democracy whilst democracy is the political expression of capitalism. The result, he suggests, has been the stoking of a "market populism" by many neoliberals, one in which it is assumed that democracy and capitalism are the political and economic systems people naturally favor. Although there are myriad examples of such a joining of rhetorics of democracy and capitalism in the neoliberal imagination, the 2005 signing by President George Bush of the Central American Free Trade Agreement (CAFTA) clearly highlighted how these are often presented as two sides of the same coin. For Bush (2005a), then, CAFTA was not just a trade bill but was "a commitment among freedom-loving nations to advance peace and prosperity throughout the region." It would "not only provide more prosperity in our hemisphere" but would "help spread democracy and peace [by] strengthen[ing] democracies," thereby "lead[ing] to greater security and stability" (Bush 2005b). The CAFTA bill was, he suggested, "pro-jobs, pro-growth, and pro-democracy."[10]

The third way in which global capitalism is frequently represented in neoliberal discourse, according to Gibson-Graham, is as a unified system or body, one that is bounded, hierarchically ordered, vitalized by a growth imperative, prone to crisis (a crisis often presented in the language of disease) but always capable of self-recovery to a higher plane of existence (as represented in the common refrain that stock market collapses are market "corrections," the implication being that they allow markets to get back "on the right track"). The use of biological discourses to talk about global capitalism – "capitalism as a unified *body*" – has a long history, both in popular and academic writing (cf. Foucault 1971). Indeed, Amariglio (1988: 585) argues that "modern economic theory in most if not all of its variants (neoclassical, Marxian, Austrian, Institutionalist, Post-Keynesian, and so forth) is grounded partly in a concept of the body" and suggests that it is possible even to "differentiate alternative positions in economic theory since [Adam] Smith and [David] Ricardo in terms of the type of body and its internal relations." Contemporary examples of such biological imagery range from the widely used expression that if "America sneezes, the rest of the world will catch a cold" to arguments that globalization represents "the cancer stage of capitalism" (Burgmann 2003). Invoking the image of bio-engineering, Ohmae (2005: 245) has even argued that if companies are to succeed in the new global economy they must be "genetically different" from the companies of the pre-global age – they must have "a different set of chromosomes." Less recent examples include Marx's (1973 [1858]: 517) comparison of the circulation of blood with the "content-filled circulation of capital" and seventeenth-century English political theorist (and physician) John Locke's explicit matching of the circulation of money through the economy to the circulation of blood through the human body (Finkelstein 2000). Other commentators, whilst not explicitly equating the economy with biological entities, have nonetheless sought to blur the boundary between the biological and the social. Hence media critic Marshall McLuhan (1964: 3) has invoked the image of humans as cyborgs (cf. Haraway 1991) in describing how, beginning with the telegraph, we "have extended our central nervous system itself in a global embrace, abolishing both space and time as far as our planet is concerned."

Certainly, not all writers who use biological metaphors to describe economic relations necessarily favor the spread of global capitalism. Nevertheless, there are several implications of using such metaphors. To begin with, such representations play a central, even if sometimes unintentional, role in making global capitalism – and hence globalization – appear normal, natural, and therefore unchallengeable. Arguably, one of the most deliberate examples of such a representation occurred in the previously mentioned (Chapter 3, footnote 3) Merrill Lynch advertisement celebrating the tenth anniversary of the Berlin Wall's fall, wherein the company suggested that although "no one ever said growing up was easy . . . for a 10-year-old, the world continues to hold great promise" – the birth and infancy of globalization, in other words, were argued to be as natural as those of a human child, presumably leading to the conclusion that as it grew and matured the globalized economy would stand on steadier legs. Going one step further, in some cases globalization and capitalism have even been represented as almost *super*natural and divinely inspired and/or animated. Thus, Buck-Morss (1995: 450) has argued that Adam Smith's use of the term "invisible hand" to describe the operation of markets was itself inspired by the long tradition of "natural theology" (as promulgated by Thomas Aquinas, amongst others), "which saw effects of the hand of God everywhere in the natural world."[11] More explicit associations of globalization and divinity are presented in Goudzwaard's (2001: 20) argument that the Christian church was always intended to be a global community and that "God is guiding or administering the course of history toward that end," with the result that "God's economy entails its own style of globalization, oriented to the coming of his Messiah King." Such associations are reinforced by the fact that the Greek word used in the Bible to describe humanity's stewardship over the Earth (*oikonomia* [οικονομία]) is also the root of our modern word "economy."

Using biological metaphors, then, opens a discursive avenue to arguments that social entities (like markets) operate according to the same rules and processes as do biological ones (e.g., evolution).[12] More recently this connection has been strengthened by the emergence of business models based upon the ideas of "viral marketing" (Rushkoff 1994) and "biomimicry" (Benyus 1997), with the latter seen as a way to produce a "natural capitalism" (Hawken et al. 1999) by developing closed-loop production systems modeled on natural systems in which output is harmless to the ecosystem or serves as inputs for other manufacturing processes.[13] Although this idea is intended to allow for more ecologically sound methods of production, it nevertheless facilitates a discourse of "Social Darwinism" in which the competition driving capitalism is seen to have its correlate in an evolutionary "survival of the fittest."[14] Indeed, such a connection has been explicitly made in at least one review of Hawken et al.'s book *Natural Capitalism*: "[W]hat makes this book worth reading," the reviewer stated, "is the fact that the authors have taken as first principles for their Utopia the harsh truths of Darwinian capitalism: individuals and companies act in their self-interest, and markets guide that impulse through prices" (*The Economist* 1999: 8).

Significantly, representations of global capitalism as a unified body also draw on highly sexualized language when describing economic processes. Hence, globalization is often seen as a decidedly gendered process, both empirically – the vast majority of workers in the world's export processing zones are women – but also epistemologically. Thus, Backer (1993: 52), in describing the creation of the famous 1970s "I'd Like to Teach the World to Sing" Coca-Cola advertising campaign (see Chapter 3), states that the words of the advert's jingle were perceived to be "more a woman's wish than a man's. It was *House and Garden*, not *Field and Stream*."[15] In similar fashion, globalization has been described by Nagahara (2000: 935) as a process whereby *Monsieur le Capital* ("Mr. Capital") and *Madame la Terre* ("Mrs. Landed Property") come together at an ever greater spatial scale so that capitalists may accumulate profit.[16] Indeed, the gendering of the relationship between capital and land is such that *Monsieur le Capital* "cannot be pregnant without *Madame la Terre*," a representation not only fraught with gendered language but also one which enforces a distinct heteronormativity to the process.[17]

Such heteronormativity of narratives about globalization has also been explored by Gibson-Graham (1996), who argues that there are parallels between much of the discourse surrounding sexual intercourse and that of globalization. For example, global capital is frequently spoken of as "penetrating" non- or pre-capitalist regions of the world, in a discourse in which men are viewed as proactive sexual actors and so are associated with those actors seen to be bringing about globalization (i.e., capital and its embodiments, TNCs), whereas women are represented as capitalism's Other, metaphorically standing in for the economies or regions of the globe which rather passively await global capital's coming.[18] Indeed, in suggesting that representations of globalization frequently create a binary of a strong global capital and a weak Other (the local community, the nation-state, workers), Freeman (2001) has even posed the rhetorical question of whether the Local/Global binary is equivalent to the Feminine/Masculine binary. In such a case, if capitalist penetration of non- or pre-capitalist territories is conducted forcibly then such a narrative can quickly become what Gibson-Graham calls a "rape script." Indeed, the image of "globalization as rape" has been a powerful one, especially when linked to the destruction by outside forces of "feminized" natural environments (Merchant 1980) like the Amazon.[19]

If capitalism is represented in much globalization discourse as omnipotent, natural, and penetrating, then a fourth way in which global capital is characterized, Gibson-Graham has argued, is as unfettered by local attachments, labor unions, or national-level regulation and unfazed by having to adapt to the myriad cultural, linguistic, legal, and other differences which characterize disparate parts of the globe. Capital is seen, as it were, as footloose and fancy free, able to conquer space and easily adapt to difference. Thus, within many representations of global capitalism (those of both neoliberals and their critics) the capitalist economy is seen as the new reality by which social possibility is dictated as all things increasingly come to be measured by how they will affect "the market" and particular locations' abilities to compete globally, rather than on the basis of other considerations (like whether they are moral). In such a discourse, capitalists' globalization typically is positioned rhetorically as a naturally more powerful force than that of labor, which is seen to be hopelessly divided by language, culture, politics, law, and tradition. The result is that any talk of workers, environmentalists, or others coming together to challenge TNCs' actions is often considered doomed from the start. This is a powerful and widespread discourse, though it is one which neglects an obvious fact: in reality, TNCs themselves also have to come to terms with different languages, laws, and customs across the globe, and sometimes do so with disastrous results, at least based upon the number of advertising mistranslations and other cultural gaffes which have plagued those operating in foreign markets (Axtell 1995; Ricks 2000).[20] Certainly, TNCs may have greater resources than do workers and others to engage in translating documents or hiring lawyers to find their way around the specificities of foreign laws and regulations, but the point is that this is an empirical phenomenon that plays out in historically and geographically specific ways, rather than something that should simply be assumed as a starting point for understanding processes of globalization – some international labor organizations, for instance, have significantly greater linguistic and legal capacities than do many firms who are just beginning the process of operating overseas.

The fifth aspect fairly common to many representations of globalization is one which portrays capitalist practices and spaces as increasingly subsuming non-capitalist ones. Thus, Gibson-Graham argues, capitalist economic relations are frequently scripted as penetrating "other" economic systems but not vice versa – capitalism affects other social systems, but is viewed as largely unaffected by them. Although there are myriad such examples, recently this representation has perhaps been most clearly seen in rhetoric concerning Eastern Europe's transition. Encapsulated within a discourse of what Burawoy (1992) has called "transitology," much of the representation of post-1989 changes has presented the countries of Eastern Europe as a fairly homogenous bloc that is moving inevitably towards a market economy with the goal of transformation being "a pluralistic and democratic political system comparable to Western European countries" (Fleissner 1994: 4). This is compounded by the representation of the period under communism as a distortion of the "correct" form of economic and political organization, such that the transition, "if not stopped by a violent return to the past," will lead the region's countries "onto the path of *normal* development" (Gordon & Klopov 1992: 30, emphasis added) as institutions "deformed" under communist influence will finally be able to operate as they "should."

Such a discourse presents a narrative in which capitalism's spread into Eastern Europe is seen to have an impact on the societies of the region, yet there is little sense that their specific histories and geographies will affect the kind of capitalism which develops – it is simply assumed the region will come to resemble Western Europe in a fairly unproblematic transition from the past through the present and into the future, an assumption that erases the region's history and geographical specificities by suggesting its experience of communism will leave no lasting marks. Yet, in reality, a particular brand of East European capitalism appears to be developing. Hence, Stark (1996: 993) has argued that in Hungary the emergence of what he calls "recombinant property" – "a form of organizational hedging in which actors respond to uncertainty by diversifying assets, redefining and recombining resources" – has heralded the materialization "of a distinctively East European capitalism that will differ as much from West European capitalisms as do contemporary Asian variants." Similarly, in many East European countries various firms appear to be bankrupt by Western standards but have nonetheless been kept running by their creditors, many of whom are banks worried that shutting down such firms will damage their own asset bases (Hawker 1993; *Prague Post* 1995; Herod 2001).

Equally, whereas in Western economic and political theory the major economic cleavages are usually seen to be between labor and capitalists, in Eastern Europe workers and employers have often come together in coalitions on the basis of former political allegiances (communist v. anti-communist) to compete for access to government resources as the privatization of state-owned enterprises occurs (Kirichenko & Koudyukin 1993). Furthermore, the transition means that unions have been placed in the awkward position of having to protect their membership from the negative consequences of privatization yet also of being agents of change against the entrenched interests of the old *nomenklatura*, many of whom have become the new capitalists.[21] Moreover, as Piore (1992: 172) has argued, the fact that the term "market economy" has different connotations in Eastern Europe than it does in the West means that, even if they represent what they are doing as "building market economies," the people in these societies may have in mind quite different things than do Western commentators.[22] The point in all this, then, is that despite the rhetoric of neoliberal transitology, the form in which capitalist social relations are taking hold in Eastern Europe is being palpably shaped by the legacy of almost half a century of communism, to the degree that what is emerging is no mere copy of Western European or North American models of capitalist economies.

A sixth and final element of much neoliberal globalization talk is a narrative presenting the nation-state as increasingly irrelevant. I shall address this issue in greater detail in Chapter 7, but it is important here at least to recognize this as a central element in many discourses concerning globalization, particularly as globalization has frequently been represented as a process that is forcing nation-states to "deregulate" markets. Whether nation-states actually have lost power in the face of globalization is, however, a complex issue – in some areas they appear to have done so, whereas in others their power seems to have grown (as with border security). Equally, how we interpret the emergence of entities like the European Union or the World Trade Organization (WTO), which are generally viewed as challenging the power of individual nation-states, also shapes how we understand the process of globalization. Hence, whereas the EU and WTO are frequently seen to be undermining national sovereignty, it should not be forgotten that both had to be brought into existence by individual nation-states who, presumably, had good reasons for doing so. Indeed, Hudson (2000) has argued that the EU's creation can be read as an effort by its individual member nation-states to come together to create a body which could serve as a counterweight to the United States and so protect their interests. Like beauty, then, loss of sovereignty may be in the eye of the beholder. The central element in all this, though, is that neoliberal discourse typically portrays the process as one in which nation-states' capacities for action *in toto* are being weakened. Importantly, this argument fuels assertions that nation-states – historically the guarantors of many public goods like education and health care – should simply accept their now-relegated political and economic positions vis-à-vis globally organized capital and supranational entities like the WTO and no longer attempt to provide such public goods, as their efforts will be doomed to failure.

Putting all of this together, then, leaves us in a situation in which the discourse of globalization – as articulated both by neoliberal but also sometimes by anti-neoliberal groups – serves, as Gibson-Graham (1996: 120–5) has put it, "as a language of domination, a tightly scripted narrative of differential power . . . [in which c]apitalism is represented as . . . naturally stronger than the forms of noncapitalist economy." As I now explore, within such a script a particular language of geographical scale – or, more precisely, geographical rescaling – has been key.

Talking Globalization

The Scales of Globalization

Globalization as a process has generally been presented in highly geographically scalar terms, variously as a means by which social life is delocalized and/or denationalized and, in the case at least of Ohmae (1995, 2005), "reregionalized." Equally, globalization has frequently elicited slogans which incorporate an implicit sense of spatial scale, such as the famous "Think globally, act locally," which has itself spawned myriad derivatives in various areas of social life – advertising (Blackwell et al. 1994), human rights (Rodman 1998), and banking (Figure 4.1), to name a few. However, the notion that it is possible to talk of processes or phenomena as being "local" or "regional" and that globalization is eradicating some scales of social life, or that we might "think globally, but act locally," requires exploring the concept of geographical scale in some detail. What, for instance, would it mean to urge a "buy local" campaign as a challenge to globalization? Would this simply mean shopping at the store closest to one's home, even if that were Wal-Mart, or would it mean buying only locally produced goods, and what would those be if the raw materials to make/grow them came from overseas? Which would be more illustrative of "buying locally" – forgoing shopping at a Wal-Mart in one's own community and traveling instead hundreds of miles to buy goods produced in the community of one's residence, or instead making purchases at the shop closest to one's home, even if what you bought was manufactured overseas? In considering the issue of scale, then, there are two important matters to ponder: first, does scale have ontological status (in other words, is scale "real"?)?; second, how are different discourses of scale constructed to get us to think about the relationships between different places in particular terms? It is these issues that I will now outline.

Figure 4.1   An advertising billboard enticing customers to "Earn globally. Bank locally."

Conceptions of geographical scale

There are three principal ways in which the geographical scales at which social life is often seen to be organized are thought of ontologically.[23] The first draws from the work of German philosopher Immanuel Kant (1943 [1781]), who argued that any structure or order humans perceive in the world results from their own cognitive functions. Order, he suggested, is imposed upon the world by our brains. With regard to geographical scale, then, this approach sees scales as entities created by our minds. Drawing upon the neo-Kantian perspective, geographer John Fraser Hart (1982: 21), for instance, has argued that regions (the spatial manifestation of the "regional scale") are not real things that have material form in the landscape but are, rather, "subjective artistic devices . . . shaped to fit the hand of the individual user." For him, they are not "real" in any ontological sense – in the case of economic regions, their boundaries are not seen to be determined by material processes but by where the analyst wishes to draw their regional limits. One person's "region," in other words, is just as good as the next.

A second common conception is one in which scales like "the local," "the regional," "the national," and "the global" are simply felt to be the most logical or natural way by which to divide the world into manageable bits for purposes of investigation and management. In this approach, it simply just makes intuitive sense to carve up the world in this way and scales are seen simply as areal circumscribers which, like boxes, contain various social and natural processes – sociologist Anthony Giddens (1985: 172), for instance, has described the nation-state as a "power container." Ontologically, scales are conceived of as existing prior to the processes they contain. Importantly, both of these approaches – the neo-Kantian and the "scales as logical units" of analysis view – have an "absolutist" conception of space which sees space as capable of being infinitely divided and/or collated together without consideration paid to the geographies of the social and natural processes playing out across the landscape. Thus, regions are typically seen as units of territory – bigger than cities but smaller than the globe – which are not organic wholes but which are capable, instead, of being endlessly divided into smaller and smaller units or amalgamated into larger and larger ones. Likewise, nation-states are not viewed as having emerged in a particular time and place and therefore as being human inventions with particular histories and geographies but, rather, are conceived of simply as natural and logical territorial entities by which to organize social life. Equally, the global scale is regarded as the natural limit – determined by geology or God – of social organization, one that has always existed.[24]

By way of contrast, a third view argues that geographical scales should be understood as social products – that is to say, they are actively created rather than simply imposed mentally or viewed as logical/natural ways for dividing up the world. There is a significant literature concerning such approaches (see Herod 2001, 2008) but one of the earliest interventions came with Taylor's (1981, 1982) argument that "the urban," "the national," and "the global" scales have occupied particular roles in capitalism's development. Hence, he suggested, the global scale is the "scale of reality," for it is the scale at which capitalism is organized, whereas the urban scale represents the "scale of experience" because it delineates the realm within which everyday life is conducted. For its part, the national scale should be considered the "scale of ideology" because, Taylor suggested, it is the scale at which the capitalist class promulgates ideologies of nationalism to divide workers.

At about the same time, Smith (1990 [1984]) was developing a different approach to understanding the scales at which capitalism is apparently organized economically. Whereas Taylor focused upon the roles he felt different scales played in the functioning of global capitalism, Smith concentrated upon how such scales are actually created. Specifically, he argued that there exist two contradictory tendencies shaping how capitalists make decisions. On the one hand, capitalists must fix their investments in the landscape so that accumulation can take place. They must, in other words, create what Harvey (1982) has called "spatial fixes" (Chapter 1). This results in a clearly visible differentiation of the landscape, as capital is fixed in particular ways in particular places. However, capitalists must also seek to keep their capital mobile to take advantage of opportunities for profit making which arise elsewhere.[25] This need for mobility is of great consequence, for through relocating investment to new places capitalists level economic space by equalizing the rate of profit across the planet's surface, a phenomenon which is the geographical expression of the tendency for competition to equalize the rate of profit.[26] For Smith, then, geographical scales are produced through the spatial negotiation of these opposing tendencies within capital: the tendency to differentiate the landscape by being fixed in it, and the tendency to level conditions across the landscape in the process of flowing over its surface, with scales being the mechanisms for differentiating between spaces which are more similar to each other and those which are less similar.

Using such a conceptualization Smith identified four primary geographical resolutions at which this negotiation unfolds. For him, then, the urban scale is delimited by the spatial coherence of labor markets and daily commuting patterns (usually represented by Travel-To-Work Areas) – it represents the geographical limit to the day-to-day travel of people from where they live to their places of paid work. Stepping up a scale, he argued that the regional scale is constituted by the spatial concentration of capital to form particular territorial divisions of labor – here a steel-making region, there a farming region – whilst the national scale results from the need of capitalists located in different parts of the globe to control particular territorially defined markets within the world economy – they cooperate to support laws domestically which favor capital accumulation and to compete with foreign capital, even as they vie with each other domestically.[27] Finally, the global scale results from capitalists' quest to universalize geographically the wage–labor relation, so that all parts of the planet will eventually come under capitalist social relations' sway. Thus, whereas the physical limits to the planet are geologically given, the emergence of an increasingly integrated economy organized at the global scale is the historical and geographical outcome of capital's expansionist nature, with "capitalism inherit[ing] the global scale in the form of the world market . . . based on exchange [but transforming it] into a world economy based on production and the universality of wage labour" (Smith 1990 [1984]: 139).

As a first cut at a theory of scale production, Smith's analysis was path-breaking, arguing that geographical scales are capable of being remade as economic and political processes are reworked. Hence, as people's capacities to travel farther to work increases, so may the urban scale be seen to expand territorially. Likewise, as economic restructuring occurs so may processes of de- and re-regionalization occur. Smith's approach, then, connected scale production to broader social processes, arguing that "not only does capital produce space in general, [but] it produces the real spatial scales that give uneven development its coherence" (Smith 1990 [1984]: xv). Nevertheless, there was also criticism of Smith's approach, not least because it tended to ignore the role of other actors – like workers – who contribute to making such scales through their actions (as when workers engage in international labor solidarity, thereby affecting how the planet's economy is integrated globally [Herod 2001]).

Significantly, a fourth perspective has recently been articulated by Marston et al. (2005), who suggest abandoning the concept of scale because, they contend, notions that the world is scaled privilege views which see it in hierarchical terms, and this tends to promote one scale (usually "the global") over others. Although they recognize that a language of scale may be invoked as part of a descriptive terminology of ordering – naming something "national" or "global" shapes radically how it is perceived – they maintain this is different from declaring that the landscape is organized into various spatial resolutions ("scales") which have ontological presence. They argue instead for what they call a "flat ontology," one in which different parts of the Earth's surface (what they call "sites") are interlinked but are not seen to be in any kind of spatially hierarchical relationship.[28] In such an ontology, "scale" may have material effects but it is, ultimately, simply a representational trope (Herod 2008) – it does not exist in any "real" sense. Although Marston et al.'s line of reasoning has been criticized by a number of commentators, who suggest they have simply misunderstood the distinction between ontology and epistemology (Hoefle 2006), that theirs is little more than a repackaged Kantianism, and that to "equate scalar hierarchies with a vertical . . . view of political action and change is misleading" (Jonas 2006: 403), it does raise the question of how scalar "verticality [often] symbolises Power" (Lefebvre 1976: 88), such that being atop a scalar hierarchy (as "the global" is frequently seen to be) is habitually to be seen to enjoy a position of power, allowing those who are global to "loo[k] down from above."[29]

There are, then, three important issues which emerge from this discussion. The first concerns whether geographical scales are seen to be "real" – in other words, do they have ontological "heft"? This is key to debates concerning the contemporary processes of economic restructuring that are taken to be "globalization." Thus, if scales are viewed in a neo-Kantian light, contemporary economic and political restructuring represents little more than a visual shift in our gaze – whereas we used to focus upon the local or national scales of life, we are now increasingly drawn to a global gaze. If, on the other hand, scales are understood to be materially produced social artifacts, then processes of "delocalization" or "denationalization" are fundamentally material as well as discursive – we have refocused our gaze, in other words, because of material transformations in how economic and political life is organized. This means that a fundamental process of rescaling is taking place.

The second issue concerns how we consider scale. Generally, scales like "the regional" or "the national" have been considered to surround areal units of territory. In other words, scales serve simply as boundaries around particular territorial realms. However, there are other ways in which scales may be considered. In particular, French theorist Bruno Latour (1996: 370) has argued that the world's complexity cannot be captured by "notions of levels, layers, territories, [and] spheres," and so it should not be thought of as being made up of discrete units of bounded spaces which fit neatly together in some overarching gestalt. Instead, Latour maintains that we should think of the world as "fibrous, thread-like, wiry, stringy, ropy, [and] capillary." His is, then, a topological rather than a topographical understanding of how places are interconnected scalarly (Castree et al. 2007), with terms like "the global" and "the local" being not opposite ends of a scalar spectrum but, rather, a terminology for contrasting shorter and less-connected networks with longer and more-connected ones (see also R. Smith 2003).

The third issue to emerge from such considerations is that of scalar discourse and the metaphors used to describe the relationship between different scales. For example, the scalar progression from "the local" through "the regional" and "the national" to "the global" is often presented in vertical terms, as if one were climbing up a ladder from "the local" to other scales (Figure 4.2). In such a representation, each scale is viewed as a rung on the ladder: "the local" is seen as the most grounded scale (just as one end of the ladder rests on the ground) and social actors climb up the scales as

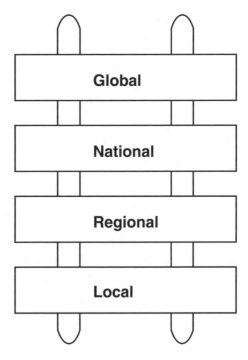

Figure 4.2 Scales as rungs on a ladder.

Figure 4.3 Scales as concentric circles.

they approach "the global." However, on other occasions the progression through the scales has been presented in more horizontal terms, with scales viewed as a series of concentric circles and with "the local" at the center (Figure 4.3). In this representation, social actors move *out* to "the national" or "the global" scale, rather than *up* to it. Certainly, these two metaphors – scale as ladder rungs and as concentric circles – share similarities. For instance, they both see scales as discrete entities (separate rungs and circles). Yet, there are also important differences – the ladder metaphor presents "the global" as *above* other scales whereas in the concentric circle metaphor it *encloses* them. This has critical implications for how we understand the relationship between, say, the global and the local and the associations we attach to each scale – viewing the global scale as "bigger" or "higher than" or "wider than" others may lead us to assume, perhaps, that it is inherently more powerful.

Equally, other metaphors leave quite different images of scales. For instance, viewing scales as highly nested, as with Russian matryoshka dolls, suggests that scales are separate entities but can only be fitted together as a coherent whole in a very particular order. Thus, there is much less sense that one could skip over, say, the regional scale in moving from the local to the national in the way one could using the ladder metaphor – in the latter a "big step" could take a social actor like a firm or labor union directly from local to national presence without first becoming a regional force (Figure 4.4). Likewise, using the metaphor of tree roots or perhaps earthworm burrows (Figures 4.5 and 4.6) to capture the fibrous, thread-like, wiry, stringy, ropy, and capillary nature of the scaled world à la Latour, a view in which terms like "the local" and "the global" "offer points of view on networks that are by nature neither local or global, but are more or less long and more or less connected" (Latour 1993: 122), is a very different way of representing the scaled nature of the world – where does one scale end and another begin and what does that mean for how we conceptualize

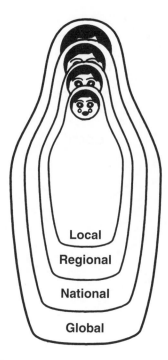

Figure 4.4 Scales as matryoshka dolls.

the relationship between, say, local and global? Does it, in fact, make more sense to say that some scales are deeper or shallower than are others, as opposed to saying they are bigger/smaller than, or above/below, each other, and what would this mean anyway?

Rhetorics of scale

If questions regarding scale's ontological status and which metaphors we use to represent scalar relationships are one element in understanding the contemporary scaling of the globe, then a second concerns how scalar rhetorics are used in constructing narratives of global transformation. There are two principal issues. The first relates to how rhetorics of scale are used to frame particular issues, thereby shaping how we understand processes or phenomena. For example, determining whether heavy pollution in a particular area is an instance of environmental racism may turn on how the scalar boundaries are drawn around a particular locale. Thus, as Kurtz (2002) has shown in the case of a petrochemical facility in Louisiana, circumscribing the boundaries of areas affected by pollution more narrowly or more widely can dramatically change the demographic make-up of the population that lives within an "impacted area," thereby making it easier or more difficult to sustain a claim of environmental racism. The way in which boundaries are drawn around processes or phenomena at

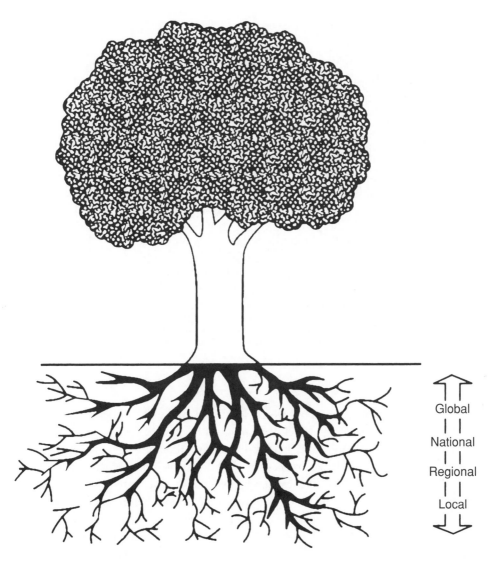

Figure 4.5 Scales as tree roots.

different spatial resolutions, thereby fixing them within particular scalar frames, then, plays a central role in what Snow and Benford (1992) call "naming, blaming, and claiming" – at one scale a noxious facility's siting may be seen as clear environmental racism whereas at another it may not. In the case of debates about globalization, framing some phenomena or processes as "local" and others as "global," with "the global" being the "ultimate" scale, the one that "really matters" (Taylor 1982: 26), can have the effect of shifting elsewhere blame for things which are initiated locally (like factory layoffs), as well as making it appear impossible to resist such things' pull – "the global" can

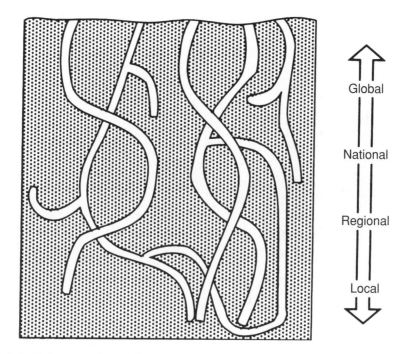

Figure 4.6   Scales as earthworm burrows.

be a convenient scapegoat upon which local and national politicians or business leaders may blame things for which they might otherwise be held accountable.

The second issue of importance here is that of how different scales are counterposed rhetorically to create particular impressions of how the world is. In this regard I want to focus on the two scales ("the local" and "the global") which are typically argued to represent scalar extremes and to explore how this local/global binary has frequently been associated with other binaries – as we have already seen, "the global" has been associated by some with masculinity and "the local" with femininity (Freeman 2001), but there are other dualisms, like "the concrete" v. "the abstract," wherein "global" activities are often perceived as somehow more abstract/less concrete than are "local" ones. Within such binary thinking, Gibson-Graham (2002) has identified several ways in which the relationship between "the local" and "the global" is viewed.

1   "The global" and "the local" are not things in and of themselves, but are viewed as interpretive frames for analyzing situations – thus, whereas from a "global perspective" a worldwide economic slowdown or a boom may appear to be taking place, viewed from the "local perspective" of particular places current economic events may look very different.

2   "The global" and "the local" each derive meaning from what they are not and through contrasting with each other – much like concepts of dark only have meaning in contrast to those of light. Drawing on Dirlik (2001: 16), Gibson-Graham suggests

that in such a discursive construction "the global" is often seen as "something more than [the] national or regional . . . [and] anything other than the local," whilst "the local" is seen as its opposite. This view, however, allows a significant flexibility in terminology, such that anything other than "the global" is sometimes seen to be "local" (for example, in much globalization rhetoric, nations and even entire extra-national regions like the European Union are referred to as "local" actors).

3 "The global" is really "local," such that "the global" does not really exist and everything "global" is really a collation of things "local." Hence, "global" processes rarely affect all parts of the globe but are felt in particular localities, such that it makes better sense to think of TNCs as "multilocal" entities rather than "global" ones.

4 "The local" is really "global," and localities are simply the nodal points in spatialized networks of social relations. Thus, "the local" is simply the location where "global" forces "touch down." "The local," in other words, is the geographical point of entry into a world of global flows encircling the planet.

5 "The global" and "the local" are not locations but processes. Put another way, all spaces are produced as "glocal" sites, such that "local" initiatives can be broadcast to the world and adopted in multiple places, whilst "global" processes always involve localization (hence, although it may operate "globally," McDonald's adapts its products to particular local tastes). "The local" and "the global," then, are not fixed entities – such that a social actor can simply jump from one scale to another – but are always in the process of being remade. However, whilst most, if not all, phenomena can be seen as both "global" and "local," according to Dirlik (2001: 30) "they are not all local and global in the same way."

The purpose of exploring how the local/global binary has been represented is twofold. Obviously, different representations leave us with different understandings of what "globalization" might be and mean – thinking of a TNC as "multi-locational" rather than "global" transforms how workers view challenging it, what they think are their chances of winning against it, and what tactics they may choose to adopt. It is therefore of little surprise that many TNCs stress their globality when faced by union organizing drives, with the aim of presenting themselves as such juggernauts that only the very dedicated – or foolhardy – will even contemplate challenging them. Equally, one can imagine the same TNC trying to emphasize its multi-locationality and localness on different occasions, perhaps when trying to suggest it is a community "insider" and not a "foreign" company (e.g., when companies like Toyota stress how many Americans they employ).

The local/global binary, though, is of interest also because many such binaries – masculine/feminine, concrete/abstract, large/small – typically imply some inequality of characteristics. In the case of the local/global dualism, Gibson-Graham (2002) suggests that neoliberal globalization discourse has overwhelmingly represented "the global" as *inherently* more powerful than "the local" – this imaginary is at the heart of the infamous TINA discourse, after all. Equally, such a representation has often been incorporated within critical perspectives on globalization – as when Marxist accounts of globalization assume a priori, rather than demonstrate, that globally organized capitalists are inherently more powerful than are locally organized ones. Certainly, "the local" has been lauded by some as a scale within/at which diversity and freedom can be expressed, but even this lauding frequently conceives of "the local" in terms of its serving as a refuge from the broader and more powerful homogenizing tendencies of global capitalism. Neoliberal portrayals of "the global" as inherently more powerful than "the local," then, are central to refocusing our scalar gaze and allowing us to imagine a world in which the global is erasing other scales of social organization to create a seamlessly integrated whole, "a global village" in which "[t]here are no boundaries" and all problems and opportunities "will become so intimate as to be one's own" (Asimov 1970: 19).

Myriad examples of such a refocusing of the scalar gaze exist, but perhaps the most explicit comes from Virilio (1995), who has suggested that in our contemporary world wherein information can travel across space virtually instantaneously "[w]hat is being effectively globalized by [such] instantaneity is time," to the extent that "[e]verything now happens within the perspective of real time [and that] henceforth we are deemed to live in a 'one-time-system.'" For Virilio, then, the establishment of a "global time" is the culmination of a conflict between the global and the local, the universal and the particular, one in which "real time [is] superseding real space [and is] making both distances and surfaces irrelevant in favor of the time-span, and an extremely short time-span at that":

> For the first time, history is going to unfold within a one-time-system: global time. Up to now, history has taken place within local times, local frames, regions and nations. But now, in a certain way, globalization and virtualization are inaugurating a global time that prefigures a new form of tyranny. If history is so rich, it is because it was local, it was thanks to the existence of spatially bounded times which overrode something that up to now occurred only in astronomy: universal time. But in the very near future, our history will happen in universal time, itself the outcome of instantaneity – and there only.[30]

For some, then, the discursive annihilation of "the local" by "the global" appears to be what Smith (2005), in a slightly different context, has called "the end game of globalization."

Placing Scale

To sum up, in their more triumphalist versions neoliberal globalization discourses frequently invoke a scalar imaginary in which the erasure of localness, regionalism, and nationalism (and the scales of social identity associated with these) is the natural outcome of stitching the planet together as an integrated whole. It is interesting to note, however, that some of the most powerful metaphors invoked to express this erasure are those of the "global village" (McLuhan 1962), wherein the enormity of the global is reimagined within the diminutiveness of the village, and the idea – drawing upon William Blake's 1803 poem *Auguries of Innocence* – that the whole world can be seen in a grain of sand (cf. Howitt 1993): in both it is things small in scale – a village, a grain of sand – which are taken to represent the contemporary scene's globality. In much globalization discourse, then, the more we become delocalized the more does the globe's vastness get expressed through reference to smaller and smaller objects, whilst the emergence of a world in which everything has become immediate and, paradoxically, local (even places on the other side of the planet) means that distance – both temporal and spatial – is eliminated and we become caught in an endless continuous present of "simultaneous happening."[31]

Furthermore, as Bartelson (2000) has argued, one outcome of the spread of the kinds of universalist rhetoric explored above is that "the global" has become an increasingly common object of thought and analysis, challenging other scales of social organization (especially the national scale); it has become the acme of scales, the scale from which there is, ostensibly, no escape. Certainly, within the TINA discourse the supposed omnipotence of "the global" is reinforced through a continual "performance" (Butler 1990) of its power and extent – TNCs, for instance, constantly reinforce their "global structure" through quite mundane things, including organizational charts and images which graphically project their global presence (Figure 4.7).[32] Such pictorial schemas of corporate structure, then, are not simply naïve images of managerial configuration but play a powerful role in policing how we envision TNCs' economic and spatial organization (Buck-Morss 1995), and whether we see them as hierarchically configured "global" entities or as something else. Such representations, in other words, serve to rework many people's identities and sense of their own place in the world relative to organizations like TNCs.

Obviously, the material practices in which economic and political actors engage – sending capital overseas, migrating internationally, building planetary communications networks – have significant impacts upon how the world economy operates. However, as illustrated above, the discourses through which "globalization" is presented – particularly those which relate to how the world and its social entities are seen to be scaled – also have a significant impact upon how we understand the material processes that are affecting the contemporary world economy. Thus, whether one views the practices whereby firms establish branches overseas as instances of such corporations becoming "global" or, instead, "multi-locational" has significant – and quite different – implications for understanding material economic and political processes. Indeed, one of the central purposes of such a globalization discourse, at least according to Gibson-Graham (2002: 35–6), is to get us to restructure collectively our personal identities so that we reimagine ourselves in relation to the geographical scales of social life within which we live – no longer can we simply concern ourselves with what happens at the scale of the nation-state but we must now be concerned with the strictures of global processes, practices, and phenomena. Increasingly, as we come to believe that we live in a global(izing) world, so will we perhaps rework our identities and perform them in an appropriately "globalized" manner.

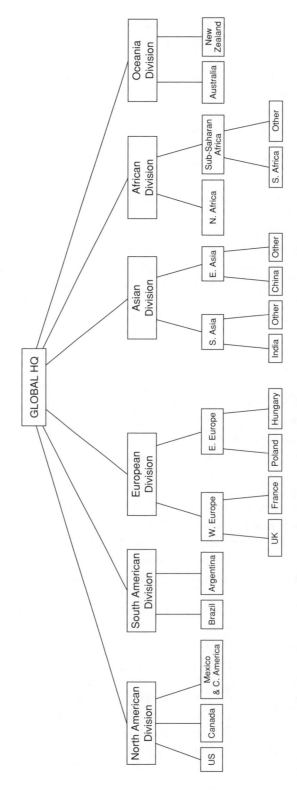

Figure 4.7 A typical corporate organization chart, emphasizing visually a corporation's global reach.

> ### Questions for Reflection
>
> - What is geographical scale and how does it help us make sense of the world?
> - How is the relationship between "the global" and "the local" manipulated for political purposes?
> - How do different representations of scales like "the global" and "the local" shape how we view social processes like globalization?

Notes

1. Gray (1998: 57) argues: "Behind all these 'meanings' of globalization is a single underlying idea which can be called *de-localization*: the uprooting of activities and relationships from local origins and cultures. It means the displacement of activities that until recently were local into networks of relationships whose reach is distant or worldwide."

2. There is significant debate concerning capitalism's geographical origins. Marx (1976 [1867]: 876) locates its origins in the early Renaissance Mediterranean. Brenner (1977) also suggests a European place of birth. Others, like Blaut (1994), argue that other regions – including parts of Africa and Asia – were experiencing similar transitions in economic and political structure at the same time as Europe, though Europe eventually emerged supreme because of its proximity to the Americas, from which wealth from gold and silver production and plantation agriculture could be appropriated, thereby facilitating the early stages of capitalist accumulation and a transition away from feudalism. My purpose here is not to argue for one side or the other but, rather, to indicate that capitalism clearly was developed in particular places and has not always been with us.

3. It is, nevertheless, a common one. Hence, in the Introduction to a book on the transition in Eastern Europe, Mandelbaum (1993: 1) has argued that, "Making markets where none existed before is an enterprise of daunting scale," the implication being that no markets existed in Eastern Europe during the communist era.

4. This is not the place to discuss whether the Soviet Union, China, Cuba, North Korea, and other "communist" nations have developed a "communist" economic system or merely "state-capitalist" ones. However, suffice it to say that if we were to adopt the latter position it would be possible to argue also that a "capitalist" society – albeit a particular form (state-capitalist) thereof – could exist in which market functions were highly restricted. For more on arguments about whether societies like the Soviet Union were examples of "state capitalism," see Mandel (1951), Cliff (1988), and Resnick and Wolff (2002).

5. At the heart of the Marxist understanding of profit lies the notion that workers are never paid the full worth of the value of commodities they produce (Wolff & Resnick 1987). For Marx, all wealth is ultimately created through the production of commodities, rather than through their purchase and sale, for although an individual may benefit by buying low and selling high, this simply means that someone else has lost out by having bought high but sold low – within the system as a whole these simply cancel each other out, and more wealth *in toto* can only be generated through the application of human labor.

Shockingly, for many neoliberal advocates, both Marx and Adam Smith agreed on this point – societal wealth emerges from production, not exchange.

6. Using such a definition, then, one of the key determinants of categorizing any mode of production is the manner in which surplus labor is secured from one class of people by another. Under capitalism, this occurs through the mechanism of the labor market, whereas under feudalism surplus labor was secured by forced appropriation of labor on the lord's lands. Whereas for peasants it was easy to see how much of their labor was being appropriated because they had to work a certain period of time on their lord's land – in other words, their manorial labor was separated from their general labor in time and space – under capitalism the extraction of surplus is less transparent, for there is no separation in either time or space between when a worker is working to cover his/her wages ("necessary labor time") and when they are generating profit for their employer ("surplus labor time") (Burawoy 1985).

7. Hegel argued that ideas drive history and that, eventually, societies reach an intellectual plateau in which all the good ideas have been thought up and society is run according to them (see Chapter 2, footnote 10, for more on Fukuyama's argument). Marx, on the other hand, argued that social struggles between classes drive history.

8. "Modernization theory" argues that to become "developed" LDCs must adopt the characteristics of Western, liberal democratic, "free market" countries. Slater (2002: 92) argues that modernization theory has been constructed around three interrelated components: "an uncritical vision of the West, largely based on a selective reading of the history of the United States and Britain; a perspective on the non-West, or traditional, societies that ignored their own histories and measured their innate value in terms of their level of Westernization; and an interpretation of the West–non-West encounter which was based on the governing assumption that the non-West could only progress, become developed, throw off its backwardness and traditions, by embracing relations with the West." Although popular in Western academic and policy circles in the 1950s and 1960s, this approach has since been subject to significant critique (Peet & Watts 1996; Willis 2005). Critiques of modernization theory have also been leveled in literature, particularly in the works of Nigerian author Chinua Achebe, most prominently in his 1958 novel *Things Fall Apart* challenging the European view that Africa had little indigenous culture of value.

9. The notion that traveling across space might allow one to travel back in time has been common in Western thought concerning modernization. Thus, Joseph Conrad's 1902 *Heart of Darkness* uses the metaphor of traveling up the Congo River to explore humanity's baser instincts – the further the protagonist Marlow proceeds into the continent's interior, the farther back in humanity's moral development he is seen to journey. Likewise, Hagen (2003) describes how eighteenth-century Western European notions of Eastern Europe as "Other" played into a post-Enlightenment representation of traveling from West to East (particularly to the Balkans) as a trip from a region of "normalcy" and "rationality" to places which were viewed as pre-Modern (and therefore further down the ladder of social evolution).

10. Such attitudes have been promulgated by those from a wide political spectrum. Hence Democratic President Bill Clinton (1996: 36) suggested that, "Just as democracy helps make the world safe for commerce, commerce helps make the world safe for democracy," whereas Daniel Griswold (2003), associate director of the libertarian Cato Institute's Center for Trade Policy Studies, has argued that, "Free trade can till the soil for democracy and respect for human rights."

11. Pictorially, this is perhaps most familiarly represented by Michelangelo's painting the "Creation of Adam" in the Sistine Chapel, wherein God's outstretched hand is just about to give humanity life. Tribe (1981: 141) has argued that, contra some later interpretations, Smith sought to use the metaphor of the invisible hand to represent "the principle of unity among a collection of divided individuals . . . not the behaviour of economic agents, be they individuals or enterprises." The idea that the "invisible hand" represents the deliberate and calculated actions of economic actors was an invention of nineteenth-century neoclassical economists.

12. The cross-over between ideas concerning biology and those concerning economic relations is even more stark when one remembers that Charles Darwin drew inspiration for some of his ideas about natural selection by reading the works of Adam Smith and Thomas Malthus, the latter famous for his theorizing on the relationship between population growth and economic development.

13. Hawken et al. (1999: 3, emphasis added) argue that, "Humankind has inherited a 3.8-billion-year store of *natural* capital." It is important to recognize, however, that this conception of "capital" is quite different from that of political economists like Marx, who argued that capital is produced only out of human labor. Viral marketing is a means of marketing that exploits consumers' social networks, thereby getting consumers themselves to spread news of a product. As such, information spreads across social networks in much the same way as viruses replicate themselves and spread throughout a corporeal body.

14. Although usually associated with Darwin, the phrase "Survival of the fittest" was actually coined by the nineteenth-century British economist Herbert Spencer, after reading Darwin's (1859) *On the Origin of Species*. Darwin himself had used the term "natural selection" in the earliest editions of *Origin*, and only began using "Survival of the fittest" after reading Spencer's (1864) work *Principles of Biology*. He subsequently rejected the term.

15. The lyrics are:

> I'd like to buy the world a home and furnish it with love,
> Grow apple trees and honey bees, and snow white turtle doves.

16. Nagahara is here adopting Marx's (1981 [1894]: 969) phraseology.

17. This language of biological reproduction is encountered in other descriptions of the emergence of new economic states (like that of globalization). Hence, Marx (1976 [1867]: 916) argues that, "Force is the midwife of every old society which is pregnant with a new one."

18. Ideas of male activity and female passivity have a long history in the Western imagination. Martin (1991) has shown how gendered stereotypes have influenced scientific understandings of human reproduction and the relative activity and passivity of sperm and egg. Significantly, however, using a different descriptor of sexual intercourse – one in which the phallus is seen not as *penetrating* another body but, perhaps, as *enclosed* by it – changes the bodily representation and, hence, the depiction of capitalist expansion. The image of outside capitalists raping the natural environment also has a long history in the Western imagination, given the associations between "Mother Earth" and femininity (Merchant 1980).

19. In an effort to query/queer such representations, Gibson-Graham outlined a discursive approach to challenge narratives portraying global capital in such rugged, dominant, and

masculinized terms – provocatively, she calls this getting the neoliberal narrative to "lose its erection" (1996: 127). Interestingly, though, much of the feminist writing in which narratives of "globalization as rape" have been articulated has drawn upon the image of heterosexual rape, with a powerful male violating a weaker female. It should not be forgotten, though, that part of discursive strategies of querying/queering the "globalization as rape" script should involve pondering how a non-heteronormative approach – one in which the act of rape is seen as male–male or female–female in nature – might challenge globalization discourses. Equally, heterosexual rape can involve involuntary forced intercourse of a male by a female, a situation which decidedly impacts on how the "globalization as rape" narrative plays out.

20. In Coca-Cola's case, for instance, "The characters chosen by the Company to translate Coca-Cola into Chinese, which are still used today, can be translated to mean 'delicious and enjoyable.' When the Company first entered China and an appropriate translation was being researched, some small shopkeepers decided to make up their own Chinese equivalents. These shopkeepers tried to roughly imitate the sound of the name Coca-Cola, with no thought to meaning. In the case of one small shopkeeper, characters were chosen that could be translated to mean 'bite the wax tadpole'" (Coca-Cola 2006).

21. Given their political connections and access to resources, many former managers have become owners of the firms they managed. In the transition, then, unions have taken on the role of advocates of marketization as a way to challenge such interests. In the Czech Republic, for instance, the power workers' union early on challenged conservative Prime Minister Vaclav Klaus over his government's failure to dismantle the energy monopoly České Energetické Závody (de Luce 1993), whilst one adviser to the national labor federation argued in the early 1990s that unemployment was too low and that "An increase would be healthy" (Rusnok 1993) – presumably because it would bring higher productivity and, possibly, higher wages for those still in employment. In Poland, Solidarność has also played a significant role in state enterprise privatization (Ost 1989, 1995).

22. He suggests that in Eastern Europe the term refers to the "dispersion of *political* power, the breakup of the state monopoly over economic resources and the destruction of its capacity for central control through a totalitarian state which those resources confer," whereas in the West "the market is a much more narrowly defined economic institution ... for coordinating and directing economic activity in which resources are controlled by private individuals who interact freely with each other in a competitive marketplace."

23. I am not here referring to scale in its technical meaning as an expression of the ratio between distance on a map and distance on the Earth's surface.

24. Such a view was reinforced in the Western imagination for several hundred years not only by lack of technology enabling humans to go beyond the Earth, but also by religious dogma – the Catholic Church's denial of the Copernican revolution and its banning of Galileo's book *Dialogue on the Great World Systems* until the nineteenth century reinforced the view that Earth is at the center of the universe. Interestingly, if somewhat bizarrely, a number of entrepreneurs have recently sought to extend things beyond the global scale to the extra-global. Hence, the Lunar Embassy Company, billing itself as "leaders of the extraterrestrial real estate market," claims the right to sell land futures on the Moon, Mars, Jupiter's moon Io, and Mercury, under Article II of the 1967 United Nations Outer Space Treaty (www.lunarembassy.com).

25. This notion that capital must negotiate two geographical conditions – one of spatial fixity, the other of spatial mobility – reflects Marx's argument that capital must cope with a

similar temporal conundrum. Thus, in his famous model of capital circulation in which money is turned into commodities and back again into (more) money (M–C–M¹), capital is either fixed in particular forms at particular moments (as commodities or money) or is in the process of transforming itself from one to the other. The spatial correlate suggests that because accumulation cannot occur on the head of a pin, even the most mobile of capital must alight somewhere, however fleetingly, if accumulation is to occur: "it must spend some time as a cocoon before it can take off as a butterfly" (Marx 1973 [1858]: 548–9).

26. This is because the situation in which some parts of the economic landscape promise high rates of return and others low will not last for long. As capital seeks out the former, profit rates will tend to equalize because labor will be in greater demand and workers can thus negotiate better conditions and wages. Equally, low profit locations that capital leaves will often begin to suffer increased unemployment, thereby reducing workers' negotiating power and providing an opportunity for capitalists subsequently to increase their profit rate. This seesawing provides stimulus to capitalists to engage in technological innovation and to reorganize the labor process so as to secure profit rates that are greater than the social average, thereby starting the process over again.

27. Although he comes at the issue from a quite different political direction, Ohmae (1995, 2005) has also argued that the territorial division of labor is the basis for regionalization, and suggests that the contemporary restructuring of the world economy is resulting in a denationalization and growing regionalization of economic activity.

28. They distinguish such a "flat ontology" from a "horizontal ontology," which they see as merely replacing the "up–down vertical imaginary" with a "radiating (out from here) spatiality."

29. However, in contrast to neo-Kantians like Hart (1982), who recognize a nested hierarchy of scales ("the regional," "the national," etc.), Marston et al. (2005: 420) argue against the notion that the world is hierarchically constituted "as a nesting of 'legal, juridical and organizational structures.'"

30. As explored in Chapter 2, the idea of a "global time" is, however, much older than Virilio's musings might lead us to believe.

31. Delocalization results in a world in which everything is local in the following sense: because people, commodities, capital, and information can more readily cross the vastness of the globe compared to previously (a process of delocalization), places across the planet are increasingly as reachable as if they were within one's local vicinity.

32. With her concept of "performativity" Butler argues that our identities are constantly reinforced through our actions. For example, through stylized repetitive acts, men and women "perform" their sexual identities – they behave in certain ways which are deemed either appropriate or inappropriate for their sex. It is important to recognize, though, that Butler does not see such identities as voluntarily performed but, rather, as located within socially regulatory mechanisms.

Further Reading

Howitt, R. (2003) Scale. In: J. Agnew, K. Mitchell, and G. Ó Tuathail (eds.), *A Companion to Political Geography*. Oxford: Blackwell, pp. 138–57.

Marston, S., Woodward, K., and Jones III, J. P. (2007) Flattening ontologies of globalization: The Nollywood case. *Globalizations* 4.1: 45–63.

Sjolander, C. T. (1996) The rhetoric of globalization: What's in a wor(l)d? *International Journal* 51: 603–16.

Tagg, J. (1991) Globalization, totalization and the discursive field. In: A. D. King (ed.), *Culture, Globalization and the World-System: Contemporary Conditions for the Representation of Identity*. Binghamton, NY: Department of Art and Art History, SUNY Binghamton, pp. 155–60.

Electronic resources

100 Mile Diet: Local Eating for Global Change (http://100milediet.org)

Global Citizenship (www.oxfam.org.uk/education/gc)

Glocal Ventures (http://glocalventures.org)

The Glocal Forum (www.glocalforum.org)

Chapter 5

Globalizing Empires

Chapter summary: This chapter shows how, beginning in the fifteenth century, empire building and growing trade linkages reworked geographical connections between different parts of the planet. Such practices can be read as an early element in what is today called globalization.

- Mercantilism and the Geographical Ties that Bind
- Nineteenth- and Twentieth-Century Imperialism as Geographical Practice
  - Britain and India
- The Geographical Structure of Non-Imperial Economic Linkages
- Geographical Legacies

Colonization is the expansive force of a nation, its power of reproduction, its dilation and multiplication across space.
  (Paul Leroy-Beaulieu, De la colonisation chez les peuples modernes, *1874)*

As long as we rule India, we are the greatest power in the world. If we lose it we shall drop straightway to a third rate power.
  (Lord Curzon, Viceroy of India, 1901)

In 2003, a man's body was recovered from a bog in Clonycavan, Ireland, near the border with Northern Ireland. He had been ritually murdered. Initially, police believed him the victim of a political killing related to "The Troubles," as the violence which had rocked the area since the late 1960s is euphemistically called. However, it soon became clear that this unfortunate being had been dumped into the bog some 2 300 years ago, where conditions preserved his corpse. Of import here, though, is not the manner of his death – brutal as it was – but how he lived, for "Clonycavan Man" was from the upper echelons of his society, having used on his hair a gel made from plant oils and resin which scientists have identified as coming from southwest France. Meanwhile, at about the same time Clonycavan Man was being lowered into his boggy sepulcher, 4 500 miles eastwards Alexander the Great was expanding trade routes from the Eastern Mediterranean into India through the time-honored medium of military conquest. Indeed, so developed did the routes he established become that by the first century BC Roman merchants were importing such quantities of Chinese silk that the Senate tried to ban the trade because of the gold outflows it entailed, whilst by the first century AD the *Periplus Maris Erythraei*, a travel and commercial guide, was detailing dense trading networks linking India, East Africa, and the Arabian peninsula, a trade which has left tangible evidence in the form of Roman artifacts recovered from archeological sites in Tanzania (Chami 1999) and Vietnam.

My purpose in recounting the linkages within which Clonycavan Man and ancient Greek, Roman, Chinese, and African traders were embroiled is to remind us that the process of connecting different parts of the globe via long-distance trade – a process seen as central to globalization – has long been in motion. For sure, at certain times in particular regions the process has run a little quicker or slower and covered longer or shorter distances. But the point is that it has been an ongoing process, one in which humans have been engaged for millennia. My objective here, then, is to provide an historical overview concerning how economic practices have linked different parts of the globe. By necessity, the narrative involves an element of periodization, a process which imposes artificial breaks on the dynamic flow of the temporal and spatial processes under examination. With this in mind, though, here I focus upon the development of global trading empires and early colonies by various European powers in the fifteenth to eighteenth centuries, and the growth of empires (primarily European, but US and Japanese also) during the nineteenth and twentieth centuries linked to industrialization in the imperial metropoles. Certainly, empire building is as old as recorded history. From the Hittites and Babylonians to Alexander and the Romans, from Chinese Emperors to the Moors, Incas, and pre-colonial Ghana and Kongo, empires have waxed and waned, in the process bringing previously separate peoples into contact and transforming extant ways of life. However, although these empires were quite different, they shared one characteristic: the territorial swaths they cut were confined to the regions within which they arose. By contrast, the imperialism of the fifteenth to twentieth centuries was quite different in geographical scope, resulting in the carving up of virtually the Earth's entire surface. Through examining such empire building, then, we can begin to understand how and why our contemporary world is linked spatially in the way it is.

Mercantilism and the Geographical Ties that Bind

Long-distance trade and the geographical extension of credit have long tied different parts of the globe together. The Greeks and Romans had sophisticated means of tendering credit over large distances, as did the Arab traders of the Middle Ages, whilst by the twelfth century the military-religious order the Knights Templar had developed a complex banking system stretching across Europe and into Asia. For their part, by the ninth century Indian merchants were trading throughout the Persian Gulf, and by the fourteenth were regularly sailing to China. In the early fifteenth century Chinese navigators sailed to Sri Lanka, India, the Persian Gulf, and East Africa (Ray 1993) and by the early seventeenth had established communities in the Philippines and Japan (Gungwu 1990). By the early eighteenth century Indian textiles, long traded with coastal East Africa for slaves, gold, and ivory – a phenomenon which spawned African communities in India and parts of Arabia (Harris 1971) – were finding their way into West Africa. By 1800 Indian traders could be found from Yemen and Zanzibar to Malaysia and Thailand, whilst communities of Indian merchants had existed in Russia for two centuries (Markovits 2000). On the other side of the Atlantic pre-Columbian traders had spread plants like maize, pumpkin, and squash from Meso-America as far north as southern New England by about AD 1000 (Chomko & Crawford 1978).

Beginning in the late fifteenth century, however, European "voyages of discovery" started to transform dramatically the way in which the world economy was being linked together. The story of such explorations is well known. Portugal and Spain largely divided up South America and parts of the Pacific, whilst the Portuguese also exerted control over parts of coastal Africa and India. Britain, France, and Spain carved out colonies in North America, with Britain also securing control over large segments of Oceania and Asia, including India, Malaysia, and Borneo. The Dutch took over Indonesia and parts of India, the Caribbean, and South and North America. Other European nations established smaller empires – the Swedes in North America, the Caribbean, and West Africa, the Danish in the Caribbean, West Africa, India, and Indonesia, whilst even the tiny Duchy of Courland (in today's Latvia) held colonies on Tobago and the Gambia River (Berkis 1969). Whereas in some cases colonies were founded as crown or government possessions, in others they were established by private companies like the Dutch United East Indian Company (*Vereenigde Oostindische Compagnie* [VOC]), the Dutch West Indian Company, and the British East India Company (EIC). Shaped by mercantilist economic thinking, which held there to be a finite amount of trade available in the world of which each nation had best get its share, European merchants and their governmental sponsors increasingly looked to the Americas, Africa, and Asia for trading possibilities and to acquire the gold and silver seen as embodiments of national wealth. Indeed, bullion exports from Spanish America – some 134 000 tonnes of silver equivalent between 1493 and 1800 (Barrett 1990: 237) – would fundamentally transform how the world was networked economically. Clearly, export affected producing regions, as *conquistadors* ransacked them and turned much of the indigenous population into slaves in the mines of places like Potosí, Bolivia. But it also dramatically impacted the European economy: silver's availability for coinage relieved liquid capital shortages created by population increase and growing consumerism; the more cheaply produced bullion encouraged reductions in gold production in both West Africa and Europe, together with silver production in the latter, with all its attendant consequences for communities' diminished influence in the wider economies in which they were implicated; and the influx of large quantities of bullion fueled inflation, initially in Spain but later across Europe, an inflation which benefited property owners and commodity producers but worsened wage-earners' positions (Hamilton 1934).

Bullion export, though, had implications for a wider geographical area than just Europe and the Americas. For instance, whereas silver imports from the Americas (coming via Manila) initially boosted the Chinese economy, China became so dependent upon this silver that when imports declined after 1640 its economy suffered significant recession. Equally, as European trade with India expanded, silver and gold from the Americas flowed into the subcontinent to pay for the cottons and spices Europeans increasingly craved. Trade for other commodities also began reworking connections between different parts of the globe. Hence, the VOC sought spices from Asia and Africa, whilst by the 1660s the EIC was increasingly dealing in Indian textiles. These new linkages had important impacts at both ends of the trade route, and at points in between. For Europeans, not only did trade bring new products (like Chinese porcelains), which fueled the development of both a consumerist class at home and several domestic industries to satisfy it (like furniture-making, which relied upon imported hardwoods from Africa, Asia, and South America), but it also spawned new tastes in cuisine and social habits – by 1731, the value of West Indian sugar imported to Britain topped £1 million per year [2005 = £17.2 billion] and would reach £2.1 million [£27.3 billion] by 1761, whilst coffee and tea imports likewise mushroomed (Ormrod 2006: 4 and Table 3). It also brought a flood of wealth into Europe. Indeed, it has been estimated that by 1700 India accounted for more than half of Britain's trade (Keay 1991: 170), whilst between 1650 and 1780 the VOC extracted some 600 million gold florins' worth of spices and other profits from Indonesia alone (Mandel 1962b: 443), such that Amsterdam soon became a – if not *the* – crucial financial center of the world economy and arguably the most important port in Europe.

For communities in Africa, Asia, and the Americas, the growing interactions with European traders also had significant impacts as the arrival of European goods disrupted long-standing trading patterns and as trading posts' establishment at places like Ghana's Elmina in 1482 encouraged shifts in regional trading patterns – whereas previously trade in salt and other commodities had linked coastal populations with the continent's interior, increasingly the geography of exchange linked the coast with Europe. Likewise, escalating European demand for spices and other commodities had an impact on agricultural land use as production was increased to meet it. Thus, beginning in the 1640s the introduction of sugar cane revolutionized West Indian economies, not only transforming agricultural patterns but also dramatically increasing demands for labor, demands which would be met first by the enslavement of the indigenous population and then by Africans. As sugar production to satiate the European and North American sweet tooth exploded, thousands of small farmers on the islands were displaced by large plantation owners, with such displacement ultimately reducing the volume of trade conducted by British and local West Indian merchants and leading to its capture by well-capitalized North Americans, who used the profits from trading West Indian sugar to pay for European consumer goods imported into places like Boston (Pares 1956).

Plantation agriculture also stimulated the transatlantic slave trade, which both increasingly linked the economies of Africa with those of the Americas and the European nations craving sugar, rum, tobacco, and, later, cotton, and transformed the geography of economic relations within Africa, as slaving parties ventured into the interior for captives. Hence, in West Africa parts of the Sahel became linked to the coast and, therefore, to the Americas through the securing of slaves for work across the Atlantic, whereas in southern Africa Angola's interior was increasingly networked to the outside world through the slave portal of Luanda. Plantation agriculture in the Americas even touched the East African hinterland, as slaves destined for Brazil and Cuba were wrenched from upland Mozambique. The transatlantic trade also transformed the geographical relationships between parts of the world less directly involved in slave transportation. Thus, whereas in the seventeenth century Indian cottons had found their way to West Africa via the East African trade, European slavers increasingly established direct routes to the Atlantic coast, trading Indian textiles for slaves. Moreover, large quantities of Indian textiles to clothe slaves were exported to the Americas, an activity which both flooded India with bullion to pay for them (Parthasarathi 2004) and devalued the gold and silver stocks of European powers like the Ottomans. Furthermore, the slave trade encouraged the spread of agricultural commodities like peanuts, introduced into West Africa from Brazil by the Portuguese in the sixteenth century as a cheap food by which to sustain slaves on the crossing. The trade's geographical implications, then, cannot be underestimated: as Klein (1990: 289) has argued, it "involved the direct participation of East Indian textile manufacturers, European ironmongers, African caravan traders, European shippers, and American planters in the purchase, transportation, and sale of the largest transoceanic migration of workers known up to that time in recorded history."

Colonial trade's importance for reshaping geographical connections between different parts of the world, then, was unprecedented. By the mid-seventeenth century the Atlantic economy was becoming increasingly integrated as West Indian sugar and rum was shipped to Europe, North American horses, lumber, and agricultural commodities were imported into the Caribbean, and European-manufactured goods were sent to the Americas. By the same token, Asia and Europe were being connected through trade in porcelain, silk, tea, textiles, and spices. Indeed, between 1664 and 1700 the EIC imported into Britain £6.4 million [£110.7 billion] worth of textiles from Asia and exported there £11.3 million [£205.8 billion] worth of goods (Chaudhuri 1978: 547, 507). Moreover, a degree of singularity was already being seen in the world market for certain goods – from about 1720 the European market for Indonesian sugar declined precipitously as imports of cheaper Brazilian sugar increased, such that many Chinese sugar traders went bankrupt. Equally, the establishment of European colonies produced a transfer in the geographical locus of power, such that the fortunes of planters in the Caribbean or of tea merchants in China were ever more linked to decisions made by European bankers in places like Amsterdam. Millions of people's destinies, then, were increasingly being decided in far-off locations.

If such trading linked different parts of the world together in new ways, its profits greatly fueled European economic expansion, providing a crucial stimulus to the eventual emergence of capitalism. Thus, Mandel (1962a: 109) records that between 1636 and 1645 the VOC sold for 300 florins a head 23 000 African slaves purchased for fewer than 50 florins each. For its part, the EIC was making 32 000 percent profits selling Indonesian nutmeg in seventeenth-century London (Keay 1991: 4) and clearing a more modest average 281 percent profit on textiles imported to Europe from Asia between 1664 and 1704 (Chaudhuri 1978: 547).[1] The profits from colonial trade, then, were enormous. Mandel (1962b: 443–5) estimates that between 1760 and 1780 alone, profits from Britain's trade with the West Indies and India doubled the capital available for investment within the British mainland. These financial injections provided sufficient economic catalyst that not only were European class relations impacted – merchants increasingly challenged the power of the landed aristocracy – but money was available for entrepreneurs and inventors to develop new production methods, methods which would presage industrial capitalism's emergence in Europe.

Nineteenth- and Twentieth-Century Imperialism as Geographical Practice

Although European empire builders of the fifteenth and sixteenth centuries saw their greatest expanse in territory come in the Americas, and those of the seventeenth and eighteenth century in Asia (often through surrogates like the EIC), the nineteenth-century European carve-up of Africa marked one of the most rapid and dramatic transformations ever in the planet's political economy. Whilst various European nations had long controlled significant swaths of coast, the 1884–5 Berlin Conference provided the major European powers (together with the United States, which had an interest particularly in Liberia) an environment in which they could carve up amongst themselves the interior. By the time they were done, colonies in Africa would contain three-quarters of the land area and more than half of the population of all colonies of the period. Whereas in 1800 Western powers controlled about 35 percent of the Earth's surface, by 1914 they controlled 85 percent, an expansion which "united [the world] into a single interacting whole as never before" (McNeill 1982: 260–1, quoted in Said 1994: 6). By 1909 the British Empire alone covered one-fifth of the Earth and contained one-quarter of its population, whilst France controlled more than a third of Africa. Such was the sense of gaiety with which this deadly serious territorial division was conducted that British Prime Minister Lord Salisbury would later state that the colonial powers had been engaged "in drawing lines upon maps where no white man's feet ever trod; we have been giving away mountains and rivers and lakes to each other ... hindered [only] by the small impediment that we never knew where exactly these mountains and rivers and lakes were" (quoted in Castellino 1999: 529). In such an enterprise, the destinies of millions would be determined by those who could control the new cartographic representations of the continent's interior, as the negotiated placement of lines on maps divided peoples or placed enemies together in new territorial configurations, frequently with disastrous results for the colonized (Bassett 1994).

In addition to its geographical magnitude, what made nineteenth-century imperialism distinctive was that whereas in prior centuries some raw materials had been brought to Europe for finishing into other goods (e.g., teak for furniture), commodities from colonies were largely imported in finished form and Europe or North America were frequently the location of their final consumption. By the nineteenth century, though, the imperial enterprise had changed significantly. Increasingly, colonies became providers of raw materials for European and US factories and markets for the goods so produced, thereby dramatically transforming the world economy and the geographical connections between places. Certainly, empire building was never solely about economic matters. Many of those involved – from planners in imperial metropoles to civil servants, colonial bureaucrats, and soldiers – believed they were engaged in a moral imperative, one of benefit to those to whom they brought "civilization." But the new geographies laid down during the nineteenth and twentieth centuries through imperialism must be understood within the context of European industrialization. Indeed, industrial expansion and imperial expansion largely went hand in hand, in several ways.

First, colonies provided raw materials: rubber from Congo, copper from Zambia, tin from Malaysia, and myriad other commodities essential to fuel industrial production in Europe, the United States, and Japan. Even commodities like tea played a role, providing an invigorating beverage that allowed tired British workers to keep laboring after a brief respite in the form of a new institution – the "tea-break." Typically, such commodities were produced using external investment and imported management – prior to World War I, for instance, the Indian jute industry (which enjoyed a virtual global monopoly) was entirely owned and managed by Europeans (Roy 2000: 166). Even where export crops were grown by local smallholders, much of the capital to finance operations was obtained from branch banks headquartered in Europe or the US. In only a small number of instances – like Kenya, where large numbers of British settlers immigrated in the early 1900s – was export production largely financed by resident capital (Greaves 1954: 6). The degree to which these raw materials were important can be shown by examining the nation with the largest empire – Britain – and its relationship with its colonies. Thus, between 1854 and 1934, the Empire's share of imports into Britain of foodstuffs and raw materials grew from 19 percent and 26 percent respectively to 42 percent and 31 percent (Schlote 1952: 99) (Table 5.1), and by 1934 the colonies provided the bulk of many raw materials (Table 5.2). Similar relationships developed elsewhere. Whereas in 1892 imports from French colonies represented 8.6 percent of all imports into France, by 1929 it was 12.0 percent, whilst the value of colonial goods imported jumped from 358 million francs [€21 billion] to 7.01 billion francs [€34 billion] in this timeframe (Southworth 1931: 60, and Table IV) (Table 5.3). By 1956 24 percent of France's imports came from colonies (Sicking 2004: 210). Between 1892 and 1896 15 percent of all imports into the Netherlands were colonial, as were 10 percent of Spanish imports, 18 percent of Portuguese, and even 1 percent of Danish imports (Flux 1899: 491). In the US's case, although it possessed less overseas territory than major European powers, colonies were also important. Whilst in the early 1890s 15–20 percent of Philippine exports (mostly agricultural products, minerals, and timber) went to the US, by 1932 that figure had risen to 87 percent (Nagano 1997).

Table 5.1 Imports of foodstuffs and raw materials to Britain from the British empire, 1854–1934 (% of total imports)

Year	Foodstuffs	Raw materials	Year	Foodstuffs	Raw materials
1854	19.1	26.2	1900	20.9	24.3
1860	18.2	23.5	1913	27.0	28.0
1870	17.5	27.4	1925	36.4	30.7
1880	17.4	31.9	1929	30.7	28.8
1890	19.1	30.7	1933	42.3	31.1

Note: Figures for 1925, 1929, and 1933 do not include the Republic of Ireland.
Source: Schlote (1952: 99).

Table 5.2 Imports of selected goods to Britain from the British empire, 1854–1934 (% of total imports)

	1854	1860	1870	1880	1890	1900	1913	1925	1929	1934
Total grain, flour	5.8	6.7	12.6	15.7	16.5	13.6	35.3	48.6	32.1	44.4
Copper ore	9.1	12.6	19.0	17.4	12.8	15.2	37.4	55.3	44.3	80.0
Tin, smelted	71.2	78.6	49.1	91.0	94.4	88.3	94.8	84.5	88.4	60.3
Zinc, smelted	0.0	0.0	0.0	0.0	0.5	0.4	0.9	11.4	26.1	79.5
Sheep's wool, lambswool	70.6	68.5	88.5	87.0	88.8	84.5	80.2	86.0	85.8	83.4
Printing paper	–	–	–	–	–	3.2	19.6	36.5	57.2	65.2

Note: Figures for 1925, 1929, and 1933 do not include the Republic of Ireland.
Source: Schlote (1952: 164–5).

Table 5.3 Imports and exports between France and French colonies, 1892–1929

Year	French exports to France's colonies as a proportion of all French exports (%)	Imports into France from French colonies as a proportion of all French imports (%)
1892	8.5	8.6
1897	9.9	10.1
1902	12.0	11.0
1907	12.3	9.8
1912	13.5	10.8
1917	15.2	8.1
1922	14.2	7.7
1927	13.9	11.4
1929	18.9	12.0

Source: Southworth (1931: Tables III and IV).

Second, colonies served as markets for products manufactured in the imperial cores. For instance, whilst in 1892 9 percent of French exports went to France's colonies, by 1929 the figure was 19 percent (Southworth 1931: Table III) and in 1956 was 31 percent (Sicking 2004: 210). Likewise, imports into the Philippines from the US jumped from 1 percent of all Philippine imports in 1892 to 74 percent by 1940 (Nagano 1997). By the mid-1890s, 33 percent (£74.8 million annually [£60.2 billion]) of British exports were going to the colonies. During the same time period, fully 24 percent of Spain's exports went to its colonies, as did 9 percent of Portugal's, 5 percent of Dutch exports, and 2 percent of Denmark's (Flux 1899: 491). Although proving beneficial to the imperial powers, such exports frequently required displacing indigenous manufacturing industries, a goal usually achieved by a combination of physically destroying local production capacity and implementing taxation policies favoring imported goods.

The third way colonies shaped the planet's rewiring in the nineteenth and early twentieth centuries was through facilitating the creation of an urban proletariat in the imperial nations by providing the metropoles food, thereby allowing agricultural laborers to migrate to new industrial regions. In Britain's case, for instance, wheat was increasingly imported from colonies like Canada, Australia, and India, as well as non-colonies like the US, whilst colonies also provided meat (especially mutton from Australia and New Zealand [Perren 1978]), fruit, butter, and cheese (Schlote 1952: Table 21). This had not only dramatic implications for international flows of commodities, but it also significantly shaped agricultural practices in producing regions, as the British working classes' ecological footprint was extended across the Empire and changing tastes in Britain affected land use in Canada, Australia, and India. Furthermore, although such food exports fed European workers, the result in the colonies was frequently famine, which occurred more regularly and was more geographically widespread than previously (Ghose 1982).[2]

Fourth, colonies served as an economic and political safety valve by providing destinations for surplus capital and labor. Thus, when Germany – experiencing both a population boom and an industrial depression – established African colonies in the 1880s it was hoped they would serve as outlets for the country's unemployed (Henderson 1938). During the nineteenth and early twentieth centuries, 1.4 million Portuguese left for Brazil, looking both for economic opportunity and escape from political repression, whilst others went to Angola and Mozambique (Segal 1993: 16). In the case of Britain, 29 percent of all emigrants (62 000 people) who left the country between 1843 and 1852 went to colonies (primarily Canada, Australia, and New Zealand), whereas by 1911–13 the figure had grown to over 65 percent (302 000 people) (Thomas 1973: Table 7). Meanwhile, by the late 1920s the market value of French colonial stocks and bonds stood at 16 billion francs [€79 billion] (Southworth 1931: 99), whilst British overseas investments grew from some £1 billion [£963 billion] in 1874 to almost £4 billion [£2 trillion] in 1913, of which just under 50 percent was invested in British colonies (Saul 1960: 66).

Colonies, though, were also providers of capital and labor. Saul (1960: 56), for instance, has calculated that in 1880 alone British investors earned some £8 million [£8 billion] in interest payments from their Indian loans, whilst between the mid-nineteenth century and the early 1910s 1 260 000 Indian laborers left for other parts of the Empire, including 240 000 for Guyana, 152 000 for South Africa, 39 000 for East Africa, 130 000 for Malaysia, 452 000 for Mauritius, and 56 000 for Fiji. Indeed, between 1850 and 1910 more than 150 000 indentured laborers every decade departed one region of the British Empire for another, such that by World War I's outbreak 14 000 from British West Africa had gone to Guyana and nearly 65 000 Chinese had migrated to South Africa. Importantly, these figures do not include non-indentured migrants, like the 540 000 Chinese who left Hong Kong between 1854 and 1880, of whom 119 000 went to Australia. Neither do they include those from British colonies who went to locations outside the Empire, like the 5 000 Hong Kong Chinese who left for Cuba between 1856 and 1858 (Northrup 1999: Tables 5.1 and 5.3). Such migrations were not confined to the British Empire. French authorities imported indentured Indians to places like Réunion, whilst between 1853 and 1870, 19 910 Africans and 41 261 Indians went to French Guiana, Guadeloupe, and Martinique (Northrup 2002). For their part, Portuguese cocoa producers in São Tomé and Príncipe imported 131 000 contract laborers from Angola between 1880 and 1910 (Ishemo 1995: 164), whilst Japan began importing Korean workers after its 1910 conquest of that nation – by 1930 300 000 had migrated (Min 1992: 10). Meanwhile, particularly within Africa, European imperialism helped establish the phenomenon of long-distance labor migration as workers came seeking employment on plantations and in mines. Such migration not only had significant impacts upon family structure – separating husbands from wives and children – but also encouraged the spread of ailments like syphilis and intestinal worms, which workers either brought with them or acquired in their destination locations.

In discussing imperialism's role in reworking trade routes and geographies of power, though, it is important to recognize that although their goals were similar – securing raw materials and markets, spreading "enlightenment" as part of what the French called the *mission civilisatrice*, and gaining political advantage over rivals – there were nevertheless significant differences in the imperial powers' relationship to their colonies. For example, although colonies were certainly important for France, they were more so for Britain. Likewise, whereas the French imagined their colonial subjects would eventually become French citizens through *assimilation* (Betts 1960), this was certainly not the attitude of the British nor most colonial powers. Equally, although all colonial powers utilized forced labor and sometimes prisoners to build infrastructure like railroads (Sene 2004 provides a Senegalese account), some regimes – especially the Belgians and the Portuguese – gained a reputation for greater brutality than others, with rubber workers' treatment in Belgian King Léopold II's infamous Congo Free State sufficiently appalling to spawn one of the first international human rights campaigns (Hochschild 1998). Furthermore, colonial relationships changed over time and as the imperial embrace tightened, individual colonies increasingly traded with their respective imperial metropoles, although the degree to which this occurred varied. Thus, as British settlers arrived in larger numbers in Canada, South Africa, Australia, and New Zealand, these colonies became increasingly important sources of raw materials and export destinations – in 1860 they provided 35 percent of all imports to Britain from the Empire and accounted for 36 percent of exports to it, whilst by 1922 the figures were 65 percent and 47 percent (Schlote 1952: 93 and Table 20b). Between 1892 and 1896 British colonies annually imported an average £167 million [£139 billion] worth of goods, of which 55 percent came from Britain. At the same time, 61 percent of the French empire's imports came from France and colonies consumed 62 percent of French exports, although the figures for individual colonies varied considerably – 36 percent of Indochina's imports came from France and 10 percent of its exports went there, whilst the figures for Réunion were 60 percent and 94 percent and for Algeria were 79 percent and 82 percent (Flux 1899). In 1960, on the eve of independence, Côte d'Ivoire purchased 65 percent of its imports from France and sent 67 percent of its exports there (Milhomme 2005). Other empires likewise exhibited tight economic links – in Dutch Surinam, by the early 1890s 55 percent of imports came from the imperial metropole and 51 percent of exports went there, whilst for Portuguese Angola the figures were 46 percent and 98 percent, for Spanish Cuba they were 49 percent and 11 percent, for German South-West Africa 75 percent and 12 percent, and for German East Africa 27 percent and 22 percent (Flux 1899).

Imperialism, then, transformed the planet's economic geography and linked various parts of the globe together in new and different ways. However, to gain a better understanding of the complexities of imperial relationships and *how* they came about, here I look at a case study of Britain's relationship with India, for the magnitude of the metamorphosis in their economic bonds illustrates not only how imperialism reworked spatial connections between places but also how it shaped the uneven geographical development which continues to contour contemporary processes.

Britain and India

British interests in India were first secured in 1617 when the Mughal emperors, holding sway over much of the subcontinent, gave the EIC trading rights. After the 1757 defeat of Indian forces at Plassey, the Company expanded its dominion over eastern India and, eventually, the subcontinent. Indeed, for the first century of its existence "British" India was actually the EIC's private concession (only after the 1857 Mutiny did the Crown take formal control). Eventually, though, India would become what Viceroy Lord Curzon called "the pivot of [the] Empire" (quoted in Das 1969: 44), "a magnificent jewelled pendant hanging from the Imperial collar" (Curzon 1909: 8). Indeed, by the 1880s British investments in India totaled £270 million [£272 billion], roughly one-fifth of Britain's entire overseas investment (Ferguson 2003: 215). Between 1892 and 1896, on average Britain yearly exported to India goods worth £37.8 million [£32.9 billion] and imported from India goods worth £22.7 million [£19.8 billion] (Flux 1899: 495). In 1906, of the £78 million [£47 billion] worth of goods imported into India (exclusive of specie), only 31 percent came from outside the Empire (Morison 1908: 424).

Although Britain's India trade involved several commodities – jute, sugar, wheat, silver, indigo – it is in the realm of cotton textiles that the colonial relationship's unfolding is most clearly illustrated. Indian merchants had long played significant roles in global trade, and during the seventeenth and eighteenth centuries India's textile industry was at a technological level at least equivalent to that of Britain, if not more advanced.[3] In the 1660s Indian cloth imports into Europe reached 10 million yards annually, and by 1684 the EIC was importing 45 million yards. Overall, Indian producers exported more than 100 million yards annually (Bronson 1982). By 1700 78 percent of all Asian imports to Britain – mostly textiles – came from India (Ormrod 2006: 2) and India was such a well-developed producer that London silk weavers rioted to protest the imports threatening their business. Indeed, fear of competition led the British government to limit Indian printed silk and calico imports, whilst later legislation banned the import into Britain (though not the colonies) of all Indian plain white calico. Government efforts to exclude Indian textiles continued throughout the century, and by 1813 Indian calicoes faced a 78 percent duty and muslins a 31 percent one (Robins 2006: 148). Unsurprisingly, this had a deleterious impact upon Indian manufacturers, and, with the British industry's expansion in the early nineteenth century, Indian producers began to lose their prominence in world trade (Parthasarathi 2001). Thus, although when the first cargo of Lancashire textiles was dispatched in 1814 Indian exports of cotton manufactures to Britain still amounted to £2 million annually [£6.85 billion] (Maddison 1972: 57), between 1828 and 1840 exports plummeted 48 percent (Chaudhuri 1966: 347). Whereas Dhaka in Bengal exported 2.85 million rupees' worth of textiles to Britain in 1753, by 1800 this had fallen to 1.36 million rupees and by 1818 exports to Britain had ceased (Robins 2006: 148). In the case of Kolkata, in 1813 the city's merchants exported £2 million worth of cotton goods but by 1830 were importing that amount (Mandel 1962a: 372).

At the same time, British efforts to create openings for Lancashire cottons by levying substantial taxes on Indian-made textiles produced for consumption within India (thereby largely pricing them out of the market) led British exports to the subcontinent to grow an average 61 percent annually between 1814 and 1820 (Farnie 2004: 399). Whereas in the mid-1810s the value of textile exports to India was relatively small – £100 000 [£333 million] – by 1850 they had risen to £5.2 million and by 1896 to £18.4 million [£11.1 billion and £14.4 billion] (Charlesworth 1982: 33). In fact, facilitated by favorable tariff and tax structures and the emergence of distance-shrinking technologies like the steamship and the 1869 opening of the Suez Canal, both of which shortened transportation times and so cheapened imports, by the early 1870s British cotton manufactures comprised 56 percent of all imports into India and 70 percent of all British exports to India (Table 5.4). In terms of volume, British exports increased from 1 million yards of piece-goods in 1814 to 13 million in 1820, 315 million in 1850, 995 million in 1870, 1.4 billion in 1880, and 2.04 billion in 1896 (Farnie 1979: 91; Bairoch 1974: 565; Bairoch 1993: 89). This latter figure represented 39 percent of all British cotton exports that year (up from 23 percent in 1850), and by the late nineteenth century India was British textile producers' single most important market. Such changes had profound impacts: between 1850 and 1880 3.6 million jobs were lost in the Indian handloom sector (Prakash n.d.: 33) and by 1896 Indian mills provided just 8 percent of the cloth consumed in the subcontinent (Maddison 1972: 57).[4] For sure, not all these losses were the result of British imperialism, as there were also local structural impediments, like poorly developed financial institutions capable of

Table 5.4 British cotton goods as a proportion of the total value of British merchandise exports to India and as a proportion of the total value of all (British and non-British) merchandise imports into India, 1820–96

	Proportion of total value of British exports to India (yearly average)	*Proportion of total value of all goods imported into India (yearly average)*
1820–4	30.5	—
1825–9	42.0	—
1830–4	54.5	—
1835–9	60.6	—
1840–4	61.7	—
1845–9	61.3	40.6
1850–4	63.0	48.5
1855–9	54.6	43.9
1860–4	63.1	45.3
1865–9	62.3	50.1
1870–4	69.5	56.4
1875–9	65.9	50.4
1880–4	66.4	49.5
1885–9	64.4	46.6
1890–6	62.9	42.6

Note: All figures as percentages; (—) = no data.
Source: Farnie (1979: 116–18).

channeling household savings into industrial development (Roy 2002). Equally, many of these weavers found work in other sectors, like railroad construction and agricultural labor (Harnetty 1971). Nevertheless, imperial policies to turn India into a market for British textiles were felt in myriad homes across the subcontinent. As William Bentick, Governor General in the 1830s, plainly put it: "The misery hardly finds parallel in the history of commerce. The bones of the cotton-weavers are bleaching the plains of India" (quoted in Marx 1976 [1867]: 558).

Meanwhile, as the export of finished textiles from India to Britain fell, the export of raw cotton increased phenomenally, growing from 34.5 million lbs in 1846, to 204 million in 1860, to a staggering 615 million lbs in 1866, 45 percent of all cotton imported into Britain that year (Marx 1976 [1867]: 579; Harnetty 1972: Table 3.1). This increase in the export of raw cotton for European (and, later, Japanese) mills had significant impacts upon land-use patterns – some land formerly used for subsistence crops or other cash crops like sugarcane (as in the Punjab) was turned into cotton fields, whilst the cotton boom sparked the clearing of previously uncultivated land for both cotton and subsistence crops. Although high cotton prices in the 1860s meant many farmers actually made some significant sums of money, the increased value of the land which such production brought led the British to demand higher taxes, and when cotton prices fell many peasants found themselves heavily indebted to local moneylenders and lost their land (Harnetty 1971).

During the eighteenth century, then, India had enjoyed a competitive advantage in production – the result of its more efficient agriculture allowing weavers to produce textiles more cheaply – and weavers had generally had higher incomes and greater financial security than their British counterparts (Parthasarathi 1998: 82). However, British "free trade" policies (which exposed Indian manufacturers to competition from British-made goods but which protected British manufacturers by imposing 70–80 percent tariffs on Indian exports to Britain) and the manipulation of taxation schedules (charging 6–18 percent duties on the inter-regional movement of Indian textiles whilst allowing British goods free movement, together with placing taxes on Indian-made textiles to the advantage of British-made imports [Bagchi 1982: 80; Lamb 1955: 468]), undercut Indian producers. Whereas before colonization textiles accounted for nearly three-quarters of India's industrial production and 60–70 percent of its exports, by the mid-nineteenth century Indian textile exports to Europe had fallen dramatically and were also dropping sharply in Africa and North America (Bairoch 1993: 88–9) – between 1814 and 1835, for instance, British cotton cloth exports to India rose 51-fold whilst finished textile imports from India fell by three-quarters (Robins 2006: 148). Indeed, British mills sealed their commercial victory in 1831 when they shipped to India 40 million yards of cottons, matching for the first time the quantity imported from India a century and a half previously (Bronson 1982).

The new relationships – geographic and economic – wrought by British imperialism, however, were complex and Indian merchants were not pushed completely aside. Rather, many shifted into producing cheaper grey cloth, which resulted in new and quite different market opportunities and production geographies (Rothermund 1988: 55). Equally, whereas Indians largely lost out on the trade between India and Europe, which came to be a virtual monopoly of large British trading houses in Kolkata, Mumbai, and Chennai, they did retain influence in the rest of Asia (especially China) and Africa (Markovits 2000). For instance, whereas between 1879 and 1884 India exported 30.1 million lbs of yarn to China (84 percent of all Indian yarn exported), between 1904 and 1909 the figure had increased to 220.7 million (89 percent) (Fukazawa 1965: 240). In fact, by the early twentieth century Indian exports to China had largely displaced British exports there – whilst British manufacturers exported 20 million lbs of cotton yarn to China in 1885, by 1913 British exports had fallen 90 percent (Saul 1960: 189). Indeed, the British were not indisposed towards Indian economic development if it augmented the Empire's markets and did not damage Britain's overall economic position or political goals (Maddison 1972: 35). Hence, whilst they pursued a (one-sided) policy of free trade when Britain was India's main supplier of textiles, when Japan emerged as a competitor after World War I the British implemented protectionist policies to shield the Indian market, even to the extent that this subjected British-produced textiles to tariffs. Equally, when Britain adopted a system of imperial preferences in 1932 in response to the Depression, this move towards protectionism had important effects, and after suffering over a century of decline the Indian industry gradually began to recover. With the growth of factory production of Indian textiles (from 422 million yards of piece-goods in 1900–1901 to 1.97 billion in 1924–5 [Fukazawa 1965: 238]), increased tariffs to protect "British" India from Japanese imports, and the Swadeshi movement's boycott of British textiles, between 1913 and 1937 British textile exports to India actually dropped 85 percent (Foreman-Peck 1983: 210).

Nevertheless, at the end of the day British policies towards India had fundamentally transformed the relationship between the two countries. Three interconnected elements in this changed relationship can be identified. First, whereas India had been a significant textile exporter in the pre-colonial period, it was turned into a net importer of such. Second, India largely became a captured market for British-manufactured goods, particularly textiles but other goods as well. Third, India was turned into a producer of raw cotton for Lancashire textile mills. The unfolding of this new economic and geographic relationship, in which British industrialists sought raw materials and markets in the colonies, was well understood by Marx (1976 [1867]: 579), who wrote at the time that, "By ruining handicraft production of finished articles in other countries [like India], machinery forcibly converts them into fields for the production of its raw material" – in this case, cotton. Only after India had been turned into a textile-importing nation were the discriminatory internal taxes placed upon Indian-made cottons repealed. Indeed, British calls in the 1840s for worldwide free trade as part of a Pax Britannica only became a mantra after much of the competition in places like India had already been destroyed by decidedly unfree trade. Furthermore, after India's conquest the structure of colonial trade was such that even if Indian-produced raw materials or finished goods eventually ended up in countries like Germany or the United States, they were frequently imported first into Britain and then re-exported. Hence, whereas India exported £8.2 million worth of raw jute to Britain in 1907, £3.0 million worth of that was re-exported for final processing (Morison 1908: 425), with the result that British importers and exporters – rather than their Indian counterparts – enjoyed the fruits of the trade.

The Geographical Structure of Non-Imperial Economic Linkages

Although imperialism brought new connections between different parts of the world and was crucial in shaping patterns of development in both the colonized and colonizing parts of the world, it is important to recognize that imperial powers also developed new economic relationships with non-colonies, relationships which transformed the nature and spatial extent of both parties' global interactions. Significantly, these relationships were often more substantial than were imperial powers' trading relationships with their colonies. Furthermore, one imperial nation's colonies frequently traded with those of another and even with other imperial nations themselves, whilst those few parts of the world not imbricated within the imperial nexus also developed new linkages. Thus, much as colonies like India played sizeable roles as providers of raw materials and markets, at no point between the mid-nineteenth and mid-twentieth centuries did Britain import more from either Africa or Asia than it did from the rest of Europe, nor did it export more to either than it did Europe (Schlote 1952: Tables 19 and 18). Equally, of the total £3.78 billion [£1.9 trillion] British overseas investment held in 1913, 53 percent was invested in non-colonies, including £755 million in the US, £760 million in Latin America, and £220 million in Europe (half of this in Russia) (Saul 1960: 67). In East Asia, Britain was the principal supplier of manufactured goods to Japan (which itself embarked upon empire building with its 1910 Korean annexation), including several of the warships which destroyed the Russian fleet in 1905, although by 1929 Britain's share of Japanese manufactured imports had fallen by three-quarters, replaced largely by the US (Foreman-Peck 1983: 210). For its part, although by the early twentieth century the US had begun to exploit its imperial relationship with colonies like the Philippines, other places were of greater import – of the $2.7 billion [$922 billion] worth of FDI held by US investors in 1914, virtually all was in Europe, Canada, the Caribbean, and Latin America (Lewis 1938: 606). The US had even established near-monopoly trading status with some European colonies – in 1894–5, for example, Cuba received 36.6 percent of its imports from, and sent 84.5 percent of its exports to, the US (Flux 1899: 511) and US dominance of Cuba's economy was highlighted by chocolate baron Milton Hershey's establishment of several company towns on the island to ensure a constant sugar stream for his Pennsylvania factories (Winpenny 1995). US merchants also played roles in expanding peanut production in Britain's West African colony of the Gambia in the 1830s and even during the US Civil War, when Northern markets could not secure peanuts from Southern states (Brooks 1975; Weil 1984).

One of the most significant cases of an imperial power developing substantial economic relations with places not formally colonies is that of Britain with South America, as Brazil, Argentina, and Uruguay became virtual British commercial captives during the nineteenth century. Domination of Brazil had its origins in the British government's helping the Portuguese royal family flee Napoléon Bonaparte, in return for which British imports were allowed to flow virtually unrestrained into Brazil. By 1808 Britain was exporting over £2 million [£6.98 billion] worth of goods to Brazil (Bethell 1970) and by 1827 was the most important supplier of Brazilian markets, outpacing even Portugal, with cotton textiles the dominant commodity imported from, and raw cotton the dominant product exported to, Britain (Manchester 1964: 312–14). This dominance continued, and in 1880 Britain supplied 51 percent of Brazil's imports and consumed 37 percent of its exports (Davis 2001: 378). Brazil's financial subservience, meanwhile, was secured through the London banking house Rothschild, which made its first Brazilian loan in 1823 and by 1855 had become the Brazilian government's exclusive London financial agent. Between 1855 and 1914 Rothschild made loans totaling £173 million [£186.6 billion] to the government, of which £37 million was for railroads linking Brazil's coast with the interior to facilitate the coffee industry's expansion, which Rothschild also bankrolled (Shaw 2005: 173). British gas companies also invested heavily, bringing light to cities like Pará and Pernambuco (Manchester 1964: 325). For its part, between 1880 and 1914 Argentina became one of Britain's most important raw materials and food suppliers, as well as a destination for capital and commodity exports (Ferns 1953). British capital facilitated the expansion of the Argentine cattle, sheep, and wheat industries, which itself required building railroads (financed by British capital) to connect coastal ports with the *estancias* of Patagonia and the Pampas (Pulley 1966). Similarly, in Uruguay British investments totaled £10 million [£9.8 billion] by 1875, with a further £25 million invested in the 1880s, mostly in railroads. By 1900 British investments exceeded £40 million [£26.0 billion], roughly comparable to Britain's investments in its West African colonies, and Uruguay sent some £2 million annually back to Britain in the form of repatriated profits (Winn 1976: 110–13). In Peru, British merchants acted as intermediaries in exporting guano, which greatly stimulated British agriculture (Miller & Greenhill 2006). As a result, between the 1820s and 1914 Britain was the pre-eminent investor in Latin America, with its holdings in the region (£1.2 billion [£607 billion]) representing one-fifth of its total foreign investment in 1913.

Even in the case of its former colony, the United States, British investors saw opportunities. By 1880 British demand for beef resulted in 156 490 head of live cattle being shipped from East Coast ports, whilst British investors began purchasing ranches in western states. Indeed, British capital was responsible for much of the American West's transformation into cattle pasture, as investors established 37 cattle companies with an initial value of over $34 million [£5.53 billion] between 1879 and 1900 (Brayer 1949: 91–2). Meanwhile, British importers bought US wheat, thereby sustaining the Great Plains agricultural economy and encouraging thousands of prairie acres to be ploughed under. The US also served as an outlet for surplus British capital and workers. Between 1843 and 1852 70 percent of all British emigrants went to the US, whilst by the 1870s capital exports to the US had become inversely synchronized with Britain's economic cycles – periods of high domestic investment were marked by low capital exports to the US, whereas periods of low British domestic investment were marked by high levels of capital exports, much of which went into building US railroads (Thomas 1973: 57, 97–8, and Fig. 23).

Other European countries also had significant investments in the Americas. From 1880 until 1903 French financiers were the largest foreign investors in Central America and were the prime initiators of building the Panama Canal, designed to shrink the globe by allowing ships to avoid the long trip around Cape Horn (Schoonover 2000: 192). For its part, the French government approved significant loans to Nicaragua in 1909 (subsequently withdrawn under US pressure), whilst financiers Crédit Mobilier and Crédit Lyonnais made major loans to El Salvador and Costa Rica just prior to World War I's outbreak. In the 1870s and 1880s, France purchased 7–13 percent of Guatemalan, Nicaraguan, Costa Rican, and Salvadoran exports and French goods constituted 11–21 percent of these countries' imports. By 1887 France was importing over 5 million francs' worth [€338 million] of goods annually, and by 1896 French investment in Central America totaled 270 million francs [€16 billion] (Schoonover 2000: 58–61). German merchants likewise had commercial interests in Central America, having exported goods to Guatemala at least since the 1830s. Although trade with the region was not particularly important for Germany's economy as a whole, for the Central American nations the German trade was very significant – Germany accounted for 53 percent, 20 percent, and 18 percent of Guatemalan, Nicaraguan, and Salvadoran coffee exports in 1913, whilst German manufacturers captured 20 percent of the Guatemalan import market, 15 percent of the Costa Rican, 11 percent of the

Salvadoran, and 9 percent of the Honduran (Schoonover 1998: 139 and Table 6). Equally, by 1906 German investors had sunk 250 million marks [€14 billion] into Guatemala, 35 million into Costa Rica, and 250 million into Nicaragua, whilst small settlements of Germans had sprung up across the region (Schoonover 1998: Tables 7 and 9).

In giving a fuller picture of how the world economy was being knitted together during this time it is also important to recognize that colonies, too, had important trading relationships with nations other than their imperial metropoles. Hence, although the bulk of India's exports went to Britain, in 1906–7 11 percent went each to Germany and China, 9 percent to the US, 6 percent to France, 4 percent each to Belgium and Japan, and 3 percent each to Italy and Austria-Hungary (de Webb 1908: 153). Likewise, although a French colony, Moroccan railroads built in the 1920s were largely funded by Swiss and Dutch loans, whilst British and Syrian bankers financed developments in French West and Equatorial Africa. By the same token, French investments in British West Africa totaled 1.6 billion francs in 1900 (White 1933: 317) whereas Chinese investors had more invested in French Indochina than did French investors and held a virtual monopoly on the rice industry (Southworth 1931: 101–2). Finally, and despite imperial tariffs, colonies' trade with their imperial metropoles could be replaced as other powers exerted their economic muscles – the proportion of manufactured goods imported into Australia, New Zealand, and South Africa from Britain declined from three-quarters in 1913 to three-fifths in 1929 (Foreman-Peck 1983: 210), the result of growing US competition.

The trading networks laid down by the early twentieth century, then, resulted in Cuban sugar being consumed in the US, Guatemalan coffee supped at Berlin breakfast tables, and Mozambiquan cotton processed in Portuguese factories. In terms of financial capital flows, London had come to sit at the center of a planetary network of criss-crossing credit and debt arrangements – by the 1870s the Atlantic cable meant New York newspapers carried information concerning the previous day's dealings in London, whilst the fact that London possessed the planet's central gold market, especially after discovery of huge quantities of the metal in South Africa, meant the Bank of England could readily shape exchange rates, interest rates, and movements of wealth across the world. As a result, the financial centers of Amsterdam, New York, Berlin, Paris, and elsewhere – and, through their own local credit arrangements, myriad other places across an ever-expanding financial horizon – became increasingly tied to, and disciplined by, London's financial markets. The outcome was the evisceration of much local financial autonomy, a point reinforced by Lord Curzon's exclamation that India's tariffs and taxes "were decided in London, not in India" (quoted in Davis 2001: 290). Equally, growing economic integration meant that prices for the world's commodities became progressively more harmonized on the basis of European and North American commodity exchange activities – cotton prices in the US, Brazil, India, China, and Japan, for instance, were increasingly shaped by Liverpool cotton exchange brokers.[5] Consequently, economic crises starting in one region of the globe were more readily transmitted elsewhere than was previously the case – the 1890 Argentine financial crisis which led to the collapse of the British Baring Brothers' banking house, for example, disrupted London capital markets and subsequently spread to Brazil, where Barings had investments, and thence across Latin America (Triner & Wandschneider 2005).

Finally, it is important to recognize that labor migration also played a crucial role in linking different parts of the globe. I have already detailed how indentured and free labor emigrated from one part of a particular empire to another, or from the imperial metropole to the colonies (and occasionally vice versa), or even how migrants left the colonies of one empire for those of another. However, of the estimated 90–100 million voluntary international migrants (Segal 1993: 16) who left their homelands for pastures greener between 1815 and 1914, a significant proportion did not fall into these categories. Thus, 5 million Italians emigrated to the US, whilst a further 2.4 million went to Argentina and 1.3 million to Brazil. Likewise, 5 million Germans left for the US and Canada, together with 300 000 for Argentina and Brazil. Some 300 000 Mexicans went to the US and Canada, as did 2.7 million Scandinavians. Approximately 11 million British migrants arrived in the US, whilst 12 million Chinese and 6 million Japanese went to other parts of Asia (Segal 1993: 16). In 1908, the first of some 189 000 Japanese arrived in Brazil, settling mainly in São Paulo and northern Paraná to work coffee plantations (Seyferth 2000–1). Large numbers of Syrians and Lebanese also came to Brazil, as well as to Uruguay and Argentina. Through their movements – often back and forth – and their sending of remittances to family members in their home communities, such migrants played crucial roles in linking different parts of the globe in new and lasting ways.

Geographical Legacies

What do the economic relationships detailed above mean for understanding the geography of the world economy? We can divide the discussion into two broad areas – how the processes described shaped patterns of development at the time and how they have continued to do so subsequently. Certainly, in the cultural realm these relationships led to the import of new institutions and concepts (like ideas concerning private property, which would subsequently impact geographies of land tenure). They also transformed people's geographical imaginations, as colonial subjects learned European languages and histories and came to see themselves as small cogs in a much larger cultural – and spatial – enterprise, even if they did not have much control over how they came to belong to this enterprise. Constantly reinforced by the imperial maps which showed their homeland's place relative to other parts of the particular empire within which they were embroiled, such new imaginative geographies were key in refocusing many colonial subjects' attentions from local and regional matters to global ones.

Without doubt, the impact upon such societies varied tremendously, depending upon their location, by whom they were colonized, and the resistance of local inhabitants. Sometimes, imperial powers sought deliberately to destroy indigenous cultural institutions and practices, either because they believed that not doing so would provide their subjects a cultural reservoir from which to continue resistance or because they genuinely found them to be barbaric and "primitive" – as in British efforts to ban *sati* in India. In other cases, imperial powers left indigenous cultural practices largely untouched, as long as their economic goals were met (as with the British in northern Nigeria). In still others, imperial powers introduced practices which the local populations undoubtedly found barbaric and primitive – as in the forced labor practiced in rubber collecting in the Congo (Hochschild 1998). In yet many other cases the cultural impacts were less deliberative but no less significant. Hence, British officials' gendered assumption that men were the key economic decision makers in households led them to deal almost exclusively with male Maasai in East Africa when negotiating cattle purchases, with the result that the household politics of a society in which women had played significant decision-making roles in the pre-colonial era were transformed (Hodgson 1997).

The social and intellectual life of the imperial nations themselves also felt imperialism's impress. Profits from colonial trade and the looting of cultural treasures provided the raw materials for much of Europe's art and museums, whilst the spirit of the imperial age was reflected in the literature it produced – without empire there would have been no *Mansfield Park*, *Jane Eyre*, *Vanity Fair*, and *Great Expectations* (Said 1994) or, in the Francophone world, no *L'Étranger* (cf. Dunwoodie 1998; Miller 1998). Empire's impact, too, can be seen through the introduction into English of words of Hindi origin (caravan, mugger, shampoo, thug, and others). Similar impacts are evident with Spanish and Portuguese incorporating words from the Americas, whilst for its part US English has adopted words from Tagalog picked up by troops occupying the Philippines, the most well-known probably being "boondocks" (from *bundok*, "mountain").

In the economic realm, imperialism and non-imperial linkages have had monumental implications for how the geography of global capitalism has been fashioned, both historically and contemporaneously. Specifically, whereas the LDCs – which, virtually without exception, were European, US, or Japanese colonies – accounted in 1750 for 73 percent of world manufacturing, by 1913 this had fallen to 7.5 percent (Table 5.5). India's share alone dropped from one-quarter to less than one-seventieth. Equally, whereas Europe produced less than one-quarter of world manufactured goods in the mid-eighteenth century, by the early twentieth it was producing over half – Britain's share alone grew from less than one-twentieth to more than one-seventh. Of course, this is not to say that all of this transformation in the planet's economic geography resulted from the practices outlined above, nor that all of the global South's contemporary problems are the result of imperialism and neo-imperialism. To adopt such an argument would be to ignore the impact of local phenomena upon global uneven development and the agency of those colonized. But it is equally untenable to believe that imperialism and neo-imperialism had no impact upon them, that the geography of development in the contemporary global economy is the result solely of local processes and phenomena.

There are, then, several issues of importance to consider when seeking to understand how imperialism and the building of commercial empires affected the geography of global development. The first concerns the historical geography of capitalism's

Table 5.5 Relative shares (%) of total world manufacturing output for selected countries, 1750–1913

	1750	*1800*	*1830*	*1860*	*1880*	*1900*	*1913*
Europe	23.2	28.1	34.2	53.2	61.3	62.0	56.6
Austria-Hungary	2.9	3.2	3.2	4.2	4.4	4.7	4.4
Belgium	0.3	0.5	0.7	1.4	1.8	1.7	1.8
France	4.0	4.2	5.2	7.9	7.8	6.8	6.1
Germany	2.9	3.5	3.5	4.9	8.5	13.2	14.8
Italy	2.4	2.5	2.3	2.5	2.5	2.5	2.4
Russia	5.0	5.6	5.6	7.0	7.6	8.8	8.2
Spain	1.2	1.5	1.5	1.8	1.8	1.6	1.2
Sweden	0.3	0.3	0.4	0.6	0.8	0.9	1.0
Switzerland	0.1	0.3	0.4	0.7	0.8	1.0	0.9
United Kingdom	1.9	4.3	9.5	19.9	22.9	18.5	13.6
Canada	–	–	0.1	0.3	0.4	0.6	0.9
United States	0.1	0.8	2.4	7.2	14.7	23.6	32.0
Japan	3.8	3.5	2.8	2.6	2.4	2.4	2.7
Less Developed Countries overall	73.0	67.7	60.5	36.6	20.9	11.0	7.5
China	32.8	33.3	29.8	19.7	12.5	6.2	3.6
India	24.5	19.7	17.6	8.6	2.8	1.7	1.4
Brazil	–	–	–	0.4	0.3	0.4	0.5
Mexico	–	–	–	0.4	0.3	0.3	0.3

Note: all figures are triennial annual averages, except 1913. Countries' geographical boundaries are appropriate for the dates given (e.g., "India" includes also the territory that is now Pakistan and Bangladesh).
Source: Adapted from Bairoch (1982: 296).

emergence and, particularly, why Europe materialized as the dominant economic region in the nineteenth century, especially given that as late as 1830 India and China alone accounted for more manufacturing output than did Europe, and throughout the eighteenth century Indian textile workers had real wages which were higher than those of their British confrères (Parthasarathi 1998). Some commentators have suggested the explanation largely lies in the internal dynamics of Europe itself. For them, Europe's emergence as the dominant capitalist region can be traced back to the transition from feudalism, whether it was labor shortages created by the fourteenth century's Black Death, which gave peasants greater economic power (thereby leading to feudalism's crumbling and capitalism's sprouting), or perhaps the growth of trade and its impact upon urbanization which eventually brought about capitalist production (Wallerstein 1974; Brenner 1977; Aston & Philpin 1985). Thus, Wallerstein (1976: 31) has maintained that world capitalism grew as a result of a "historically unique combination of events [that] first crystallized in Europe in the sixteenth century" and that its "boundaries slowly expanded to include the entire world" – in other words, that capitalism diffused out from Europe in a largely one-way, internally driven movement.

Others have argued that this perspective ignores the influx of imperial wealth on Europe and fails to consider how the geography of the emerging world economy was being transformed, such that events happening in one part of the globe were increasingly affecting places thousands of miles distant. Such critics suggest that although there were clearly important processes of transition occurring in Europe in the late Middle Ages, similar processes were taking place in Asia and Africa – the use of money, the growth of commercial manufactures, and the rise of more bureaucratized state forms had all occurred in Asia and West Africa by 1700, for instance (Perlin 1983; Abu-Lughod 1989; Goody 1996). Given that there was a degree of equivalence in levels of economic development, such commentators argue it must have been exogenous factors that were responsible for European economic advancement. Thus for Blaut (1992), the key explanatory factors are imperial conquest and geographic location – Europe's physical proximity to the Americas allowed the Iberian powers to acquire huge quantities of bullion and profits from Caribbean plantations, which gave rise to a proto-capitalist class in Europe that began both to dissolve feudalism and to destroy proto-capitalist communities elsewhere.

In articulating such arguments, though, both sides rely upon a reification of the geographical boundaries of various nation-states and/or regions and view them as little more than spatial containers by which "internal" and "external" processes can be distinguished. Thus they both view Europe as a relatively discrete geographic unit which came into contact with regions beyond its shores sometime towards the end of the Middle Ages, and then argue that this contact either had very little impact or served as the "Big Bang" of European economic development. However, as we have seen, contacts between Europe and the outside world pre-date the Middle Ages and European voyages of discovery by centuries, thereby rendering meaningless the concept of a discrete "European economy," one uncontaminated by contact with the outside world. If we recognize that both the endogenous and exogenous camps rely upon the discursive geographical fiction that national or regional economies have ever been discrete spatial entities (Mitchell 1998; Herod 2007a), then the issue is no longer whether European (and, later, US and Japanese) imperialism/neo-imperialism is either the *deus ex machina* which explains particular countries' economic rise or is simply the context within which it occurred. Rather, it becomes one of *how* different parts of the globe were linked together and how this shaped patterns of uneven development. Put another way, the language of analysis changes from the topographical to the topological (Castree et al. 2007), from a discussion of endogenous and exogenous processes as opposites that only make sense if economies are viewed as discrete units of bounded spaces to a terminology for contrasting shorter and less-connected networks with longer and more-connected ones (see p. 96).

Following from this, the second set of issues concerns what the tying together of different economies meant for the transfer of wealth in both directions and how differential abilities to shape such flows contoured patterns of development. In other words, how did structural transformations in the world economy shape the context within which local actors could make decisions about their own economic futures, decisions through which they could reshape, to a greater or lesser degree, the structural relationships within which they found themselves? Thus, the economies of India and Britain did not evolve independently but were increasingly linked through the trading connections bringing textiles and other goods to Europe and taking back silver. Indeed, Parthasarathi (1998) has argued that the British textile industry's mechanization was carried out primarily *in response to* competition from higher-quality and cheaper Indian imports – in other words, the mechanization of the industry which served as the basis for Britain's industrial revolution was a retort to processes originating half a world away. However, as the narrative above demonstrates, this was not a trading relationship in which both parties were equal. Although mechanization and protectionist policies allowed British textile factories to develop, it was only through conquest that many of the market advantages which gave Lancashire cottons their world dominance could be secured. The degree to which this was understood by those seeking to ensure such advantages for British interests was clearly revealed by one witness before the 1830 Parliamentary Committee on East Indian Affairs: "India requires capital to bring forth her resources; but the fittest capital for this purpose would be one of native growth *and such capital would be created if our* [British] *institutions did not obstruct it*" (quoted in Lamb 1955: 471, emphasis added).

The example of British rule in India is fairly typical in terms of its impact upon indigenous manufacturing and raises important issues for how imperial and commercial policies and practices shaped the broader geography of global uneven development. Whereas neoliberal interpretations see trade as an interaction between equals in the marketplace, from which both parties are equally free to disengage, the reality is that the economic relationships laid down in the fifteenth to twentieth centuries were frequently unequal in nature, with trade backed up, when necessary, with gunboats. Indeed, the speed with which many colonies became economic captives of imperial metropoles suggests either that local populations suddenly decided to break out of trading relationships within which they had been embroiled for centuries, or that their economic relationships were being forcibly reoriented. In the case of India, the fact that in the early twentieth century fewer than one-third of goods imported into the subcontinent came from outside the British Empire but that only 27 percent of India's exports went to Britain (Morison 1908: 425) is illustrative of the dynamics (and imbalance) in the imperial relationship: Britain needed India more than India needed Britain, a fact recognized by many on both extremes of the ideological spectrum – whereas arch-imperialist Lord Curzon proclaimed that if Britain lost "any portion of the Dominions of the Queen" it would still "survive as an Empire" though "if we lost India . . . our sun would sink to its setting" (quoted in Das 1969: 44–5), Russian revolutionary Leon Trotsky (1970 [1930/1906]: 152–3) would equally proclaim that "Britain's dependence upon India naturally bears a qualitatively different character from India's dependence upon Britain . . . India is a colony; Britain, a metropolis. But if Britain were subjected today to an economic blockade, it would perish sooner than would India under a similar blockade."

In considering how imperialism shaped flows of wealth between different parts of the globe, some (Southworth 1931; Nemmers 1956) have argued that colonialism was, by and large, economically unprofitable, costing the imperial powers more than it brought in. Others (Hobson 1948 [1902]; Lenin 1939 [1916]) have suggested that it was, in fact, highly profitable, providing essential capital and raw materials as the basis of industrialization. Still others suggest that even in individual cases where colonies cost more to run than they brought in, profits were available to particular sets of interests – merchants, military supply firms – who could shape national policy to their advantage, even as the ordinary taxpayer picked up the tab. For his part, however, Nowell (2002/3) has argued that imperialism has to be seen not on a colony-by-colony basis, nor even an empire-by-empire one, but within the totality of the world economy. Specifically, he contends that even colonies that were, on their face, unprofitable played an important economic role: preserving the asset values of the metropoles' investors. Such preservation could be accomplished through one of two ways. On the one hand, by ensuring access to particular markets and raw materials, colonies helped reduce investors' exposure to market uncertainty – because of its monopoly trade with India, for instance, for years EIC shareholders were guaranteed 8 percent per annum dividends for very little risk (Robins 2006: 31). Investors, then, could use the predictable returns from India to balance their portfolios as they engaged in riskier ventures elsewhere. By the same token, whereas independent countries might unilaterally change the terms of investment, colonies could not. Money invested in a colony, in other words, was safer.

On the other hand, by establishing cartels and limiting the development of regions across the globe whose products might flood the world market, imperial conquest and neo-imperial trading practices provided a mechanism to keep commodity prices from falling – investors in South American copper, for instance, would not be keen to see new mines open in Africa which might threaten their existing copper stocks' value, so conquest of African copper regions by one power could keep another from flooding the market. Consequently, imperial countries would formally and informally acquire control of territories "because maximizing profit somewhere [usually] requires that investment in another place be stopped or delayed" (Nowell 2002/3: 322). What this suggests is that rather than calculating the economic costs and benefits of empire on the basis of single colonies or even all an empire's colonies combined, it is important to see how the imperial project unfolded with an eye towards the global. Incorporating one part of the planet into a political or commercial empire, then, was often less to do with the specifics of that place and more about maintaining the asset bases of investments elsewhere. In the process, though, conquering territories to keep them from developing simply as a way to protect such investments had momentous implications for patterns of economic development and helps explain to a significant degree the geography of the global economy, both then and now.

Equally, colonies could allow imperial powers to bring pressure to bear on their competitors. For example, the US, British, and Indian economies had become so connected by the early nineteenth century that whenever US cotton was overpriced, Indian supplies could be increased to "rescue the British manufacturer from inflationary pressures" (Webster 1990: 410). Indeed, Siddiqi (1981) has argued that Britain used

India to counter inflationary pressures caused by US cotton price increases as early as the Napoleonic Wars. Likewise, colonies could rescue imperial nations whenever they might have difficulty securing raw materials from their usual suppliers. In 1861, for instance, immediately prior to the US Civil War's outbreak, India contributed 34 percent of the raw cotton imported into Britain. However, as a result of the blockade of Southern ports and British mills' inability to secure US cotton, by 1862 India's contribution had jumped to 94 percent (Logan 1958). Although Indian cotton was considered of inferior quality for the articles British manufacturers produced (underwear and summer clothes) and many British manufacturers returned to using US cotton after 1865, this switch to sourcing cotton from India is rich testament to how integrated the world economy had become by the mid-nineteenth century and how colonies gave imperial powers a degree of geographical flexibility they otherwise would not have had.

Certainly, in considering how empire shaped the geography of the world economy it is important to recognize that some in the colonies benefited from the new relationships laid down. Thus, membership of the British Empire gave many Indian investors access to British capital markets at interest rates lower than had they been independent (Ferguson & Schularick 2006). By the same token, some Indian cultivators made significant amounts of money as gold and silver to pay for cotton poured into the subcontinent. Nevertheless, the increased demand for Indian fields to produce the raw materials necessary for British textile manufacturers had dramatic impacts upon other segments of Indian society. As cotton and other commercial crops became important sources of the cash required to pay colonial taxes, many farmers progressively switched away from producing subsistence crops, with the result that grain output declined steadily until the middle of the twentieth century. This had widespread effects. For instance, life expectancy fell 20 percent between 1871 and 1921, the result of dramatic increases in death rates, especially amongst infants (Kumar 1983: 501, 502).[6] Equally, despite the money made from selling cotton, in the last half of the nineteenth century India's national income actually fell 50 percent. As Davis (2001: 311) soberly concludes: "If the history of British rule in India were to be condensed into a single fact, it is this: there was no increase in India's per capita income from 1757 to 1947."

By the mid-twentieth century, then, the world had been tied together in ways that cannot be understood without examining the political and economic geographies of imperialism and the creation of various commercial empires – the 667 percent increase in exports from India to New South Wales (Australia) between 1828 and 1840 (Chaudhuri 1966: 351), for instance, is hard to explain without recognizing how British imperial interests shaped emergent trade patterns in the southern hemisphere. Equally, importation of rice from Indochina into French West Africa to combat famines caused by an overemphasis on peanut production can only really be explained by the linkages created by French colonialism, as can the fact that West African soldiers fought for France in Indochina in the early twentieth century. Clearly, as imperial power networked the globe together in new ways, the geography of global uneven development was indelibly marked. Yet the consequences of this were shared unevenly, and wealth created in one part of the globe was increasingly enjoyed in others. Thus, Bagchi (1982: 81) estimates that between 1813 and 1822 5–6 percent of resources which would otherwise have been available for investment in places like Bengal were siphoned off by the EIC and private traders. Given that 7–8 percent of Britain's national income was invested into economic development during its industrial revolution, this "missing investment" represents a significant lost potential for India that had significant impacts upon the economic geography both of India but also the wider world within which it became embroiled. Similar examples can be found with other colonies and imperial powers. At the same time, non-imperial linkages – as with Britain's insatiable demand for US beef and wheat to feed its industrial labor force – also networked the planet together in novel ways, with all the attendant consequences for transnational transfers of wealth.

Finally, colonialism's legacy can be seen in the fact that the overwhelming number of international boundaries laid down in the global South during the colonial period remain intact. Today's African borders are virtually identical to those designated at the 1884–5 Berlin Conference. Indeed, Europe's own borders have changed to a much greater degree in the past century than have those of its former colonies. Simultaneously, colonialism transformed nations' internal political geographies, frequently shifting the geography of power within areas colonized – in India, by the late nineteenth century power had shifted from the traditional heartland of the Ganges Plain to the maritime provinces focused upon the ports of Kolkata, Mumbai, and Chennai (Charlesworth 1982). Thus, it is no accident that most of the major urban areas of Latin America, Africa, Australia, and Asia are coastal, with many having begun as colonial entrepôts, linking the hinterland with Europe. Moreover, much physical infrastructure, like railroads, continues to link places as it did during the colonial era – virtually the only railroad built in Africa in the post-colonial period has been the 1 100-mile TanZam Railroad linking Zambian copper mines to Dar es Salaam, the result being that the geography of transportation inherited from the colonial era has made it difficult for many countries to break out of the economic relationships in which colonialism placed them.[7] Similarly, the division of the world by various empires into specialized agricultural zones – one for groundnuts, one for tea, one for bananas, and so forth – not only had significant impacts upon local ecosystems (like the conversion of Sri Lankan rainforests into tea plantations) but it also created a new international division of labor which continues to influence patterns of trade, investment, and development today. Appropriating Marx (1963 [1852]), then, it is clear that the landscapes of the imperial era have weighed like a nightmare on those which have come after them.

Questions for Reflection

- How have different empires linked the globe together in different ways and with what effect?
- What legacies are there today for the world economy of such empire building?
- How have non-imperial trade practices linked the globe together and with what effect?

Notes

1. Marx (1976 [1867]) called wealth accumulated by buying goods low and selling high "primitive accumulation," whereas Adam Smith (1961 [1776]) called it "previous accumulation." Profits from such trade would eventually be used to purchase labor power for commodity production – the beginning of capitalist production.

2. Moreover, the new technology of the telegraph accelerated famine's spread by ensuring grain price hikes "were coordinated in a thousand towns at once, regardless of local supply trends" (Davis 2001: 26).

3. Even the 1916–18 British Royal Industrial Commission confirmed that Indian industrial development was "not inferior to that of the ... European nations" when European merchants first arrived (cited in Chomsky 1993: 13).

4. This proportion increased during the twentieth century as the British government imposed tariffs to fend off Japanese competition and locally produced textiles replaced imports: by 1913 20 percent of textiles consumed in India were produced domestically, by 1936 62 percent were, and by 1945 76 percent were (Maddison 1972: 57).

5. In 1870 the price of wheat was 58 percent higher in Liverpool than in Chicago, but by 1913 was only 16 percent. Similar global convergences can be seen with other commodities (O'Rourke & Williamson 1999: 43, 53). Typically, price convergence followed the telegraph's arrival, which allowed more rapid price transmittal across space.

6. Kumar (1983: 502) suggests that although the 1876–7 and 1896–1900 famines caused a significant increase in mortality (the latter killed 17–20 million), other famines' effects on mortality were mitigated by activities like canal building, which brought irrigation to drought-prone areas (although such activities themselves helped spread diseases like cholera, influenza, and malaria into previously isolated areas). Nevertheless, between 1875 and the 1920s, Indian population growth came to a virtual standstill, whilst between 1875 and 1900, when the worst famines occurred, Indian grain exports increased from 3 million tons a year to 10 million (Davis 2001: 299). Put another way, increased mortality did not result from a Malthusian "overpopulation" but from transformations in the agricultural economy. In fact, the yearly rate of population growth for each decade from 1871 to 1941 was minimal: 1871–81, 0.20 percent; 1881–91, 0.89 percent; 1891–1901, 0.11 percent; 1901–11, 0.65 percent; 1911–21, 0.09 percent; 1921–31, 1.05 percent; and 1931–41, 1.41 percent. According to the Indian Census, infant mortality jumped from an annual average of 263 per 1 000 live births between 1871 and 1881 to 290 by 1920, whilst overall death rates increased from 40.7 per 1 000 population to 49.8, with the result that in some regions and eras – western India in the 1870s, the 1890s, and the 1910s, central India in the 1890s and 1910s, northern India in the 1910s, and southern India in the 1870s – population growth was actually negative (Kumar 1983: 490).

7. One difficulty African nations face is that European powers used different gauge railroads, making it difficult for nations colonized by different powers to send each other goods via train (Siddall 1969). This is a clear example of how the landscape's physical construction can affect regions long after the social context within which the infrastructure was laid down has changed.

Further Reading

Franke, R. W. and Chasin, B. H. (1980) *Seeds of Famine: Ecological Destruction and the Development Dilemma in the West African Sahel*. Montclair, NJ: Allanheld, Osmun.
Hardt, M. and Negri, A. (2000) *Empire*. Cambridge, MA: Harvard University Press.
Knox, P., Agnew, J., and McCarthy, L. (2003) Pre-industrial foundations. In: P. Knox, J. Agnew, and L. McCarthy, *The Geography of the World Economy*. London: Edward Arnold, pp. 119–42.

Electronic resources

Index of Possessions and Colonies (www.worldstatesmen.org/COLONIES.html)
List of largest empires (http://en.wikipedia.org/wiki/List_of_largest_empires)

Chapter 6

Manufacturing Globalization

Chapter summary: This chapter examines the geography of foreign direct investment, stretching back several hundred years, and looks at how FDI has linked the world together geographically. The chapter contrasts contemporary FDI patterns with historical geographies thereof.

- FDI in the Era of High Imperialism
- FDI between the Wars
- Post-World War II FDI
- Transnational Firms as Geographical Actors

Colonies . . . do not cease to be colonies because they are independent.
(Benjamin Disraeli, 1804–81)

Though the wound is hidden, the blood does not cease to flow.
(Asadullah Khan Ghalib, 1797–1869)

As Chapter 5 showed, the relationships laid down during the era of formal imperialism tied the world together in new ways. They have left indelible marks on its economic geography. For instance, although between 1980 and 2000 some 125 free trade agreements were negotiated, the Soviet Union collapsed, the European Union expanded twice, and the volume of world trade grew 175 percent, for former colonies international trading patterns remained remarkably stable – in 2000, half a century after a great wave of decolonization began reshaping the planet's political geography, having been a colony was the second most important factor explaining why one country was the leading exporter to another (only geographic proximity was more significant) (Cassing & Husted 2004). Thus, in 2003 France still supplied 26 percent of Tunisia's imports and accounted for 31 percent of its exports and was Niger's most important trading partner, consuming 48 percent of its exports and providing 15 percent of its imports. Likewise, Britain remains the second most important export market for Ghanaian and Kenyan goods (CIA 2006) and Portugal is Angola's second largest provider of imports. In the case of the Philippines, the US remains that nation's most important trading partner some six decades after its 1946 independence.[1]

The colonial legacy continues to shape development patterns in other ways. In the quarter-century after gaining independence, France's former African colonies, tied by historical links to French manufacturers, paid on average 20–30 percent more for iron and steel than had they bought it from coming from France than did other nations, with such adverse prices costing them $2 billion by 1987 (Yeats 1990). Equally, British Commonwealth members were required to keep their gold and foreign exchange reserves in London and to maintain sterling balances in British banks, a policy which shaped global financial markets in the post-colonial period. The imperial legacy can also be seen in the migration of Algerians to work in French automobile factories and of South Asians to Britain. Similarly, non-imperial linkages developed in the nineteenth century have continued to shape the world economy's geography. Whereas thousands of Italians migrated to Argentina in the nineteenth century, in the late twentieth many of their descendants sought to escape Argentina's economic problems by migrating to Italy, where their ancestral ties allowed them to claim citizenship. Historical linkages with colonies and non-colonies has likewise shaped migration to the United States, with US overseas investment and military intervention often mir-rored by subsequent immigration from these locations (Sassen 1988). The persistence of such linkages, then, appears to challenge what many neoliberal commentators have argued should happen in an age of "globalization" – that "the individual, consumers, corporations, and regions [become freed] from the legacy of the nation-state in which they belong" (Ohmae 2005: 122).

Understanding this persistence is important because it has had a critical impact on the twentieth-century evolution of global manufacturing's geography and the behavior of TNCs, many of whom have preferred to invest in parts of the globe which are culturally familiar. For instance, the UK is now the overwhelming favorite destina-tion for US TNCs looking to locate in Europe, hosting 38 percent of their European headquarters (UK Trade and Investment 2006: 20). Most of the petroleum produced by Elf Aquitaine, France's largest oil company, comes from former French colonies in the Gulf of Guinea, whereas Italy's ENI secures much of its crude from former Italian colony Libya. Former British colonies in Africa still trade more with Britain than they do with other African countries, a situation mirrored by France and its colonies. Likewise, the UK is the third largest investor in India and the top destination for Indian FDI. Indeed, in 2006 Indian companies were the third largest investors in the UK, with London being their gateway to Europe – 60 percent of Indian investment in Europe now comes via the UK (Darling 2006). Two Indian companies are amongst the 40 largest in the UK and more than 500 Indian companies (60 percent in the IT sector) have a UK base (UK Trade and Investment 2006: 34; Darling 2006).

In this chapter, then, I explore how TNCs shape how the world economy is spatially connected. This is important for understanding global development patterns, for if the age of mercantilism was marked by a tying together of various parts of the globe largely via trade, and the nineteenth century marked a growing connectivity via international flows of speculative financial capital (facilitated by the telegraph), then the twentieth century was perhaps most distinctively marked by the growth of TNCs, especially – but not only – those engaged in manufacturing. Certainly, this is not to ignore how new trading patterns established between independent producers have also reshaped the world economy's geography. However, the growth in the number of firms with foreign direct investments, particularly those who have stretched their manufacturing processes across national boundaries, has become an important aspect of how the world economy functions, for it is this stretching that, for many, truly marks our contemporary period as different from previous ones – as an era of "globalization."

In exploring TNCs' growth as major economic and geographical players, though, it is important to keep in mind that, contra neoliberal claims that globalization's advent marks a distinct break with the past, there are, in fact, deep connections between the present world economy and the economic geography of the imperial era. First, a not-insignificant number of TNCs either began life as colonial trading companies or were deeply implicated in the practices of empire. Hence, by World War I numerous imperial trading houses had transmogrified into global financial, mining, and manufacturing operations, whilst many firms that did not begin life as imperial trading houses nevertheless grew as a result of imperial connections – between 1880 and 1914, for instance, the number of overseas branches of major European joint-stock banks jumped from 525 to 1 610, as many established colonial divisions (Battilossi 2006). Second, many US TNCs played a role in encouraging decolonization after World War II, although not necessarily out of any great antipathy toward imperialism nor any great sense of altruism toward colonial subjects: they were, rather, largely driven by the desire to access colonial markets to which they were previously much excluded by imperial tariff and preference systems. Third, all TNCs have had to engage with the geography of the world economy as it was structured during the colonial era. The pattern of global development which this era left – a largely industrialized global North and non-industrialized global South – has had signal impacts upon patterns of global manufacturing FDI in the half-century since decolonization began in earnest, particularly as TNCs have sought new locations for labor-intensive operations.

Organizationally, this chapter is divided into four sections. In the first three, I survey the historical geography of TNCs' investments during the twentieth century, especially as they relate to changes in manufacturing, for such changes have been crucial in reworking patterns of uneven development created in the colonial period. Thus, whereas in 1980 fewer than 10 percent of world manufacturing exports came from the global South, by the early twenty-first century that figure had grown to 30 percent, with predictions that by 2030 it will be 50 percent (Sainsbury 2005). Certainly, not all this manufacturing is conducted by TNCs. However, TNCs do have significant influence on the world economy – they account for two-fifths of all exports from global South countries, whilst one-third of all US exports and 45 percent of US imports are intra-firm sales from one TNC division to another (United Nations 1999: 86). This is significant, for whilst in 1973 only 22 percent of global South exports were manufactured goods, by 1999 70 percent were (often technology-intensive products like electronics) (UNCTAD 2002: Chart 3.2). Having explored TNCs' FDI activities, the final section examines how they act as geographical agents and what this means for arguments about globalization. Certainly, in focusing upon FDI I am not suggesting other investment forms are unimportant. But, given their multi-locationality and ability to transform dramatically the geography of power and control within the world economy through their investment and transference of technological, managerial, and organizational skills, thereby providing knowledge and goods perhaps not available locally, TNCs are some of the more far-reaching geographical agents of the contemporary era. Furthermore, because they are seen within much neoliberal discourse as the acme of economic organization, TNCs have a significant impact on imagining processes of globalization.

FDI in the Era of High Imperialism

In suggesting that what largely distinguishes the twentieth century from earlier eras is the rise of TNCs, it is important not to forget that, as with other aspects of what has come to be called globalization, the history of firms operating beyond their national borders is much longer than merely the recent past. For instance, in the Middle Ages the Italian Bardi and Peruzzi banking houses set up in Britain to export wool and made loans from myriad offices across Europe and the Middle East (Wilson 1974; Hunt 1994). In fifteenth-century Europe family firms frequently conducted long-distance trade, with some family members going abroad to secure trading opportunities whilst others remained behind to manage the firm's import houses (Mathers 1988). Meanwhile, sixteenth- and seventeenth-century trading companies shared important characteristics with contemporary TNCs – the growth of a managerial hierarchy, systems of control exerted over significant distances, and cross-border transactions through sales or production branches in two or more countries – and so could be considered early examples of the kinds of firms many neoliberals view as globalization's avatars (Carlos & Nicholas 1988). Thus by 1610 the British East India Company (EIC) had built its first trading post and warehouse at Machilipatnam on India's east coast. Before long Company representatives secured rights to build posts in Surat, Chennai, Mumbai, and Kolkata, and by 1647 had 23 Indian posts. Ultimately, the EIC would develop an extensive network throughout South and East Asia, including in China, Taiwan, Sumatra, Vietnam, and Java. Likewise, the Danish East India Company built warehouses at Machilipatnam and Tharangambadi on India's Coromandel coast, others at Pipli and Balasore in Bengal, and at Banten, Sukadana, and Makassar in Indonesia (Macau 1972; Subrahmanyam 1989). At the same time, the Dutch United East Indian Company founded warehouses in Jakarta, Sumatra, and Nagasaki. By 1641 it had built a saltpeter refining plant in Bengal, by 1651 a print works for textiles, and by 1717 was employing more than 4 000 silk spinners in Kaimbazar (Carlos & Nicholas 1988: 399). For its part, the French East India Company established warehouses at Puduchcheri, Machilipatnam, Karaikal, Mahi, Surat, and Chandannagar (Haudrère 2006). Meanwhile, across the Atlantic the Virginia Company, chartered in 1606, soon established Jamestown to export tobacco to Britain (Craven 1957). A few years later, English investors created the Newfoundland Company to import fish into Europe (Cell 1982). Meanwhile, in 1632 two Dutch merchants founded the first foreign-owned foundry in Russia and by 1700 foreigners owned 60 percent of all the large factories in the country (de Goey 1999: 38). These activities suggest, then, that the difference between nineteenth- and twentieth-century firms' overseas operations and earlier trading companies was "one of degree, not of kind" (Carlos & Nicholas 1988: 399).

FDI in manufacturing, then, clearly has a long history. In Britain's case, many companies had become quite transnational prior to 1914, including British American Tobacco, Lever Brothers (soap), J&P Coats (cotton thread), Courtaulds (synthetic textiles), and The Gramophone Company (later EMI) (Stopford 1974). Unilever, for instance, began building and buying soap factories overseas in the 1890s, and by the 1920s had plants across Europe and in Australia, South Africa, India, Indonesia, Sri Lanka, China, Thailand, Nigeria, the Belgian Congo, and Latin America. Whilst most such FDI was geared towards producing goods for local consumption, in 1931 the company took over the United African Company – itself created out of several pre-existing companies dating to the eighteenth century (Pedler 1974) – for the purpose of purchasing and exporting tropical commodities and importing and distributing trade goods (Fieldhouse 1978). British companies also built Brazilian factories in the 1880s to refine sugar (Galloway 1968; de Andrade 1989), whilst within three years of its 1889 Dublin founding, the Dunlop rubber goods manufacturing company had set up factories in France and Germany. Facilities in the US followed in 1893 and in Japan in 1909, whilst in Canada, Australia, and Russia the company entered into licensing agreements with local manufacturers (Jones 1984). By 1914 J&P Coats had become the most active British TNC, with 53 overseas operations in 15 countries on four continents (Kim 2005). Although disrupted by World War I, the growth of British manufacturing FDI quickly picked up again in the 1920s – the chocolate manufacturer Cadbury's, for instance, opened a factory in Hobart, Tasmania, in 1922.

Across the Atlantic US investors established a paper mill in Quebec in 1804, whilst by 1835 others had set up a tannery in Ontario and soon expanded that to shoe manufacturing (Lewis 1938: 293). These, though, were relatively small operations. By way of contrast, one of the first US firms to build overseas manufacturing plants in large number was the sewing machine manufacturer Singer. Soon after its 1851 founding, Singer began exporting machines to Europe (Davies 1976; Carstensen 1984). Then, in 1867, the company opened its first overseas assembly operation in Glasgow, to supply the European market. As UK sales grew from 30 000 in 1875 to 90 000 in 1884, Singer's European operations increasingly subsidized its loss-leading US expansion. In fact, by 1879 Singer was selling more machines abroad than in the US and had branches in Australia, New Zealand, South Africa, and India. In Russia, Singer first sold machines in 1859 and opened a factory in 1902 (Bissell 1999: 224), bringing its total direct Russian investment to $1.26 million [$649 million modern equivalent] (Davies 1976: 260). In 1914 85 percent of Singer sewing machines were purchased outside the US, in places as far afield as the Philippines, Japan, the Ottoman Empire, Scandinavia, India, and South Africa (Godley 2006). Other US corporations emulated Singer, and by 1900 Westinghouse, General Electric, Western Electric, Eastman Kodak, and Standard Oil all had numerous overseas facilities (Hambleton 1987) – in 1911 the International Harvester company alone had subsidiaries in 13 countries, manufacturing plants in eight, and its overseas sales constituted almost 40 percent of its net sales (Kobayashi 1974: 208).

Continental European firms likewise established overseas manufacturing and mining operations early. Belgian steelmaker Cockerill built a Prussian plant in 1815 (Franko 1975), whilst Dutch company Billiton was established in 1860 to mine Indonesian tin and would later mine bauxite in Suriname. Between 1864 and 1889 the Dutch potato flour manufacturer W. A. Scholten founded factories in Germany, Poland, Russia, and Austria. By 1880 Dutch margarine producer Anton Jurgens had four factories in Russia and by 1900 the Nederlandsche Gist-en Spiritusfabriek had a yeast factory and gin distillery in Belgium, a cream-processing plant in Germany, and owned a peanut plantation in Egypt. The electronics company Philips set up a joint venture in the US before World War I and by the late 1920s a third of the company's personnel were working outside the Netherlands, including in Spain, Poland, Belgium, Germany, Italy, Sweden, Brazil, and China (de Goey 1999). For its part the Italian tire company Pirelli built Italy's first overseas factory in 1902, a cable plant near Barcelona (Viesti 1988), and by 1920 the company had facilities in Britain, Brazil, Greece, Turkey, Argentina, and Germany. French and German banks also dramatically increased their overseas branches between 1880 and 1913, from a combined total of 43 to 311 – French banks expanded operations particularly in the Ottoman Empire, Egypt, China, and Brazil, whilst German banks focused upon Romania, the Ottoman Empire, Egypt, China, Brazil, Argentina, Chile, and Peru. Although they were still significantly behind Britain, whose banks' overseas branches mushroomed from 482 to 1 302, mostly in parts of the Empire, together with China and Latin America, French and German banks did challenge British supremacy in East Asia and Latin America, operating between them some 30–45 percent of foreign branches in these regions (Battilossi 2006: 367).

Firms from other parts of the world similarly had significantly transnationalized by the early twentieth century. The Tamil Nadu Nattukottai Chettiars established overseas financial houses in Singapore and Penang in 1825, in Mawlamyine in 1852, Yangon in 1854, and Mandalay in 1855 (Rudner 1994: 54) and by 1900 had become the region's pre-eminent bankers, with a network of interdependent family firms stretching from India to Sri Lanka, Myanmar, Vietnam, Sumatra, Java, Singapore, and Malaysia. Between 1890 and 1940 craft goods merchants from the Sind region of modern-day Pakistan set up branches across the globe, from Melbourne, Sydney, Kobe, and Yokohama, to Cape Town, Marrakech, Buenos Aires, Bermuda, and Honolulu (Markovits 2000: 112–13). The Japanese firm Mitsui founded its first overseas office in 1877 in Shanghai to exploit an export monopoly on coal. By 1900 Mitsui had developed a network of offices in China to export raw cotton and soybeans and import

Japanese textiles, and by 1914 had 46 branches across Asia, with a further five in Europe, two in the US, and one in Australia. Other Japanese firms also invested in China prior to World War I, exporting glass, paper, beer, matches, and textiles, whilst importing agricultural products and natural resources. Textile companies Chubei Itoh and Kanematsu Shoten opened branches in Korea, and by 1890 the latter operated in Sydney exporting Australian wool. Other firms established offices in Taiwan, Indonesia, and the Philippines, although little of this FDI was in manufacturing, with most being in shipping, banking, and railroads – investment in the South Manchuria Railway Company, for instance, aimed to facilitate Japanese export penetration of Asian markets and ease imports of rice and raw materials necessary for Japan's booming economy (Mason 1999). In the case of Europe, Japanese trading, banking, shipping, and insurance companies established affiliates to facilitate export to Japan of machinery and other capital goods in exchange for Japanese silk products. One of the earliest to do so was, in fact, Mitsui, which opened its first European branch in Paris in 1878, following with branches in London, Milan, Lyon, and Hamburg. Other Japanese firms followed suit, including Okura Gumi, Nihon Menka, the Yokohama Specie Bank (which established a London branch in 1881), and the Tokio Marine Insurance Company (Mason 1994).

Clearly, then, one of the elements taken as evidence of globalization – firm transnationalization – has a longer provenance than much neoliberal rhetoric might lead us to believe. However, in discussing nineteenth-century firm transnationalization it is important not simply to focus upon FDI by firms with headquarters in one country and operations in another (the archetypical TNC), for there were also myriad free-standing expatriate firms which, though established overseas as independent entities by emigrants to both colonies and non-colonies, had financial and ownership lines stretching back to various imperial metropoles. Their importance in linking parts of the globe cannot be underestimated, as prior to 1914 thousands of such firms were founded (particularly in extractive industries and agriculture) by emigrants from Britain, France, Portugal, the US, and elsewhere, with these emigrants starting businesses locally but raising the venture capital to do so in their imperial metropole and then either manufacturing for local markets or exporting to their founders' countries of origin.[2] Though not TNCs in the classic sense, then, free-standing firms' geographic preponderance is important for two reasons: (1) through their acquisition they would often provide a convenient means of entry into overseas markets for other manufacturers from expatriates' countries of origin; (2) many would eventually develop into TNCs, either as they invested back in their founders' nation of origin or as they spread elsewhere (as in the case of Glaxo, established by Londoner Joseph Nathan after emigrating first to Australia and then New Zealand, which began producing dried-milk products for sale in Britain in 1904, transferred its management center to Britain in 1914, and a decade later began manufacturing pharmaceuticals there).

In terms of the broad regional patterns of TNCs' overseas investments, British firms were by far the dominant global investors in the early twentieth century, accounting for 45 percent of world FDI by 1914 (Jones 2005: 22). For their part, US firms accounted for 14–19 percent of FDI totals, German firms 11–14 percent, Dutch firms 5 percent,

French firms 11–12 percent, other Western European nations' firms 5 percent, Japan 0.1 percent, and the rest of the world some 6 percent (Feis 1930; Woodruff 1966; Jones 1996: 30–1; Bornschier 2000). In France's case, in the four decades prior to World War I most overseas investment went not to colonies but to independent nations like Russia, Turkey, and Argentina, largely for strategic reasons. Consequently, in 1900 only about 6 percent of the 28 billion francs of French overseas investment was in French colonies, whilst in 1914 the figure was 9 percent of 45 billion francs (Berger 2003), compared to 20–30 percent for Britain. Significantly, though, between 1887 and 1913 French net capital exports (not all of which was in the form of FDI) equaled 3.5 percent of French national income, a proportion higher than in the early twenty-first century. Although a relative latecomer to overseas investment, for its part by 1914 Germany had significant interests in Russia, Turkey, the Balkans, and Latin America. However, whilst some firms, like Siemens, AEG, and Mannesmann, had fairly large overseas investments, most were domestically focused and when they did contemplate serving international markets they tended to supply them via exports rather than establishing overseas operations (Hannah 1996). By way of contrast, Dutch firms invested particularly in the US in the early twentieth century, with Dutch FDI and portfolio investment rising from $8 million in 1900 to $135 million [$72 billion] by 1914 and $380 million [$76 billion] by 1938, by which date Dutch firms were the third largest investors in the US (de Goey 1999: 46). Meanwhile, by 1914 some 160 Swiss firms controlled 265 foreign production subsidiaries, of which two-thirds were in Germany and a fifth in France (Jones 1994: 6). Other European nations' TNCs – like those of Sweden – tended also to invest primarily in neighboring countries. In the case of some European TNCs, though, a predisposition towards licensing independent overseas manufacturers meant that these companies' FDI levels were relatively low – Belgian chemical manufacturer Solvay, for instance, preferred to license local operations in the US, Britain, and Germany than to establish its own plants (Hannah 1996). As a result of these various strategies, the character of countries' FDI was often quite different in terms of quantity, geographical location, sectoral distribution, and how TNCs' different business strategies tied the planet together – licensing a local producer to make one's product and building overseas plants have quite distinct implications for the exercise of corporate power, for instance.

In the case of US TNCs, most focused upon those regions where raw materials were most readily accessible and where it was relatively easy to carve out new market opportunities. Consequently, they largely established operations in Canada, Europe, and Mexico. By the early 1880s US firms had established at least 32 branch factories in Canada, mostly in metal and textile manufacturing (Lewis 1938: 294). In 1897 Mexico was the largest recipient of US FDI (at 31 percent of US totals, mostly in railroads, mining, and smelting), followed by Canada (25 percent), Europe (21 percent), the Caribbean (11 percent), South America (6 percent), and Asia (3.6 percent), with Africa (0.2 percent) and Oceania (0.3 percent) negligible recipients. By 1914, though, these proportions had begun to change: whilst the share of US FDI going to Canada, Europe, and the Caribbean stayed roughly the same (23 percent, 22 percent, and 11 percent), FDI to South America had increased to 12 percent of the total whilst that to Mexico had

Table 6.1 US manufacturing FDI overseas (in then-current $ millions)

Location	1897	1908	1914	1919	1924	1929	1935
Europe	35	100	200	280	450	636.6	640
Canada & Newfoundland							
Paper & pulp	20	55	74	100	180	278.9	290
Other mfg.	35	100	147	300	420	540.6	550
Cuba and W. Indies	3	18	20	26	30	47.1	45
Mexico	–	10	10	8	7	6.3	6
Central America	–	–	–	–	–	7.2	7
South America	–	2	7	50	90	170.4	200
Africa	–	–	–	–	3	6.7	7
Asia	–	5	10	15	46	77.4	75
Oceania	0.5	6	10	16	26	49.8	50
Total	93.5	296	478	795	1 252	1 821.0	1 870

Source: Lewis (1938: 595).

decreased to 22 percent (Wilkins 1970: 110). Equally, FDI structure had begun to change. Whereas in 1897 petroleum accounted for 42 percent of US FDI in Europe and manufacturing only 27 percent, by 1914 the figures were 24 percent and 35 percent (Tolentino 2000: 48–9). Lewis (1938: 595) has estimated that whilst in 1897 Europe was the destination for 37 percent and Canada 59 percent of the $94 million [$72 billion] in US manufacturing FDI, by 1914 the figures were 42 percent and 46 percent, with Mexico and the Caribbean accounting for 6 percent, South America 1 percent, and Asia and Oceania 2 percent each (Table 6.1). Although World War I and its aftermath caused several firms to write down their European investments – International Harvester discounted its European investment by nearly $14 million, $10 million of which represented assets nationalized by the Bolsheviks (Lewis 1938: 295) – by 1929 US manufacturers were investing more heavily overseas, though the regional destinations had changed somewhat: 35 percent of total manufacturing FDI was now invested in Europe, 45 percent in Canada, 3 percent in Central America and the Caribbean, 9 percent in South America, 4 percent in Asia, 3 percent in Oceania, and 0.4 percent in Africa. Such estimates gel to some degree with those of Vaupel and Curhan (1969) who, though undoubtedly undercounting, nevertheless estimated there to be 47 overseas manufacturing subsidiaries of major US TNCs in 1901 (13 percent of which were located in Canada and 79 percent in Europe), 116 by 1913 (with 26 percent and 62 percent in Canada and Europe), 180 by 1919, and 715 by 1939, with the proportion in Europe and Canada having declined and that in Latin America and the "Southern Dominions" having increased since World War I (Table 6.2).[3] Nevertheless, US TNCs were generally less internationally oriented than were European ones prior to World War I. Although an analysis of the largest 100 mining and manufacturing firms of 1912 shows that 56 were US whilst only 16 were British and 15 German, and whereas several US firms (Phelps Dodge, General Electric,

Table 6.2 Growth in the number of, and changes in the geographical distribution of, overseas manufacturing subsidiaries of major US TNCs, 1901–45

Host country/region	1901 No.	1901 %	1913 No.	1913 %	1919 No.	1919 %	1929 No.	1929 %	1939 No.	1939 %	1945 No.	1945 %
Canada	6	12.8	30	25.9	61	33.9	137	29.3	169	23.6	192	23.8
Europe	37	78.7	72	62.1	84	46.7	226	48.4	335	46.9	324	40.1
UK	13	27.7	23	19.8	28	15.6	78	16.7	128	17.9	129	16.0
France	8	17.0	12	10.3	12	6.7	36	7.7	52	7.3	47	5.8
Germany	10	21.3	15	12.9	18	10.0	43	9.2	50	7.0	40	5.0
Scandinavia	1	2.1	8	6.9	11	6.1	19	4.1	24	3.4	23	2.9
Italy	2	4.3	7	6.0	7	3.9	13	2.8	18	2.5	17	2.1
Latin America	3	6.4	10	8.6	20	11.1	56	12.0	114	15.9	182	22.6
Mexico	0	0.0	3	2.6	4	2.2	11	2.4	27	3.8	41	5.1
Argentina	0	0.0	2	1.7	5	2.8	12	2.6	27	3.8	39	4.8
Brazil	0	0.0	0	0.0	4	2.2	12	2.6	20	2.8	32	4.0
Southern Dominions[1]	1	2.1	3	2.6	8	4.4	25	5.4	69	9.7	83	10.3
Asia & Africa[2]	0	0.0	1	0.9	7	3.9	23	4.9	28	3.9	26	3.2
Total	47	100.0	116	100.0	180	100.0	467	100.0	715	100.0	807	100.0

Notes:
1. Australia, New Zealand, South Africa, Southern Rhodesia (Zimbabwe).
2. Excluding Southern Dominions.
Source: Vaupel and Curhan (1969: 125).

Westinghouse, Kodak) had significant overseas investments, for the most part US firms were largely domestically focused. By contrast, about a third of the largest British firms had a majority of their assets or production abroad. Moreover, even domestically based US firms tended to be less internationally oriented than their European competitors – such firms exported about 5 percent of their product compared to 30 percent for Germany and 40 percent for Britain (Hannah 1996: 147).

In terms of FDI's geographical destination just prior to World War I, Latin America was the largest recipient, absorbing one-third of global stock. Such FDI took the form of ownership of manufacturing facilities, mines, plantations, and financial institutions – by 1914 there were 100 or so branches of foreign banks across the region, and in countries like Brazil foreign banks controlled nearly half the market share (Briones & Villela 2006: 329). Meanwhile, Asia hosted 21 percent of global FDI, much of this British investments in India and China, whilst Western and Eastern Europe received 8 percent and 10 percent respectively, and Africa 6 percent (mostly imperial nations' TNCs investing in colonies) (Jones 1996: 31). With regard to individual nations, by 1914 the US was the world's leading FDI recipient, with around 10 percent of global totals, followed by Russia, Canada, Argentina, and Brazil (Wilkins 1994). By far, most FDI was invested in natural resources – 55 percent compared to 30 percent in services like banking and only 15 percent in manufacturing (Jones 1996: 32). This breakdown is highly suggestive, illustrating how lines of corporate power and control were being established to secure raw materials for the European, US, and Japanese industrializations then dramatically shaping the world economy.

FDI between the Wars

World War I dramatically affected the nature and geographical origins and destinations of global FDI. In particular, the conflagration provided significant openings for US TNCs in Latin America as local populations, no longer able to secure many European imports, turned to the US, with the result that US firms increasingly exported to the region and established subsidiaries there, displacing European capital in countries like Brazil (Manchester 1964: 332–6). Likewise, as British demand for Cuban sugar exploded to replace the sugar no longer available from beets grown in France, Belgium, and Germany, US banks established branches in Havana to fund the crop's production and movement. Consequently, although in the late 1930s Britain remained the world's leading FDI supplier (at 40 percent of planetary totals), the US had risen to second place, at 28 percent. Moreover, manufacturing began to play a bigger role in US FDI totals. Whereas in 1897 manufacturing accounted for 15 percent of US FDI and 18 percent in 1914, by 1929 it was the dominant element (Tolentino 2000: 59), with US firms having an estimated 1 236 overseas manufacturing and assembling plants. Significantly, 524 of these were in Canada (US Senate 1931), a geography of foreign investment that gave US manufacturers both greater access to the Canadian market but also allowed them to more easily export to other parts of the British Empire (due to imperial trading preferences) than did exporting directly from the US, another example of how the imperial legacy continued shaping global investments into the twentieth century. In Europe, US manufacturing FDI was principally in the electrical and telephone equipment sectors, followed by the automotive sector, agricultural implements production, and other assorted industrial fabricating like typewriters. Within Latin America, US manufacturing FDI concentrated in Cuba, Argentina, and Brazil, in the automotive, agricultural machinery, chemical, and foodstuffs industries (like meat-packing), whilst in Asia most US manufacturing FDI was in Japan (a wide range of commodities) and India (principally jute products), though there was also some investment in textile and embroidery plants in the US colony of the Philippines.

In the case of the Netherlands, by mid-century Dutch TNCs constituted 10 percent of global FDI, with 60 percent invested in the Dutch East Indies (mostly in oil) and 14 percent in the US (de Goey 1999: 49). Neutrality during World War I meant the Netherlands emerged with its FDI stock largely intact, whilst its colonial possessions provided Dutch TNCs continued opportunities to expand overseas within a relatively protected set of geographical spaces. Thus Indonesia was virtually Royal Dutch Shell's "private oil reserve" (Reed 1958: 312) and by the 1920s had become one of the largest crude producers outside the US. Although its oil's high quality made Indonesia an area of interest for other nations' oil companies, like US-based Standard Oil, the fact that the Dutch government was keen to keep foreign capital out of its colonies meant direct ownership by non-Dutch companies was difficult. Consequently, as a way to access the fields, in 1912 Standard Oil created the Nederlandsche Koloniale Petroleum Maatschappij as a Dutch firm with Dutch directors, to be owned not by Standard directly but by its Dutch marketing affiliate, the American Petroleum Company (a solution which begs the familiarly contemporary question of where one nation's FDI ends and another's begins). Initially the Dutch government refused to grant NKPM prospecting licenses, although later relented, and by 1939 US interests drilled 28 percent of Indonesia's oil (Reed 1958: 337).

World War I also disrupted European trade with Asia, and after 1914 Japanese firms increasingly challenged European dominance in the region. Although during the war Japanese textile exports to China replaced many of those from Britain, after 1919 Chinese tariff increases led manufacturers to establish Chinese operations, and by the early 1920s no fewer than seven of Japan's 10 largest cotton spinning companies had done so. Japanese FDI in China doubled between 1914 and 1919 and doubled again by 1930 (for a total of $763–$874 million [$524–$601 billion], depending on sources). By 1930 Japan was the second largest investor in China, although Japanese FDI still paled in comparison to Britain's. FDI to Japan's colonial possessions (Taiwan and Korea) also grew dramatically (Mizoguchi & Yamamoto 1984). Continuing pre-1914 patterns, investment was primarily in non-manufacturing activities, though Japanese-owned factories producing textiles, foodstuffs, beer, metals, machinery, rubber products, chemicals, and glassware were increasingly established across East Asia. In 1937 Nissan moved its headquarters to Manchuria whilst Toyota soon thereafter set up factories in Tianjin and Shanghai. Several Japanese coal, iron, and gold mining operations were also established. Although some Japanese firms initiated operations in the Dutch East Indies (particularly in rubber, fishing, tea, fruit, oil, and general trading), the vast majority of investments were in areas not part of European empires – by 1945 94 percent of Japan's overseas Asian assets were in Manchuria, Korea, China, and Taiwan (Mason 1999: 21–8).

War experience also had an impact on other nations' firms' investment strategies. Thus, Schröter (1988) argues that whereas prior to 1914 German TNCs had invested overseas in much the same way as had many of their rivals, expropriation of their foreign investments – like the US government's seizure of pharmaceutical company Bayer's plants – led them to be more cautious in the inter-war period than other nations' firms when it came to investing abroad. For German firms in the post-1918 era, then, engaging in FDI was unattractive both because it tied up capital for lengthy periods at a time when many firms were severely short of liquidity (the result of the economic catastrophe of the 1920s), and because FDI was subject to loss through nationalization. The result was that German firms tended instead to use international cartels and contracts more widely than did other nations' firms, which, though offering less direct control, were more easily adaptable to new economic conditions in the international arena. Furthermore, they tended to engage in regions where they felt their investments would be safest rather than where they could secure the highest profits. Consequently, German industry made major efforts to capture increased market shares in Scandinavia, both because the region had important raw materials like iron ore and presented market opportunities, but also because Scandinavian neutrality in World War I meant German investments had not been lost during the war. This strategy was reinforced when the Nazis came to power, as the idea of incorporating "Aryan" Scandinavia into a broader Germanic *Großraumwirtschaft* ("greater economic space") took on an ideological, as well as an economic, rationale.

In terms of recipient regions, in 1914 the US, Russia, Canada, Argentina, Brazil, and South Africa were the leading FDI destinations. Total foreign investment in the US was equivalent to 20 percent of GDP (in 2005 it was 13 percent [UNCTAD 2006: 308]) and consisted of $1.3 billion [$443 billion] in inward FDI and a further $5.8 billion in portfolio investment. Meanwhile, US firms had about $2.6 billion invested overseas. World War I's effects, though, meant that FDI stocks in the US fell to $900 million by 1919 (Graham & Marchick 2006: 3–8) and by 1929 Canada had become the world's leading FDI destination, followed by the US, India, Cuba, and Mexico (Wilkins 1994: 20–1). Most global FDI flow continued to be in the primary sector. Indeed, for much of the first half of the twentieth century it was foreign firms that controlled the export in many parts of the globe of primary materials like oil (in Iran, Iraq, Saudi Arabia, Libya, Indonesia), copper (Chile, Zambia), bauxite (Guyana, Suriname, Guinea, Jamaica), and tin (Malaysia). In 1928 nearly half the oil produced in Latin America was drilled by two foreign TNCs – Royal Dutch Shell and Standard Oil of New Jersey – whilst in Canada US firms owned 32 percent of the country's mining and smelting assets. TNCs also increasingly integrated the growing and harvesting of agricultural commodities into their overall operations – as with food manufacturers and even the Ford Motor Company, with its rubber plantation in Fordlândia, Brazil – and played a central role in encouraging new agricultural exports. Between the world wars, however, FDI became increasingly important in the manufacturing sector – for instance, in 1926 30 percent of Canadian manufacturing was owned by US TNCs (Wilkins 1994: 37). Generally speaking, though, because of TNCs' interests in securing access to primary materials, the bulk of FDI prior to World War II continued to go to the global South.

Despite, however, the large quantities of FDI invested in primary materials and, increasingly, in manufacturing and services, between 1914 and the early post-World War II era TNCs' international activities were generally reduced as two world wars, a global depression, protectionist policies (most industrialized nations' tariffs for manufactures were far higher in the 1930s than they were before World War I [Findlay & O'Rourke 2003: 51]), and, after 1945, intensifying geographical divisions between East and West all promoted a closing of the world economy. Although some companies expanded their overseas operations in the 1930s to avoid growing tariff barriers, others simply closed their overseas factories or sold them to local investors as restrictions placed on currency convertibility by various governments made it difficult to repatriate profits. Thus, whereas FDI from the US rose from $2.6 billion in 1914 to $8 billion by 1930 (Graham & Marchick 2006: 9), the stock of capital invested abroad by the UK, France, and Germany was greater in 1914 than in 1938. Overall, whilst FDI stocks worldwide equaled almost 20 percent of world GDP in 1900, by 1945 they had fallen to just 5 percent (Denis et al. 2006: 22).

Post-World War II FDI

In some regards the FDI patterns laid down before World War II have remained largely unchanged, whilst in others they have been transformed dramatically. For instance, whereas British firms' control of world FDI stocks has shrunk significantly since the early twentieth century, US firms controlled roughly the same proportion in 2005 as in 1914, though they had held significantly higher proportions in the 1960s and 1970s (Table 6.3). In general terms, though, three features have characterized post-World War II FDI. First, as newly independent colonies engaged in protectionism, import substitution industrialization, and nationalized many investments held by foreign TNCs, and as the global North's manufacturing industries became more capital-intensive, FDI

Table 6.3 Percentage share of accumulated stock of outward FDI held by the UK, US, and Japan, 1914–2005

	1914	1938	1960	1970	1980	1990	2005
UK	45	40	16	14	15	15	11
US	18	28	49	48	40	26	19
Japan	<1	3	1	3	7	12	4

Source: Jones (1994) for years 1914 to 1990; UNCTAD (2006) for 2005.

Table 6.4   Outward FDI stock, of all types, from "developed countries" to the rest of the world, 1990–2004 (percentage located in various parts of the globe)

	Developed countries	Developing countries	South-East Europe and the CIS	Unspecified
1990	82.2	16.6	0.1	1.1
1995	79.6	17.8	0.2	2.4
2000	81.4	16.7	0.5	1.4
2004	75.4	20.5	0.6	3.5

Source: UNCTAD (2006: 90).

increasingly flowed between global North nations and was focused in manufacturing and, latterly, services.[4] Whereas FDI in the nineteenth century was largely driven by industrialized nations seeking raw materials and agricultural products, together with some manufacturing to serve local populations overseas, industrialized nations' FDI post-1945 has been largely market-oriented, seeking out new markets in places like Western Europe. Thus, whilst in 1914 only about a quarter of world FDI stock was invested in Western Europe and North America, by 1980 almost two-thirds was (Jones 1996: 31, 48). Consequently, Latin America, Asia, and particularly Africa have declined sharply in importance as host regions – Africa saw its share of global FDI inflows fall from 10 percent in 1970 to less than 1 percent in 2000 (UNCTAD 2006: 41), whilst in 1980 there was virtually no FDI in either China or India (Jones 2005), both nations where foreign assets were nationalized after World War II. In fact, although fast-growing nations like China and India are now becoming significant FDI destinations – with inflows of $108 billion and $7 billion respectively in 2005 alone (UNCTAD 2006: 301) – the proportion of world FDI in LDCs is still nowhere near pre-World War I levels (in 2004 "developing countries" hosted a little over one-fifth of FDI from "developed countries," compared to over 60 percent in 1914) (Table 6.4).[5]

The second distinguishing aspect of post-1945 FDI flows has been the growth of investments by global South TNCs, with such FDI going both to other LDCs but also to global North countries. Certainly, the amount of this FDI is still fairly miniscule – whereas in 1990 TNCs from LDCs owned 3.5 percent of world outward stocks of manufacturing FDI and 3.0 percent of all outward FDI stocks, by 2004 they accounted for only 4.4 percent of outward manufacturing FDI and 8.7 percent of total outward FDI.[6] Nevertheless, FDI flows from LDCs have been growing, reaching $117 billion by 2005 (compared to $646 billion from "developed" economies) (UNCTAD 2006: 299). FDI outflows from China alone were $11 billion in 2005, the fourth largest from developing and transition economies after Hong Kong, the British Virgin Islands, and Russia (UNCTAD 2006: 114–15).[7] In India's case, the first post-World War II manufacturing FDI involved the Birla group building a textile mill in Ethiopia in 1956. After a slow start, Indian firms began increasing their FDI and by 1983 had 140 operations overseas and another 88 in various stages of development (Lall 1986: 13). More recently, in October 2006 the Indian firm Tata purchased the Anglo-Dutch steelmaker Corus, creating the world's fifth largest steel company.[8] Headquartered in Mumbai, Tata has become highly diversified over the past 140 years, with operations in tea, steel, chemicals, watches, vehicle manufacturing (it is the world's fifth largest producer of commercial vehicles), software, telecommunications, publishing, and hotels. Beginning in 2000, it went on a global buying spree, spending $3 billion acquiring 19 companies on five continents, including the Eight O'clock Coffee Co. in the US, Tetley Tea in Britain, and Daewoo Commercial Vehicle in South Korea (*Daily Times*, October 21, 2006). More broadly, in 2006 Indian companies announced overseas acquisitions worth $20 billion, twice the level of acquisitions in India by overseas firms.

Third, contrary to neoliberal proclamations that geographies of development will become more even as globalization unfolds, FDI patterns have remained remarkably consistent in many regards during the past century and a half. Indeed, although "developed nations" are investing a greater proportion of their FDI in LDCs than two decades ago, at the rate at which this trend unfolded between 1990 and 2004 it would be somewhere in the early twenty-second century before their FDI was evenly invested between the global North and the LDCs. Thus, whereas there has been much talk of FDI creating a new international division of labor, in 2005 it was actually the UK which was the largest recipient of FDI flows and, at $817 billion, the second largest holder of inward FDI, after the US. Moreover, the UK remained the world's second largest overseas direct investor, with stock valued at $1.2 trillion (UNCTAD 2006: xvii, 83–7). Overall, in 2006 more of the world's 100 largest TNCs had invested in the US than anywhere else, followed by the UK and the Netherlands. The US was also the leading location for affiliates established by TNCs from LDCs, followed by Hong Kong and the UK. In fact, virtually the only difference between the investment patterns of TNCs from the industrialized countries and those from the LDCs is that the latter have tended to invest in Asia to a greater relative degree, which is where most originated (Table 6.5).

Despite many continuities with pre-World War II FDI patterns, though, there have been important changes in the post-1945 era, perhaps the most significant being the US's emergence as the world's largest FDI provider, accounting for 85 percent of all new FDI flows worldwide between 1945 and the mid-1960s (Jones 2000: 117). With returns on overseas manufacturing substantially greater than those on domestic manufacturing (Aliber & Click 1993: 86), in the 1950s US firms increasingly established operations abroad under the Pax Americana – whereas in 1929 they had slightly over 1 200 foreign manufacturing subsidiaries (US Senate 1931), by 1957 there were 3 481 (1 165 in Canada). Although overseas portfolio investments amounted to only $7.9 billion [$224 billion], FDI totaled $22.1 billion [$629 billion], with 54 percent of portfolio investment and 34 percent of FDI in Canada, 10 percent and 33 percent, respectively, in Latin America, and 22 percent and 16 percent in Europe (US Department of Commerce 1960). These figures are significant for two reasons. First, they show that Latin America in the mid-1950s was still a more important locus for US FDI than was Europe. Second, the fact that the ratio of FDI to portfolio investment was 1.8 to 1 for Canada, 2.0 to 1 for Europe, but 9.2 to 1 for Latin America suggests there were few Latin American firms in which US companies were interested in investing, compared

Table 6.5 Most-favored locations of largest TNCs

For largest world TNCs		For largest developing-country TNCs	
Economy	*Location intensity*	*Economy*	*Location intensity*
United States	92.0	United States	50.0
United Kingdom	91.0	Hong Kong (China)	33.9
Netherlands	89.6	United Kingdom	33.7
Germany	87.4	China	30.0
France	83.5	Singapore	26.4
Italy	81.4	Netherlands	25.0
Brazil	81.0	Japan	22.5
Belgium	80.0	Malaysia	20.3
Switzerland	79.4	Canada	16.2
Mexico	78.0	Australia	15.0
Canada	77.3	Germany	15.0
Spain	76.4	Cayman Islands	13.7
Singapore	73.7	Taiwan	13.2
Poland	72.0	Virgin Islands (UK)	12.5
Japan	70.3	Bermuda	11.2
Czech Republic	70.0	France	11.2
Australia	69.7	Brazil	10.4
Argentina	68.0	Belgium	10.0
China	66.0	Mexico	9.5
Hong Kong (China)	65.6	Poland	8.8
Austria	64.0	Czech Republic	7.5
Portugal	64.0	Italy	7.5

Note: "Location intensity" is a measure of the geographical reach of TNCs. It is designed to remove the influence of TNCs investing in their home countries – as when Dutch TNCs have subsidiaries in the Netherlands, which is the destination most frequently chosen by the largest TNCs for their overseas affiliates (as of 2006, 86 of the 100 largest TNCs have at least one affiliate there). The figure is calculated as follows: the total number of TNCs having at least one affiliate in the host country, divided by 100, minus the number of TNCs from this country listed in the top 100.
Source: UNCTAD (2006: 35)

to Canada and Europe – direct ownership, in other words, was a more significant strategy in Latin America than in Canada or Europe.

Though US TNCs continued largely low-end, labor-intensive manufacturing in Latin America – with the result that in 1966 US manufacturing subsidiaries accounted for 41 percent of Latin America's manufactured exports but less than 10 percent of its gross manufacturing value added (Diebold 1973: 84) – in the 1950s US manufacturing FDI increasingly began to flow to Western Europe, Australia, New Zealand, and Canada. In the case of Europe, this shift came on the heels of the Marshall Plan, designed to rebuild Western Europe's economies to, amongst other things, serve as markets for US goods.[9] By 1994 50 percent ($115 billion [$202 billion]) of US FDI invested in the

EU was in manufacturing, with Germany the leading recipient (the UK was the leading overall recipient), followed by the UK and France, a pattern of investing in these three countries which continued trends traceable to at least 1901 (Barrell & Pain 1997). Notably, very little US FDI went either to Japan or the communist world. However, the destruction of Europe's industrial infrastructure – especially that of Britain, the largest pre-war investor in the US – meant that only modest levels of FDI entered the US in the early post-war period: whereas in 1946 $2.5 billion of FDI was located in the US, by 1956 the number had only grown to $4.5 billion, although $8.8 billion of portfolio investment was located in the country (US Department of Commerce 1958: 181).[10] With the reconstruction of the Japanese and, especially, Western European economies, however, FDI in the US began growing – whereas in 1980 there was $83 billion [$370 billion] of inward stock relative to the $220 billion invested by US TNCs overseas (i.e., inward stock was 38 percent of outward stock), by 1994 the figures were $504 billion [$886 billion] and $610 billion (which put inward stock's value at 83 percent of that of outward stock) (UNCTAD 1995: 401, 407). Put another way, whilst US FDI overseas grew 13 percent per annum, inward investment grew 36 percent. As a result, by 2004 FDI stock in the US had increased to $1.3 trillion whilst US FDI overseas grew more slowly, although at $1.5 trillion was still greater in absolute terms (Kurian 2001: 377, 378).

By the early twenty-first century, though, trends had changed again, and outflows outpaced inflows – in 2004 US outflows totaled $222 billion, whereas inflows were $122 billion (UNCTAD 2006: 299). Consequently, foreign firms presently account for 13 percent of US manufacturing employment, a proportion slightly lower than the 19 percent share of US manufacturing GDP they generate, the result of their concentration in more capital-intensive production (Graham & Marchick 2006: 19–26). Despite popular perceptions that much US FDI is relocating to places like China and Mexico, in fact the vast bulk flows to the European Union – in 2003 60 percent of the $152 billion outflow went to the EU, whilst 2 percent went to China (including Hong Kong) and 4 percent to Mexico. However, in terms of growth, whereas total US FDI stock in the EU grew almost fourfold between 1992 and 2003, that in Mexico grew four-and-a-half-fold and that in China sixfold, with much of this in manufacturing. Indeed, in 2000 the US was the largest holder of FDI in Mexico ($55.0 billion) (OECD 2004). Meanwhile, US FDI stock in China grew to $51 billion by 2005, after that from Hong Kong ($260 billion, much of which is controlled by non-Hong Kong investors, like those from Taiwan) and Japan ($53.3 billion) and followed by that from the British Virgin Islands ($46 billion, investment largely redirected from elsewhere, the result of its tax haven status), Taiwan ($42 billion), and South Korea ($31 billion) (Morrison 2006: 6).

For their part, in many regards European TNCs simply picked up in 1945 where they left off in 1939. Thus, maintenance of British imperial preferences until 1973 continued shaping firms' investment decisions in much the same way as they had prior to World War II. Nevertheless, the war's impacts, including the break-up of empire beginning in the 1940s, did dramatically transform the globe's political and economic geography. For instance, to pay war debts Britain sold off overseas assets worth £5 billion net (£618 billion) (Hatch 1975: 75), whilst at home and abroad the fact that industrial production had been diverted into war materiel significantly affected investment patterns. Such developments influenced firms' behavior (both those engaging in FDI and those not) in important ways, as did nationalization of TNCs' investments in former colonies. Hence, Kleiman (1978) suggests that by 1972 the loss of colonies – which had previously absorbed many European exports – had reduced British overseas sales by 9 percent, whilst for France the figure was 13 percent, although others (e.g., Foreman-Peck 1983: 311) have suggested the impact was somewhat less. Furthermore, the reduced trading barriers brought by the European Community's creation encouraged many firms to focus upon gaining market share within Europe. As a result, imports from colonies and former colonies declined from 26 percent of all imports into Western Europe in 1935 to 13 percent by 1998, whereas exports to such locations declined from 23 percent to 14 percent. Meanwhile, imports from other Western European countries increased from 46 percent to 71 percent of all Western European nations' imports, whilst exports to other Western European countries grew from 57 percent to 69 percent of all exports (United Nations 2000: 24). At the same time, imports from Japan increased, whereas those from North America dropped (Table 6.6).

In terms of FDI, although European TNCs' overseas direct investment in the 1950s was fairly minimal, by the 1960s many had once again begun looking abroad, particularly to the US, with its high consumer spending and large market (Jones & Wren 2006: 14). Globally, whereas in 1980 the EU-15 had overseas stocks valued at $213 billion (42 percent of global totals) (UNCTAD 1995: 407), by 2005 the EU-25 controlled $5.5 trillion (51 percent) of outward FDI stock, with the biggest overseas investors being the UK ($1.2 trillion), Germany ($967 billion), France ($853 billion), the Netherlands ($641 billion), Belgium ($386 billion), and Spain ($381 billion) (UNCTAD 2005: 303). Meanwhile, Europe's growing political and economic integration encouraged FDI flow between Western European countries, from a total of $186 billion in the 1980s to $662 billion between 1990 and 1997 (United Nations 2000: 27). At the same time, the EU enjoyed substantial FDI influx, such that whilst in 1980 the EU-15 hosted $184 billion of stock (38 percent of global totals) (UNCTAD 1995: 401), by 2005 the EU-25 was hosting $4.5 trillion (44 percent of global totals), with the biggest FDI receivers being the same six countries – the UK ($817 billion), France ($601 billion), Germany ($503 billion), Belgium ($492 billion), the Netherlands ($463 billion), and Spain ($368 billion) (UNCTAD 2005: 303). Concomitantly, the Soviet Union's collapse and associated changes in Eastern Europe encouraged intra-continental investment by European firms – whereas from 1980 to 1989 only $600 million of Western European FDI flowed to Eastern Europe and the Soviet Union, between 1990 and 1997 $41.7 billion did. Indeed, although their FDI inflows are still fairly small, several former Soviet bloc countries have become favored destinations – between 2003 and 2005 the Czech Republic received $18 billion and Poland $25 billion, more than longer-standing EU members like Greece ($4 billion) and Portugal ($14 billion), a development which suggests that, through their investment practices, TNCs are beginning to reshape the EU space-economy in dramatic ways (UNCTAD 2006: 299).

Table 6.6 Exports from, and imports to, Western Europe, 1928–98 (%)

	Western Europe[a]	Eastern Europe and former Soviet Union[b]	North America	Japan[c]	Other "developed" nations[d]	Developing countries
Exports by destination						
1928	55.2	7.7	8.6	1.2	4.8	22.5
1935	56.5	6.4	7.4	1.1	5.7	22.9
1938	54.3	8.3	6.6	0.9	6.5	23.4
1963	63.7	4.4	8.9	1.0	3.8	18.2
1970	67.2	4.7	9.2	1.2	2.9	14.8
1973	69.0	4.9	8.6	1.4	2.2	13.9
1979	68.0	4.9	6.7	1.2	1.6	17.6
1981	63.6	4.8	7.3	1.1	2.0	21.2
1988	71.3	3.7	8.9	1.9	1.4	12.8
1995	69.1	4.4	7.3	2.1	0.8	16.3
1998	68.9	5.7	8.9	1.7	0.7	14.2
Imports by source						
1928	46.4	7.4	17.1	0.5	4.6	24.0
1935	45.6	9.4	12.3	0.8	5.8	26.1
1938	43.2	8.9	14.8	0.7	6.3	26.0
1963	58.8	4.5	13.9	1.0	2.4	19.4
1970	64.1	4.5	12.0	2.0	1.9	15.6
1973	66.4	4.7	9.4	2.4	1.9	15.2
1979	64.5	5.3	8.4	2.2	1.2	18.4
1981	60.3	5.8	9.3	3.0	1.0	20.5
1988	71.3	4.4	7.8	4.5	1.1	10.9
1995	72.4	4.4	7.0	3.5	0.4	12.2
1998	70.8	4.9	7.7	3.4	0.6	12.6

Notes:
(a) For 1928–38, includes Albania.
(b) For 1928–38 comprises Czechoslovakia, Bulgaria, Estonia, Latvia, Lithuania, Hungary, Poland-Danzig, Romania, Yugoslavia, and the USSR. Includes the DDR from 1963 to 1988.
(c) For 1928–38 includes Korea and Taiwan.
(d) Australia, New Zealand, and South Africa, except for 1995 and 1998 when South Africa is included in "Developing countries."
Source: United Nations (2000: 24).

In the case of the UK – the largest EU FDI recipient and provider – both its outward and inward positions have changed in recent years. Whereas in 1987 40 percent of outward FDI was in manufacturing and 41 percent in services, in 2002 the figures were 29 percent and 62 percent respectively. Concomitantly, in 1987 34 percent of inward

FDI was in manufacturing and 48 percent in services, whilst by 2002 the figures were 25 percent and 64 percent – services, in other words, have become a more significant element both of FDI from the UK but also FDI invested in it (OECD 1999, 2004). With regard to inward FDI's origins, the bulk is from the EU (£247 billion in 2005), the US (£136 billion), Australia (£34 billion), Canada and Switzerland (£14 billion each), and Japan (£13 billion), whilst UK overseas investment was primarily in the EU (£379 billion), the US (£161 billion), Australia (£24 billion), Hong Kong (£21 billion), and Canada (£12 billion). Overall, Latin America held £18 billion of British FDI, whilst Africa held £25 billion, 66 percent of which was in South Africa (OECD 2007).

As with the UK, so with France and other EU member states, with outward FDI being concentrated largely in two regions: the EU (with a total French stock of €465 billion in 2005) and the US (€143 billion). Indeed, French FDI flows into other EU countries began increasing dramatically in the 1990s, from about 48 billion francs in 1988 to 104 billion by 1998, increasing the proportion of French FDI stocks located in other EU countries from 47 percent to 50 percent. Whereas in 1989 50 percent of French FDI was in manufacturing, by the late 1990s that had dropped to 38 percent, and the lead had been taken by services (56 percent of outward stock). In terms of the reverse, the bulk of FDI invested in France came from other EU countries, a trend which also increased during the 1990s – inflows jumped from 15 billion francs in 1987 to 132 billion by 1998, increasing the proportion of FDI stocks coming from other EU countries from 57 percent to 67 percent (OECD 2004). A noticeable trend, though, has been the growth of French investment in China (including Hong Kong), from flows of 758 million francs (equivalent to €116 million) in 1987 to €853 million by 2002, leaving French TNCs with a total investment of €4 billion (OECD 1999, 2004). By 2000 French firms accounted for 1.3 percent of all FDI in China, though 17 percent of total EU FDI (Van Den Bulcke et al. 2003: 39). Nevertheless, whilst French companies have increasingly explored new locations for investment (like China), it is important to recognize that colonial ties continue to bind them – of the €13.5 billion of French FDI in Africa in 2004, 47 percent was located in the former French colonies just of Algeria, Morocco, and Tunisia (OECD 2007).

Turning to Asia, whereas Japanese FDI remained fairly low immediately after World War II, by 1980 it had reached 7 percent of global outward FDI, although FDI into Japan was less than 1 percent of world totals (Jones 2005: 33). Although prior to the war Japanese TNCs had invested heavily in neighboring countries and colonies, by 1952 the entire stock of Japan's pre-war FDI in East Asia had been confiscated and the assets given away by the Allied powers (Mason 1999). Consequently, as the 1950s progressed and as Japan's economy was rebuilt with substantial imports of US investment as part of a Japanese "Marshall Plan," Japanese TNCs again began looking outward for investment opportunities, both in places with which they were familiar but also in new locations (investing, for instance, in Indian glass-making and manufacturing refrigeration facilities [Joseph 1990]). In the 1960s they invested significantly in iron ore and copper mining in Malaysia and the Philippines, in natural gas extraction in Brunei, in oil in Indonesia, and in the agricultural, forestry, and fisheries industries – in other words, in the economic sectors which dominated pre-World War II FDI. By the 1970s, growing import restrictions across the region, the Japanese government's lifting of many prohibitions on outward FDI, and rising Japanese wage rates were encouraging manufacturers to establish facilities in Indonesia, the Philippines, Malaysia, Thailand, and elsewhere (though not in China nor North Korea, where government opposition meant Japanese FDI was virtually banned, nor in South Korea, where anti-Japanese feeling ran high). Consequently, whereas Japanese FDI outflows to Asia totaled less than $800 million between 1950 and 1970, during the 1970s they reached $9 billion [$31 billion] (Mason 1999: 33). In the 1980s and 1990s FDI continued to climb, and by 2003 there was some ¥10.0 trillion ($86.3 billion) invested in Asia. With improved Sino-Japanese relations, stocks of Japanese FDI in China (including Hong Kong) alone increased 13 percent between 1992 and 2003 to ¥2.2 trillion ($19.0 billion), whilst those to India jumped from ¥41.4 billion ($327 million) to ¥161.2 billion ($1.4 billion) in the same time period, although Japan's ongoing economic crisis has meant that flows to Asia have declined in recent years, from an annual high of ¥1.5 trillion ($12.4 billion) in 1997 to ¥1.3 trillion ($11.2 billion) in 2003 (OECD 2004).

Although Asia, then, represents an important location for Japanese FDI, Japanese TNCs have been looking elsewhere too. North America (particularly the US) is now the dominant destination for Japanese FDI, and has been so since the early 1980s. In 2003 ¥1.2 trillion ($10.4 billion) of FDI flowed to the US, bringing Japanese stocks to ¥14.9 trillion ($128.5 billion) (OECD 2004). In more recent years Japanese firms have also looked to Mexico, seeking cheap manufacturing sites through which to supply the US and Canadian markets, courtesy of the North American Free Trade Agreement. Whereas in 1992 annual FDI flows to Mexico were only ¥7.6 billion ($60 million), by 2003 they reached ¥15.8 billion ($136 million), having hit a high of ¥165.5 billion ($1.5 billion) in 1999. By 2003 Japanese TNCs had invested ¥307 billion ($2.6 billion) in Mexico (OECD 2004), whilst as of January 2006 Japanese TNCs were three of the six largest employers of *maquiladora* workers in the country (Maquila Portal 2006).[11] Africa, on the other hand, has remained largely untouched, with total stock of only ¥220 billion ($1.9 billion), of which 52 percent is in South Africa (OECD 2004). Concomitant with such geographical shifts has been a sectoral shift. Whereas in 1960 more than half of Japanese FDI globally was in the primary sector (mostly mining), 27 percent was in manufacturing, and only 22 percent was in services, by 1994 5 percent was in the primary sector, 28 percent in manufacturing, and 51 percent in services, although these percentages varied significantly by location – in Europe investment has tended to be in service sector activities like banking whilst in Australia and Asia it has been in mining (Dicken 1988; OECD 2004).

Whilst major economic powers like the G-3 nations both send and receive the bulk of world FDI flows, an important recent development has been the growth of overseas activities by TNCs from LDCs, with such TNCs themselves a growing phenomenon on the world stage.[12] In this regard Hong Kong is the largest source of FDI from the global South and the sixth largest FDI supplier globally, mostly in services related to offshore financial centers in the Caribbean and the rest of China – as of 2004, half of all FDI from Hong Kong was invested in the British Virgin Islands and Bermuda and almost 40 percent in China (UNCTAD 2006: 112). In similar fashion, TNCs from other parts of Asia have begun increasing their FDI levels. For instance, although Taiwanese TNCs contributed only 0.7 percent of global outward FDI in 1990, a proportion roughly equivalent to Japan in 1960, between 1980 and 1998 Taiwanese outward stock grew by an annual average of nearly 40 percent, reaching $38 billion. This made Taiwan one of the fastest-growing FDI sources, responsible for almost 10 percent of outward FDI coming from developing countries in the late 1990s. Similarly, whereas South Korean TNCs did not really begin to engage in FDI until the late 1950s – the earliest recorded investments were acquisition of a commercial building in New York City and a timber operation in Malaysia – South Korean FDI levels grew as the government allowed more companies to invest overseas after 1986's economic liberalizations. In the late 1970s three-quarters of Korean FDI had gone to LDCs, the most important regions being Southeast Asia (43 percent) and Africa (21 percent), whilst virtually all the rest was in North America, with small amounts in Europe. However, by 1989 LDCs accounted for only 46 percent of South Korean FDI, the result of a shift from securing raw materials to engaging in manufacturing and mining operations in North America, Oceania, and Western Europe (Tolentino 2000).

In the case of Brazil, South America's nascent industrial colossus, TNCs' FDI practices stretch only to about 1970. Although as a proportion of global totals Brazil's FDI is small (0.7 percent of world outward FDI in 2005), by 2003 the country had the largest stock of outward FDI in Latin America and was the fourth largest "developing world" provider, after Hong Kong, Singapore, and Taiwan.[13] Over 1 000 Brazilian firms had direct investments abroad, although only three (the state-owned oil company Petrobrás, the mining company Companhia Vale Do Rio Doce, and steel manufacturer Gerdau) were among the hundred largest non-financial TNCs from LDCs (UNCTAD 2006: 283–5). Nevertheless, the $5 billion merger between Brazilian drinks group Ambev and the Belgian-based brewer Interbrew in 2004 resulted in a significant presence within Europe. Overall, in 2005 Brazilian companies had $72 billion of FDI, most of which was in offshore financial centers like the Cayman Islands for tax avoidance purposes, whilst the remainder has largely been in trading companies, mining, construction, petrochemicals, autoparts, and telecommunications (UNCTAD 2006: 113–14). Much as with Indian TNCs now investing in Britain, several Brazilian TNCs have begun investing in Portugal, a phenomenon that highlights how colonial linkages continue to shape how the world economy is networked in the twenty-first century. Indeed, a third of the $5.4 billion represented by the 20 largest new overseas investments since 2002 has gone to Portugal, such that in 2003 Portugal was the ninth most important destination for Brazilian FDI, after tax havens like the Caymans and Bahamas, neighbors like Uruguay and Argentina, and major economies like the US (UNCTAD 2004a: 5–8). Given Portugal's membership in the EU, such an investment strategy has also provided Brazilian TNCs access to Europe as a whole – a sort of inverse imperial preference scheme.

To summarize the post-1945 FDI experience, then, several broad trends can be identified. First, there has clearly been a growth in FDI's importance as a tool for linking together different parts of the global space-economy. Not only have FDI flows grown in absolute terms – from $13 billion [$156 billion] in 1970 to $916 billion in

2005 – but they have also grown relative to nations' GDPs. Thus, whereas in 1990 global FDI stocks equaled 9 percent of world GDP, by 2005 they equaled 23 percent (UNCTAD 2006: 307). This mirrors the broader transnationalization of assets in recent years. Hence, although in 1900 the total planetary stock of foreign assets was valued at 18 percent of world GDP yet by 1945 had dropped to 5 percent, by 1980 such assets had regained their 1900 level, and by the late 1990s had exploded, reaching almost 80 percent of the value of global GDP (Denis et al. 2006: 22). Such figures, then, indicate that TNCs' activities have become ever more significant as we have moved into the twenty-first century and that they are playing an increasingly influential role in shaping patterns of world economic development. It also shows that there was a significant "deglobalization" of the world economy – at least as measured by FDI – in the middle of the twentieth century, compared to its beginning years, and that only by the early 1980s had an equivalent level of global integration been reached.

Second, whereas FDI patterns prior to World War I largely reflected efforts to secure raw materials from the global South (as with investments in Latin America and Africa), to break into overseas markets (as with US manufacturers' operations in Europe and Latin America or British firms' building of factories in colonies like South Africa and Australia), and to find investment opportunities which could soak up surplus capital held by European and North American investors (like railroads in Latin America), by the early twenty-first century most of the world's FDI was flowing to the industrialized global North – in 2005 47 percent of world FDI inflows of $916 billion went to Europe and 15 percent to Canada and the US (UNCTAD 2006: 299). Moreover, of all the FDI flowing in 2005 to what the United Nations Conference on Trade and Development (2006: 299–302) calls "developing economies," 67 percent went to just six countries, with 32 percent going to China alone.

Third, this geographical shift in FDI stock location and global FDI flow destinations is undoubtedly linked, at least in part, to the transformation in the nature of FDI, which has become increasingly oriented towards services – whereas in 1989–91 services accounted for 55 percent of world inward FDI flows, by 2002–4 this had climbed to 64 percent. Although the sectoral proportions obviously vary greatly by country, even many LDCs have become primarily importers of service-sector FDI – during this period, the proportion of inward FDI going to LDCs that was service-oriented grew from 36 percent to 48 percent of the total, whilst that for "developed countries" jumped from 60 percent to 72 percent (UNCTAD 2006: 268).

Fourth, although they are still relatively rare, the past few decades have witnessed a growing number of TNCs emerge from the global South. By 2005 annual FDI outflows from developing and "transition" countries (e.g., Russia) reached $133 billion (17 percent of global FDI outflow totals) and outward stock was $1.4 trillion (13 percent of world totals), with some of the largest investors being Hong Kong-based Hutchison Whampoa, PETRONAS (Malaysia), SingTel (Singapore), Samsung (South Korea), CITIC (China), and CEMEX, S.A. (Mexico) (UNCTAD 2006: 31–2, 105, 283). Although their specific investment strategies may vary, depending upon in which region of the globe they originate (Mexican TNCs have overwhelmingly invested in the Americas, whereas Malaysian firms have invested heavily in Southeast Asia and

Table 6.7   Comparison of overseas assets, sales, and employment of the world's largest 100 TNCs and the 50 largest TNCs from LDCs, 2004

	World's largest 100 TNCs	*50 largest TNCs from LDCs*
Assets ($ billion)		
Foreign	4 728.0	336.9
Total	8 852.0	1 073.2
Foreign as % of total	53.4	31.4
Sales ($ billion)		
Foreign	3 407.0	323.0
Total	6 102.0	738.2
Foreign as % of total	55.8	43.8
Employment (000s)		
Foreign	7 379.0	1 109.0
Total	14 850.0	3 364.0
Foreign as % of total	49.7	33.0

Source: UNCTAD (2006: 31, 32).

Europe [UNCTAD 2006: 35]), generally such corporations follow patterns of investment similar to those of US, Canadian, European, and Japanese TNCs – investing in industrialized countries in search of consumers and in LDCs in search of cheap labor. Thus, although TNCs from the industrialized countries tend to be more transnationally focused than do those from LDCs, this is more a matter of degree than kind – in 2004, for instance, whereas 50 percent of the employees of the world's 100 largest TNCs worked overseas from the firm's country of origin, for the 50 largest from LDCs the equivalent figure was 33 percent (Table 6.7).

Fifth, TNCs which have expanded into LDCs in search of cheap labor have often been attracted to Export Processing Zones (EPZs), locations in which bureaucratic requirements are substantially reduced and tariffs and taxes fairly low, if not abolished outright. Typically, the rationale for Zones' establishment is that they will attract international investment and so reduce poverty, although the record suggests this often does not occur (Klein 2000). The number and importance of EPZs, however, has been growing – whereas in 1975 there were 79 worldwide, by 1999 there were 845, and by 2002 over 3 000 had been established in over a hundred different countries, mostly in Asia.[14] Such growth has implications not only for the geography of the world economy but also for nation-states' abilities to ensure their own space-economies' territorial integrity, given that EPZs effectively become "offshore" economic enclaves wherein various "international" activities occur but which are often somewhat sealed off from the broader national economy. Moreover, whereas in 1999 EPZs employed 27 million workers, by far the majority women (United Nations 1999: 86), by 2003 the total was more than 43 million, most of whom were working in China's 2 000 or so Special Economic Zones (ICFTU 2003: 8). This growth is significant because firms locating in EPZs typically import much of their raw materials and export the finished product back onto the world market. Given that such production tends to be labor-intensive and often low-tech, there is often little technological know-how transfer to the host economy. Furthermore, whereas in the 1970s TNCs typically located operations in an EPZ to reduce production costs when serving their home market – as with US TNCs relocating to Mexican *maquiladoras* and exporting back to the US – today TNCs commonly use such Zones to supply third-party countries. Hence, in the early 2000s several Taiwanese and Korean firms located in Nicaraguan EPZs to export to the US under the provisions of the Central American textile quota system. EPZs' establishment by many global South governments, and their increasing use by TNCs, then, is having significant impacts upon the geography of the contemporary world economy and what that means for the millions of workers toiling to make the commodities which drive it.

Transnational Firms as Geographical Actors

As the above analysis highlights, the twentieth century has witnessed growing industrialization in several regions of the globe beyond those of the nineteenth-century imperial core. This has had dramatic implications for how the planet is wired together geographically and has resulted in, amongst other things, a significant increase in the share of the world's manufacturing exports coming from LDCs (from 11 percent in 1980 to 30 percent in 2000), with much of these going to the global North (UNCTAD 2004b: 11). Certainly, much of this industrialization has been indigenously led and has deep roots. For example, by the late 1930s India's Tata Iron and Steel company was producing 1 100 tons of steel a day at its Jamshedpur blast furnace (an amount bested in the US by only two furnaces) and India was the world's twelfth largest steel producer (Keenan 1943: 1). Nevertheless, much has been stimulated by TNCs' FDI activities – Ford and GM, for instance, essentially kick-started the Brazilian automobile industry when they established assembly plants there after World War I (Shapiro 1991). TNCs' activities in linking different parts of the globe together through flows of investment, commodities, and power, then, have been paramount in shaping global uneven development. What is particularly significant in all this, however, is that although the products coming from the global South today may be different from those of the nineteenth century – increasingly, manufacturing goods instead of raw materials – many are being produced by subsidiaries of US, British, French, Japanese, and other global North nations' TNCs located in former colonies. This raises telling questions about just how much the planetary economic geography of the "era of globalization" differs from that which preceded it. Although some LDCs are beginning to break out of colonial-era trading patterns and are trading more with each other than in the past (whereas in 1975 23 percent of LDC exports went to other LDCs, by 1995 that had increased to 40 percent, though it declined subsequently [UNCTAD 2002: 52]), there are still considerable similarities between the trading patterns of the early twentieth century and the early twenty-first.

Unquestionably, in contemplating how TNCs have networked the globe together it is important to recognize that different nations' corporations have behaved in different manners and that FDI has played different roles in various nations' economic development. Thus, during the twentieth century FDI was a more important aspect of British firms' activities than it was for those of any other industrialized nation, the result perhaps of Britain's relatively small domestic market and its longer history of imperialism (which encouraged many firms to establish colonial subsidiaries). Likewise, whereas US firms typically established themselves domestically and then expanded overseas with directly owned subsidiaries, many Belgian and French firms (particularly prior to 1914) invested overseas through holding companies and financial intermediaries, making it difficult to establish where managerial control actually lay (Jones 1994: 8). Meanwhile, British FDI was often in the form of free-standing companies funded and headquartered mainly in London (Kim 2005: 70; also Wilkins 1988). Equally, although US TNCs' expansion into foreign markets was frequently characterized by the production of new, labor-saving, highly income-elastic products aimed at middle-class consumers, Continental European firms tended to lead with synthetics and goods aimed either at working-class consumers or those in the upper echelons of society (Franko 1975: 42). There have also been significant differences in the more contemporary period – Japanese and US TNCs have generally been less transnationalized than have European ones (Hannah 1996; UNCTAD 2005: xix–xx).

Regardless of the different models, however, ultimately the goal of investing transnationally has been the same: to secure geographical flexibility by operating over a greater area than firms would enjoy were they to remain bound within their home countries. Hence, by expanding their spatial reach, firms can access otherwise unattainable labor markets, consumers, and resources. They can also avoid tariffs designed to close off territory, a spatial strategy recognized early on by William Lever, who declared he was building overseas manufacturing plants because "the French tariff . . . has now reached the point where it will pay us better to have a factory inside France itself . . . If all the world were free trade, I could make all my soap in England and export it. But the whole world is not free trade, and it is the tariff which compels me to . . . go multinational" (quoted in Wilson 1974: 296). In similar vein, the Singer company was able to use its geographical flexibility to service markets it could not supply locally – when its Podolsk plant could not keep up with orders, Singer simply increased production at its Glasgow plant and exported the surplus to Russia (Davies 1976: 266). Concomitantly, transnationally organized firms may take advantage of currency fluctuations, may have access to sources of overseas capital denied their non-transnational rivals, and may engage in transfer pricing wherein goods and services are priced to avoid local taxes. Through their geographical flexibility, then, TNCs can gain significant advantages.

As a result of such corporate strategies, between one-third and two-thirds of global trade is now intra-firm trade between TNCs' different subsidiaries (Sauvant 1996; Ruggiero 1997). This has transformed the nature of the world economy as trade is increasingly not between independent entities located in separate nation-states but between various branches of firms straddling national boundaries. In exploring the

Table 6.8 Outward stock of FDI as a percentage of GDP for selected countries, 1913–95

	1913	1938	1950	1960	1971	1980	1995
Canada	6	14	6	6	7	9	20
France	23	21	–	6	5	–	25
Germany	11	1	–	1	3	4	10
Japan	11	21	–	1	2	2	5
Netherlands	82	91	–	89	35	25	47
UK	49	38	9	15	17	15	28
US	7	8	4	6	8	8	18

Source: Twomey (2000, Table 3.4).

historical geography of TNCs' FDI practices as they relate to the supposed emergence of a globalizing world economy, though, it is important to recognize two points. First, the fact that FDI was a smaller portion of world output in the 1990s than it had been in 1913 (Richter 2003) questions the supposed unidirectionality of globalization. Moreover, it suggests that whatever globalization may be occurring is a much more historically and geographically complex process than that portrayed by simplistic TINAesque statements concerning how "irresistible" (Ohmae 2005: 18) the process is. Hence, whereas nations like the US are more FDI-oriented at the beginning of the twenty-first century than they were at the beginning of the twentieth, both in terms of the proportion of global FDI they control and their level of FDI relative to their economies' overall size, others like the UK are much less so (Tables 6.3 and 6.8). Moreover, even in the case of countries which do appear to be more transnationalized at the end of the twentieth century than at its beginning, such countries have often traced a complex path to get where they are today, a path which has seen declines in their shares of FDI as well as increases – although the share of worldwide FDI owned by US companies at the end of the twentieth century was greater than at the beginning, it was still less than half what it had been in mid-century. These trajectories raise important questions about the nature of globalization, just exactly how global it is, and what is the connection between discourses of globalization and TNCs' investment practices.

Second, there is an important difference between TNCs investing overseas to serve local markets, TNCs investing overseas to produce whole goods for export to their home country, and TNCs investing overseas to produce parts to be used in their domestic assembly operations. Certainly, all three situations represent growing spatial connections between places. Yet, they are quite different in what they mean for TNCs' geographic organization and the world economy's spatial structure. Thus, investing overseas to serve local markets can have significant impacts upon local producers in the host countries, who may now face greater competition due to foreign firms' presence, but it generally has little impact upon the citizens of a TNC's home country. Indeed, if such overseas operations increase a TNC's profitability then this type of investment may actually benefit the home nation's citizens as profits are repatriated. On the other hand, investment overseas to produce goods for export back to the TNC's home country typically represents a significant challenge to that home country's workers and firms, even as it may benefit overseas workers (who gain employment) and presents less of a challenge to overseas firms (who have less to fear from foreign TNCs' products flooding local markets). This distinction is all the more important given that this latter phenomenon has become increasingly common – whereas in the 1950s and 1960s TNCs' subsidiaries frequently did not export to their country of origin to avoid competing with their parent company, many companies now serve their home markets through importing from overseas subsidiaries. Finally, stretching production chains across national boundaries represents an even greater degree of integration – production in one country must become increasingly synchronized with that in a second if the manufacturing process is not to grind to a halt due to unavailability of components from the overseas subsidiary. These distinctions, each with diverse implications, mean that discourses which present TNCs' establishment of manufacturing subsidiaries overseas as evidence of globalization must be parsed carefully, especially because it is, I rather suspect, the growing presence of commodities made overseas and the frequent inability to tell precisely from whence is a commodity – is a car assembled in the US using Mexican parts "really" a "US car"? – and not the growth of fairly intangible lines of transborder corporate control and ownership that are creating the sense that something fundamental has occurred in how the world economy is being reorganized.

Questions for Reflection

- How have TNCs linked places together and to what effect?
- How have FDI patterns differed at different historical periods?
- How do TNCs' investment practices shape the geography of the world economy and how are they shaped by it?

Notes

1. Philippine "independence" included maintenance of US bases, pegging the Philippine peso to the dollar, prohibitions on manufacturing goods that competed with US-made ones, allowing US citizens and corporations equal access to Philippine natural resources, and allowing US-made goods tariff-free access to the country until 1954.
2. Such investments raise methodological questions concerning the distinction between overseas portfolio investment and FDI. Wilkins (1988) has argued that the traditional distinction between portfolio investment (arm's-length investment in foreign companies without significant control) and FDI (the establishment of directly controlled operations overseas) is not particularly helpful in many situations: firms established overseas by

expatriates that rely upon financing from the imperial metropole do not fit neatly into either category – they are not directly controlled from, say, London (and so might appear to be portfolio investment) but, equally, were founded with the intent of conducting and managing business operations abroad. In such a situation, if such expatriates never returned to Britain their firms would *not* constitute FDI, since there would be no obligation to anyone in Britain. On the other hand, should the expatriate return home after many years and retain an interest in the overseas establishment, then such firms would suddenly become "foreign investments." Further complicating the picture is the fact that expatriates frequently registered their overseas firms in their country of origin so as to tap local capital markets, with the result that such overseas direct investment did not emerge from the operations of firms located in the imperial metropole in the way in which FDI is typically seen to do. The point is that understanding overseas investment patterns relies heavily upon definitional questions. Wilkins, then, suggests the category of the "free standing company" – firms established overseas by expatriates but registered in Britain, France, the US, etc. and not growing out of the domestic operations of firms from such expatriates' home countries.

3. As Jones (1996: 60) points out, Vaupel and Curhan's data collection method – they identified 187 large US TNCs operating in at least six countries in the mid-1960s and then traced their growth backwards – surely undercounts the number of overseas manufacturing subsidiaries by missing firms no longer in existence when they conducted their study.

4. Williams (1975) estimated that between 1956 and 1972 $10.1 billion worth of FDI – almost 25 percent of all assets owned by foreign firms – was nationalized in LDCs, with only 41 percent of that being compensated.

5. The Chinese figure includes $35.9 billion flowing to Hong Kong.

6. Figures calculated by author, based on data in UNCTAD (2006: Table A.I.3).

7. It is important to note that because many large mergers and acquisitions by Chinese companies are financed outside the country, this figure may significantly underestimate Chinese FDI. The British Virgin Islands' high number is largely accounted for by the fact that they are a major offshore financial center.

8. This acquisition was particularly symbolic, for one of Corus's constituent elements was British Steel, whose Port Talbot mill provided much of the steel to build India's railways during the nineteenth century. Predictably, the British press called this an instance in which "The Empire Strikes Back" (*Observer*, October 22, 2006).

9. As a way to do so, nations receiving Plan funds had to reduce tariffs, form an economic union, and not discriminate against US firms operating in Europe, thereby ensuring Western Europe was turned into a singular, integrated economic space for the relatively free circulation of US goods (rather than maintained as a series of economically separate nation-states). Recipient nations also had to agree to subcontract much of the rebuilding work to US firms. Equally, much aid was in the form of US-produced goods (like coal) which were exported to Europe, thereby supporting US industries. European nations could pay for such goods using the loans made them by the US government (providing a stimulus to the US financial system), by sending as barter goods (either from the recipient nation or one of its colonies) raw materials needed by US industry, or by directly transferring ownership of assets (typically in their colonies) to US interests. Recipients also had to agree to stabilize their currencies by linking them to the US dollar and ensuring their convertibility, and to rework to US satisfaction policies relating to taxation, budgeting, finance, and labor markets. This was designed to prevent establishment of economic

models in Europe based on strong national planning, seen as anathema to US visions of capitalism. The Plan's ultimate economic goal was perhaps most openly stated by US Under-Secretary for Economic Affairs William Clayton, who stated: "Let us admit right off that our objective has as its background the needs and interests of the people of the United States. We need markets – big markets – on which to buy and sell" (quoted in Tabb 2004: 114).

10. Significantly, in real terms the figure for 1956 (at its 2005 equivalent of $128 billion) was lower than that for 1946 (worth $140 billion in 2005 dollars). This, of course, raises questions about the supposed inevitability of the global economy's growing integration.

11. The six largest companies were: Delphi Automotive Systems (US, 66 000 workers in 51 plants); Lear Corporation (US, 34 000 in 8 plants); Yazaki North America (Japanese, 33 400 in 41 plants); Alcoa Fujikura Ltd (Japanese, 23 000 in 26 plants); General Electric (US, 20 700 in 30 plants); and Takata (Japanese, 15 800 in 10 plants).

12. In 2005 47 TNCs listed in the *Fortune 500* were from developing and transition economies, compared with 19 in 1990 (UNCTAD 2006: 122).

13. Proportion calculated using 2005 figures of $10.672 trillion of global outward FDI stock and $71.6 billion of Brazilian outward stock (UNCTAD 2006: 9, 115).

14. In 2001, 69 percent of EPZs were in Asia, 23 percent in the Americas, 4 percent in Europe, and 2 percent each in Africa and the Middle East (SOLIDAR 2001: 3).

Further Reading

Dicken, P. (2003) *Global Shift: Reshaping the Global Economic Map in the 21st Century*. New York: Guilford Press.

Jenkins, R. (1987) *Transnational Corporations and Uneven Development*. London: Methuen.

Jones, G. (2005) Multinationals from the 1930s to the 1980s. In: A. D. Chandler, Jr. and B. Mazlish (eds.), *Leviathans: Multinational Corporations and the New Global History*. Cambridge: Cambridge University Press, pp. 81–103.

Electronic resources

The Economist (www.economist.com)

FDI Magazine (www.fdimagazine.com)

Forbes Global 2000 (www.forbes.com/lists)

Chapter 7

Governing Globalization

Chapter summary: This chapter considers the nation-state's status as an economic regulator and the alleged undermining of its sovereignty by globalization. Through examining the Bretton Woods system and its forerunners it shows that the nation-state has not ever been as strong historically, nor is as weak today, as much neoliberal discourse suggests. The chapter also examines what is meant by "sovereignty" and contrasts government and governance of the world economy.

- Economic Sovereignty in the Pre-Bretton Woods World
- The Institutional Architecture of the "Bretton Woods Economy"
- The Nation-State in the Post-Bretton Woods Era
- Globalization versus the Nation-State or Globalization of the Nation-State?
- Sovereignty and Governance

[T]raditional nation states have become unnatural, even impossible, business units in a global economy.　　　(Kenichi Ohmae, The End of the Nation State, p. 5)

There was nothing natural about laissez-faire; free markets could never have come into being merely by allowing things to take their course . . . [L]aissez-faire itself was enforced by the state . . . Laissez-faire was planned.
　　　　　(Karl Polanyi, The Great Transformation, pp. 139–41)

Globalization is not destined, it is chosen . . . But if integration is a deliberate choice, rather than an ineluctable destiny, it cannot render states impotent. Their potency lies in the choices they make.
　　　　　(Martin Wolf, "Will the nation-state survive globalization?", pp. 182–3)

I would not use the language of the adversary. It isn't free trade. If it was free trade why do they need hundreds of pages of rules and regulations? If it was free trade they could do it in two sentences.

(Ralph Nader, Chatroom debate with Patrick Buchanan, 1999)

A central neoliberal contention is that globalization is undermining the nation-state's economic and political sovereignty. Thus, Ohmae (1995) claims globalization is bringing "The End of the Nation State," whilst Bryan and Farrell (1996: 10) maintain that "it will be progressively less possible for an individual government to pursue policies under the assumption that it can directly control its own domestic financial market." For his part, Wriston (1992) has argued that we are witnessing the "twilight of sovereignty."[1] In such narratives the rise of organizations like the EU and the World Trade Organization (WTO), together with TNCs' ability to send huge quantities of money around the globe in the twinkling of an eye and the growing phenomenon of "dollarization," are all represented as "clear" evidence that national sovereignty is being undermined.[2] This viewpoint has been reinforced by some candid proclamations by those involved. For example, Renato Ruggiero, former WTO Director-General, has declared that, "We have created a global trade architecture which is . . . a seamless web of interlocking interests and responsibilities, interdependent and indivisible . . . We are no longer writing the rules of interaction among separate national economies. We are writing the constitution of a single global economy" (quoted in Raghavan 1998).

In arguing that globalization is undermining the nation-state's power, people like Ohmae (2005: 82) are certainly correct in one regard – the nation-state is not "immutable." Given that it is a social construction, it could not ever be so. Rather, its rise as a regulator of economies is historically and geographically specific, so it should come as no surprise that it is, perhaps, seeing its power undermined by contemporary processes. Indeed, if the birth of the modern system of powerful nation-states is typically dated to the 1648 Peace of Westphalia, through which was established the principle of nation-state sovereignty (a principle later exported globally via colonial conquest), then globalization is frequently presented as a process whereby a post-Westphalian world is emerging, one in which nation-states' sovereignty and ability to engage in economic and political self-determination is eroding. Generally, such erosion is seen to date from the early 1970s and the collapse of the post-World War II Bretton Woods system in which nation-states' power was, arguably, most exemplified through their ability to control their currencies' value via international agreement.

Accepting for one moment that the nation-state *is* being undermined by contemporary developments (a contention to which we shall return), there are several questions which logically follow. For example, what kind of world is superseding the pre-globalized one in which nation-states were, by implication, more powerful than now? Is it a world of supranational entities and "world government" wherein nation-states simply become elements in some larger political structure? Is it one in which the nation-state's powers are devolved to subnational entities – or is it one which has elements of both these developments? Furthermore, if political supranationalism is, indeed, economic globalization's corollary, then it is important to recognize that different actors view this in quite varied terms and, through their actions, will shape in diverse ways how political supranationalism unfolds. Hence, whereas US financier David Rockefeller is alleged to have stated that the "world is now more sophisticated and prepared to march towards a world government . . . [in which the] supranational sovereignty of an intellectual elite and world bankers is surely more preferable to the national autodetermination practiced in past centuries" (Jasper 1999), for many nationalists world government represents a decidedly undesirable future. Likewise, whereas for some the emergence of supranational government is an inevitability – Asimov (1970: 19) has declared that "world government will become a fact even if no one . . . particularly wants it" – for others it is a goal to be achieved or an end to be avoided.

As outlined in Chapter 1, a key aspect of thinking about the world economy until recently has been the tendency to view nation-states as territorial units whose boundaries circumscribe fairly discrete "national" economies. Indeed, the reason for many people's contemporary anxiety seems to be that the planet's old territorial ordering is being shaken up, such that it is increasingly difficult to say where one national economy ends and another begins. This sense that something has changed fundamentally with the system of nationally constituted economies is, in fact, the leitmotif of practically all neoliberal narratives (and, to be fair, many on the left): whereas the "pre-global" world was one in which powerful nation-states had virtual monopoly regulatory sway over what went on within their borders, now globalization is forcing them to withdraw from proactive management of their economies, allowing both "the market" to become the pre-eminent determiner of how economies operate and for economies to become "unbound" (Bryan & Farrell 1996: 17) from regulatory constraints. Such a depiction of nation-states as formerly practically omnipotent is important, for neoliberal narratives which argue that globalization is undermining the nation-state's power only really work rhetorically if we accept the existence of an all-powerful nation-state as our starting point – the economy, in other words, cannot be viewed as becoming unbound unless it is viewed as having first been bound. However, as with many things, this representation begins to crumble when exposed to further scrutiny. Whereas representations of the Westphalian system have tended to imply that the formal recognition of nation-states in 1648 marked the beginning of a long period wherein they exercised growing economic sovereignty within their borders, upon closer inspection it becomes evident that such power was only really secured fairly recently, and then only partially.

Given this discussion, then, in this chapter I do several things. Specifically, if we are to explore how TNCs and entities like the International Monetary Fund (IMF), the World Bank, the WTO, and the EU are supposedly undermining the nation-state's sovereignty and power, then we must place nation-states' abilities to regulate their national economies in historical and geographical perspective. Consequently, I investigate several elements of the nation-state's regulatory role in the nineteenth and twentieth centuries, examining particularly the ability to regulate currency markets, which are generally seen as being on the leading edge of contemporary globalization processes. I also outline how supranational organizations, which are often taken as evidence of globalization – the World Bank, the IMF, the WTO, the EU (with its supranational currency, the euro), and others – have played a significant role in the world economy's evolution in the twentieth century. Second, I examine some of the debates concerning how the nation-state is being impacted by globalization, particularly arguments averring that the creation of supranational organizations and the criss-crossing of the globe by flows of money, information, commodities, and people in apparently ever-greater quantities are making it increasingly irrelevant as a regulatory entity. In particular, I survey how, rather than being victims, some nation-states have been active drivers of globalization, even as they have become caught up in complex and contradictory relationships with transnational capital and supranational organizations. I do this through an examination initially of the nation-state's economic sovereignty in the "pre-Bretton Woods" and the "Bretton Woods" eras, for both provide benchmarks for comparison with the allegedly globalizing "post-Bretton Woods" era. Finally, I contrast issues of "governing" and "governance," for it is important to recognize that even if nation-states and their capacities to regulate economies *are* being undermined, this does not mean the world economy has entered a period where there are no rules, where it is "unruly" (Herod et al. 1998). Indeed, rather than the neoliberal call for market "deregulation" being about creating a world in which markets are increasingly freed from regulation, I argue that neoliberal agendas are more about changing *how* the world economy is regulated than dispensing with rules.

Economic Sovereignty in the Pre-Bretton Woods World

When examining the pre-Bretton Woods era to provide a benchmark for comparison with more recent transformations, three important issues emerge, all of which relate to nation-states' exercise of currency control. First, whereas the establishment of transnational currencies like the euro and talk in some quarters of creating other such currencies have often been viewed as fairly new developments in response to growing transnational economic integration, in fact tendencies towards currency supranationalism have a long history.[3] Indeed, pointing to the use across Europe in the pre-medieval and medieval periods of Roman, Islamic, and Byzantine coins as de facto universal currencies, Mundell (2005: 466) has rather provocatively gone so far as to argue that what has made the past three decades "unique in the history of civilization" is not that there have been moves towards such transnationalism but that, rather, whereas "[n]early every previous age over the past 3000 years has had something that could pass for 'international money,'" since the collapse of the Bretton Woods "gold dollar" standard in the early 1970s (see below) there has "been no money that could even approximately be called a universal currency." Furthermore, whereas Roman, Islamic, and Byzantine coins' use beyond these empires' realms represented the emergence of informal universal currencies based upon such coins' precious metal content (i.e., it was the coin's weight, not its statutory value, that counted), the nineteenth century saw several efforts to create *de jure* universal currencies. Though Napoléon Bonaparte was one of the earliest to advocate monetary uniformity across Europe, one of the most far-reaching plans, proposed at the 1867 International Monetary Conference in

Paris attended by European, Russian, US, and Ottoman representatives (Cohen 1998: 69–70; Einaudi 2000), was for the development of an international currency based on French, British, and US gold coins.[4]

Although nationalist divisions ultimately doomed the 1867 plan, other steps were made towards currency supranationalism, particularly the formation of several monetary unions between various countries. Whilst a short-lived monetary union operated amongst several German states in the 1830s, arguably the two most successful unions were the Latin Monetary Union (LMU), established in 1865, and the Scandinavian Monetary Union (SMU), established in 1873. The LMU resulted largely from French government desire to standardize French, Belgian, Italian, and Swiss gold and silver coinages, with the Bank of France serving as the Union's central bank (Cohen 1998). Rather than issuing a single currency, the LMU made each member state's currency legal tender throughout the Union, so that travelers could "pay their way in the same coin . . . from Antwerp to Brindisi . . . without any of the risk or inconvenience of national exchanges" (quoted in Einaudi 2000: 289). For countries like France, the LMU gave greater influence over European financial matters and opened the door more widely to free trade (other governments would be harder pressed to use currency controls to limit goods' movement across national boundaries), whereas for weaker states adopting the French franc and renaming it according to local custom (the lira, the drachma, etc.) would provide a degree of monetary stability and greater access to capital markets in Paris and London. Other nations subsequently expressed interest in LMU membership (including Portugal, the Netherlands, Denmark, and the US), and Greece formally joined in 1876, such that by 1880 18 countries were associated with the Union (Bartel 1974). The SMU was likewise designed to standardize monetary systems. Created through a Swedish–Danish agreement to allow the gold coins of one nation to circulate as legal tender in the other, the SMU was subsequently enlarged through Norway's membership. By 1885 the free intercirculation of all paper currency had been achieved and exchange-rate quotations between the three member states had disappeared as the SMU increasingly operated as a single currency area (Cohen 1998).

Although both the LMU and SMU were abandoned soon after World War I, other efforts at monetary union quickly thereafter emerged. In 1920 Liechtenstein began using the Swiss franc as its official currency, and in 1922 Belgium and Luxembourg agreed to merge their monetary systems. Meanwhile, in 1945 the French government established the *franc des colonies françaises d'Afrique* (CFA franc), replacing the various monies then used in its African colonies. Upon independence, this CFA franc was replaced with two new currencies, the Central African CFA franc (currently used by six nations, including one non-former French colony [Equatorial Guinea]) and the West African CFA franc (currently used by eight nations, including one non-former French colony [Guinea-Bissau]). Similar monetary unions existed in the British Empire. In the early 1900s the West African pound was introduced throughout Britain's West African colonies to replace the multitude of extant currencies (Uche 1999), whilst in 1919 the East African shilling was established for Tanganyika, Kenya, and Uganda. In the early 1950s the British government initiated a process which would create the East

Caribbean dollar, which currently serves as legal tender in several nations (Cohen 1998). Recalling such monetary unions and efforts to establish universal currencies suggests that the kind of contemporary currency transnationalization heralded as evidence of globalization and the undermining of nation-states' sovereignty is not, then, that new. Furthermore, the fact that such arrangements had to be instituted by nation-states raises important questions about whether currency transnationalization represents the nation-state's weakening or, rather, a strategy by which particular nation-states pursue in the international arena their economic objectives (as with France, Italy, and Greece in the LMU).

If the fact that efforts to transnationalize currencies have a long history is the first issue to consider when evaluating claims about globalization's undermining of the nation-state, a second concerns the degree to which nation-states have ever had control over their own currencies or the currencies circulating within their territories. This is important because the power of narratives regarding the nation-state's loss of economic sovereignty, particularly as represented through a loss of control over its currency and the concomitant geographical spread of dollarization, relies upon an account which sees there having long been a strong correspondence between the territorial extent of the nation-state and the geographical domain within which its currency holds sway – what Cohen (1998: 27) has called the "One Nation/One Money myth." This myth, Cohen suggests, has been a powerful one and has pertinence for discussing globalization's impact on the nation-state, for by "insisting on an imaginary landscape populated by mutually exclusive sovereign moneys, the conventional [One Nation/One Money] approach in effect privileges the power of national governments over all other actors." However, an historical examination of currency sovereignty shows that not only is the nation-state's ability to enforce a single monetary standard within its territory fairly new, but it has also never been total. Thus, whilst it is often assumed that the nation-state has been inextricably tied to national currencies since the Peace of Westphalia, nation-state monopolies over currencies circulating within their territorial bounds are actually fairly recent, generally dating to the nineteenth century or later – for instance, foreign coins were legal tender in the US as late as 1857 (Schuler 2000), whereas the Bank of England only gained legal monopoly over issuing bank notes in England and Wales in 1921, after the last private notes were issued by the Somerset bank Fox, Fowler and Co. (Bank of England 2007). Indeed, even into the twentieth century coins often circulated in many parts of the world without regard to national frontiers, such that "[f]oreign coins could be used interchangeably with local money, and restrictions were only rarely imposed on what could be treated as legal tender," with the result that "currencies were effectively deterritorialized, and cross-border competition was the rule not the exception" (Cohen 1998: 28).

Recent dollarizations in Ecuador, El Salvador, and East Timor (which all replaced their currencies with the US dollar), the exploration of this possibility by Argentina in the late 1990s, and the taking up by Montenegro of the German mark and, more recently, the euro, also seem to point to a contemporary undermining of nation-states' abilities to reinforce the link between their territorial expanse and national economic sovereignty.[5] However, here too it should not be forgotten that practices of dollarization

– whether formally, as in Panama and Liberia (which began using US dollars as official currency in 1904 and the 1890s, respectively), or less formally, as in myriad LDCs where hard currencies have long served as media of exchange – pre-date considerably what many argue is the beginning of the contemporary era of globalization (the early 1970s).[6] Indeed, the fact that in 1900 some 40 percent (Eichengreen 1996: 23) of world foreign exchange was held in sterling (£) suggests that dollarization was fairly widespread over a century ago. Equally, the practice of pegging a nation's currency to another's, which could be taken as evidence of its inability to exercise economic sovereignty, is also quite old – Liberia first pegged its currency to the US dollar in 1847. The nation-state's power to enforce currency monopoly, then, appears to have varied significantly, both historically and geographically. Given that there is no clear historical convergence between, on the one hand, the apparent advent of globalization and, on the other, the breakdown of currency sovereignty and the emergence of widespread dollarization, the rhetorical use of the latter as evidence of the former is suspect.

The third issue concerning nation-states' ability to exercise economic sovereignty in the nineteenth century relates to the establishment of a different kind of inter-national standard – the gold standard. Although not an effort to form an international currency, the gold standard nevertheless served as a currency regulator and so, much like the euro, is often read to have undermined nation-states' economic sovereignty. Whereas prior to the nineteenth century several nations had adopted gold (and silver) standards wherein paper money could be converted into precious metal at a fixed price, it was not until the 1870s that what many historians consider the modern gold standard arrived. "Neither invented nor consciously planned" (Scammell 1965: 32), the modern standard emerged as an effort to counter world price instability through limiting governments' ability to print money without linking it to tangible goods like gold. The result of individual countries adopting the standard was the creation of an international system in which different parts of the world were linked together via the medium of relatively fixed exchange rates – although one in which nations with gold-producing regions or colonies (like the US and Britain) could expand their currency supply more easily than could those without ready access to gold.[7] As with monetary unions so with the gold standard: many countries in weak economic posi-tions joined to guarantee London, Paris, and New York capital markets that they were fiscally and monetarily responsible and thus credit-worthy (Bordo & Rockoff 1996), doing so even when such adherence cost them significantly in other regards, like the lost ability to devalue their currencies to make their exports more competitive on the world market and the fact that when the Bank of England – which lay at the center of the global financial nexus – increased interest rates, capital was readily drawn from the world periphery to the core (Officer 1996).

The gold standard (really, a gold-sterling standard, given Britain's financial dominance), then, served as a means through which the world economy could be governed and disciplined. However, by 1914 it had begun to break up as war and a con-sequent run on sterling led the Bank of England to impose severe exchange controls which, though not legally suspending convertibility, nonetheless meant the standard became non-functional (Officer 1996). As other European nations likewise effectively

withdrew from the standard, followed by the United States in 1917, a system of floating exchange rates emerged. When peace resumed, though, efforts were made to
re-establish a gold standard to support the international trading regimes then re-
emerging. In 1925, Britain returned to gold and others soon followed, although this
post-war standard operated slightly differently than had the pre-war one (Scammell
1965): whereas the pre-war standard functioned in a system dominated by a single
currency (sterling) which was very strong, in the post-war era sterling was weaker
and other gold-based currencies, of varying strength, began to operate as reserve
currencies. This development bred instability into the system as the 1930s "saw the
emergence of clearly differentiated groups [of nations] between whose currencies there
was no longer any fixed relationship" (Crick 1948: 18). The trauma of the Great
Depression, though, soon encouraged countries to re-abandon the standard, such that
by 1937 it had again effectively broken up.

In discussing how the gold standard served to discipline various nation-states concerning fiscal and monetary matters and what this meant for economic sovereignty
in the late nineteenth and early twentieth centuries, though, it is important to recognize several things which have lessons for interpreting nation-states' present economic
sovereignty. Certainly, the standard can be read as a means whereby individual
nation-states had to subordinate their economic sovereignty to that of global financial markets. However, evaluating the relationship between the nation-state and such
markets is complex and can be read in different ways. Specifically, it is important to
recognize that the standard itself did not enter the stage of world history pre-formed.
Rather, it had to be agreed to by various nation-states, who themselves did so for
different reasons. For Britain, the standard's guaranteeing of currency convertibility
ensured that the free trade regime of the Pax Britannica was maintained (de Cecco
1984). Other nations, especially those wherein the state was fairly weak, likewise gained
advantages. As Helleiner (2003: 215) has put it, whereas before the gold standard's
introduction such "countries usually had rather heterogeneous and often quite
chaotic monetary systems over which the state exercised only partial control,"
moving onto the standard "was often seen as the key monetary reform that could lead
to a more unified and homogeneous monetary order controlled by the state, a
project that appealed to nationalists for a variety of reasons." In other words, rather
than giving up economic sovereignty, adopting the gold standard actually provided a
means whereby such countries could exercise greater discipline over their respective
economies.

Furthermore, although the gold standard did not operate under the aegis of a transnational organization like the WTO, its specificities needed to be agreed upon by
various nation-states, each of which frequently pursued different monetary policies
and so left their distinctive mark on how the standard performed as an international
economic regulator of individual nation-states. For instance, notes from the Bank of
France were only convertible into gold or silver coin at the option of the authorities,
whilst central banks used "gold devices" (like extending interest-free loans to gold
importers) to encourage gold inflows and discourage outflows, thereby manipulating
how the standard functioned (Eichengreen 1996: 20–1). Moreover, as they adopted

various domestic policies, governments shaped the standard. Hence, the US government's virtual abandonment of bimetallism in the 1870s – the result of political conflicts between creditors (principally Eastern financiers) and debtors (largely Western farmers) – required the US Treasury to accumulate gold to back bank notes, which not only limited the amount of gold available to send to the London gold market and thus the amount other countries could acquire to back their own currencies (de Cecco 1984: 115–16) but often led the Bank of England to send gold to the US to satisfy the latter's need for paper money (Officer 1996). Additionally, changes in domestic economic objectives – like efforts to depreciate national currencies in the 1930s to provide a competitive advantage on international markets and so solve national unemployment during the Depression – meant that it was individual nation-states, pursuing largely self-serving economic policies, that helped bring the gold standard crashing down (Eichengreen 1996: 89). Indeed, as a League of Nations (1944: 230) report put it: after World War I international monetary policy increasingly "conform[ed] to domestic and social policy and not the other way round." Thus, the US government's 1934 decision to fix a new parity between the dollar and gold (devaluing the dollar by changing the paper dollar/gold rate from $20 per ounce to $35) reshaped how the standard worked and encouraged at least three currency blocs to emerge – one centered on sterling, the other on the dollar, with other currencies following their own course – which, though they had points of contact, no longer had the kind of fixed relationship that existed during the standard's zenith (Crick 1948: 18).

Exploring nineteenth-century efforts to develop international currencies and monetary unions, together with nation-states' adoption of the gold standard, then, highlights two matters of interest for considering globalization's impact on the contemporary nation-state. First, that there is a degree of similarity between late nineteenth-/early twentieth-century events and those of today suggests that tendencies towards the kinds of transnationalism which contemporary observers posit as evidence of globalization and the nation-state's supplication before global forces – the gold standard in the nineteenth century, "the global market" today – are not as novel as we might sometimes assume. The history of the nation-state is clearly not one of a unidirectional increase in power since 1648 but, rather, is one in which its economic sovereignty has waxed and waned. Second, how such transnationalism evolved in the nineteenth century was clearly shaped by the actions of various nation-states themselves, who not only were not necessarily its prisoners but, in fact, were frequently driving the process so as to secure – and even enhance – their own economic sovereignty. These precedents raise questions about the events of the twentieth and twenty-first centuries and their impact upon the nation-state, issues to which we now turn.

The Institutional Architecture of the "Bretton Woods Economy"

If the pre-Bretton Woods period provides one benchmark against which to measure contemporary events, the Bretton Woods era provides a second. Resulting from a 1944 conference attended by Allied representatives – including the Soviets, who subsequently balked at its specifics – to map out how they wished the post-World War II world to function, the Bretton Woods Agreement was sold as a way in which to "build a durable peace, and to achieve the stable world economy and democratic procedures that make a durable peace possible" (Newcomer 1944: 7). The conference's publicly proclaimed goal – stated by US Treasury Secretary Henry Morgenthau – was to end the kinds of economic nationalism that had dominated the pre-war era and to reinforce the belief that "the only enlightened form of national self-interest lies in international accord" (quoted in Van Dormael 1978: 222). Although many delegates hoped it would, as one Brazilian representative put it, forestall a return to the "drama of monetary chaos, of restrictions of all sorts of international trade, of blocked currencies, of economic isolationism, of competition instead of cooperation among central banks, and of general unemployment" (quoted in Van Dormael 1978: 1), it was clear that, because the US would emerge from World War II as the world's pre-eminent manufacturing and dominant Western military power, US interests would significantly shape the accord and thus the post-war international economic system. Whilst there were noteworthy differences of opinion – US and British delegations clashed over how much exchange-rate flexibility there should be – US concerns that being denied access to raw materials held in European colonies would strangle US manufacturing, together with fears that European powers would use trade restrictions to keep US goods out of their home markets and colonies, saw US negotiators work to break down tariff barriers and "open up" global economic spaces to US capital through "free trade."

Significantly, many of the ideas expressed at Bretton Woods were not particularly new. US governments at least since Woodrow Wilson had sought to impose a global Monroe Doctrine in which US interests of economic liberalism were presented as universal interests (N. Smith 2003), whilst a multilateral economic order with relatively free trade, stable currencies whose exchange rates could be adjusted through negotiation, government regulation of cross-border capital flows, and an international agency to coordinate this (i.e., a prototype for what would be institutionalized at Bretton Woods) had been proposed in the mid-1930s in the context of US policy towards Latin America (Helleiner 2006). However, recent experiences meant that by 1944 there was added US urgency to implement such a system. This system would embody what Ruggie (1982) has famously called an "embedded liberal" view of economic management, one in which tariffs and other trade barriers would be reduced (economic liberalism) but in which capital controls and exchange rate adjustments would sufficiently insulate national financial markets that governments could intervene domestically in the economic arena without fear of significant capital outflows (the market, then, was to be governmentally embedded). As a result of the changed economic and geopolitical context and of the growing dominance of new economic ideas – like those of British economist John Maynard Keynes, who played a pivotal role at the conference – the system which emerged after the war would be different from the gold standard system in three fundamental ways: exchange rates would be determined by international agreement but could be adjusted under certain conditions by government action; governments could limit international flows of capital; and a set of new institutions was established to monitor national economic policies, to provide funding to nations with balance-of-payments problems, to regulate trade, and to stimulate "economic development" in the global South (Eichengreen 1996). The Bretton Woods Agreement, then, would play a pivotal role in establishing the Pax Americana of the post-1945 world. It rested on several key elements.

First, whereas sterling served as the world's currency of last resort during the nineteenth century – between 1860 and 1924 about 60 percent of world trade was invoiced and settled in sterling (Williams 1968: 268) – by the mid-twentieth century the US was in ascendancy. This raised the question of which currency would serve as the international reserve currency in the post-war era. Prior to the conference both British and US policy makers had put forward plans for an international currency unit as a means to stabilize markets – the British proposed a sterling-based currency (the "bancor"), the Americans a dollar-based one (the "unitas") – though inability to agree meant these plans eventually came to naught.[8] At the conference itself considerable differences remained: US negotiators wanted to see the dollar, "the one great currency in whose strength there is universal confidence," become the "cornerstone of the postwar structure of stable currencies" (senior US negotiator Harry White, quoted in Van Dormael 1978: 200), whereas the British delegation, led by Keynes, did not wish to see either the gold standard re-stablished or the dollar become the central currency in the new system. Nevertheless, US interests held sway and a limited form of gold standard was implemented wherein the dollar's value would be set by pegging it to gold at $35 per ounce, whilst other major currencies would be pegged to the dollar. However, unlike under the old gold standard, in the new system exchange rates would be adjustable through government action (to eliminate balance-of-payments deficits) whilst exchange controls were allowed (so as to limit capital outflows) (Eichengreen 1996). The resulting system, wherein the dollar would sit at the center of a planetary financial network, gave the US immense power to shape the world economy (consciously or not), as fixed exchange rates meant that any changes in policy or conditions within the US – like rising interest rates or the growth of inflation – would be transmitted globally. Furthermore, it allowed the US to transfer vast quantities of wealth from overseas as inflation-depreciated dollars were exchanged for imports, whilst the fact that large quantities of world currency reserves were held in dollar-denominated assets allowed the US government and private individuals to borrow from the rest of the world at low interest rates, so fuelling US economic growth.[9]

Second, Bretton Woods laid the groundwork for several organizations designed to manage the post-war international economy, these being the IMF, the International Bank for Reconstruction and Development (IBRD, also known as the "World Bank"), the International Trade Organization (ITO), and the General Agreement on Tariffs and Trade (GATT). Although they worked together to encourage economic liberalism and "modernization," each had a different function. Hence, the IMF would establish currency stability by avoiding "competitive exchange depreciation" (International Monetary Fund 1944: 1). It would do this by lending countries funds sufficient to cover short-term balance-of-payments deficits and, should such deficits become more permanent, could insist that a member state reorganize its internal fiscal policy through devaluing its currency, privatizing state assets, or reducing social welfare expenditures (thereby providing more government revenue for paying creditors) before agreeing to further loans – practices known today as "structural adjustment policies." For its part, the IBRD focused upon loans and grants to facilitate "development" (defined as projects following "free market" principles) in middle-income countries – an early loan was to France for post-war reconstruction. Four associate agencies were subsequently established as part of the "World Bank Group": the International Finance Corporation (IFC), founded in 1956 to promote private-sector investment in LDCs; the International Development Association (IDA), founded in 1960 to provide long-term loans to the world's poorest nations; the International Centre for Settlement of Investment Disputes (ICSID), founded in 1966 to mediate disputes between private investors and governments; and the Multilateral Investment Guarantee Agency (MIGA), founded in 1988 to encourage FDI in LDCs by providing insurance to investors against political risks (such as asset nationalization). Meanwhile, the ITO was conceived to establish international trade rules. However, the ITO charter's rejection by the US Congress signaled the Organization's end before it had even got started, with the result that the GATT, separately negotiated in Geneva and initially designed to facilitate reductions in tariffs and other trade barriers on manufactured goods, became the primary means of regulating world trade in the early post-war period.[10] Unlike the intended ITO, though, the GATT did not have a formal organizational structure and was administered through the United Nations.

In considering the Bretton Woods system, two issues of importance arise. First, it very much relied upon a discursive representation of the world economy as an entity made up of discrete national economies. Indeed, the system could only be seen to work if this rhetorical position were maintained: only if the nation-state were imagined as a powerful delineator and discipliner of economic spaces – determining where "domestic" economic space ended and the space of "the international" realm began – could a distinction be drawn which would allow it to claim sovereignty over economic policy in the former but permit the multilateral institutions of the IMF, World Bank, and others to predominate in the latter. However, given what we have seen in Chapters 5 and 6, the degree to which economies actually were fairly discrete, nationally defined entities, even in the early post-war period, is questionable. Certainly, they may have become more spatially distinct during the protectionism of the 1930s than they had been at the turn of the twentieth century, but the world economy by the late 1940s had already been significantly knitted together through thousands of firms' FDI and cross-border portfolio investment strategies. Whereas the representation of the world economy as constituted by a group of independent nation-states may have been, then, something of a fiction, it was nevertheless an important fiction for allowing the agreement worked out at Bretton Woods to operate as a mechanism for managing the post-war economy. Furthermore, such discursive considerations again raise questions about whether the nation-state was ever as strong during the Bretton Woods era as it is frequently portrayed to have been and, if not, then what does that mean for arguments that it is being undermined by globalization, that the "modern nation state itself – that artifact of the 18th and 19th centuries – has begun to crumble" (Ohmae 1995: 7)?

Second, and relatedly, when talking about "the Bretton Woods era" it is important to recognize that the nature of the arrangements negotiated in 1944, and the operation of the organizations that came out of the Bretton Woods conference, did not remain subsequently unchanged. For example, whereas the IMF was initially formed to address problems faced by European and North American nation-states, especially problems concerning exchange rates and balance of payments (Peet 2003: 63), its focus later changed primarily to that of the "developing world." Equally, although many US Bretton Woods negotiators supported state-led "developmentalist" policies in Latin America and elsewhere, this was quickly forgotten in the post-war era as the needs of reconstructing the European and Japanese economies to provide a bulwark against Soviet expansion came to predominate and as developmentalist policies were portrayed during the McCarthyite period as smacking of communism (Helleiner 2006). Meanwhile, the Bretton Woods system's central element – the convertibility of major currencies into dollars and, consequently, gold – only really existed after 1958 when the major European countries finally allowed their currencies' free convertibility. Put another way, if we calculate the life of the Bretton Woods era as having stretched from its formal 1944 birth to its acknowledged 1971 death (when President Nixon suspended full dollar-to-gold convertibility), then for nearly half of that existence it was not fully operational in the manner frequently represented – as a system in which nation-states tightly managed their currencies' values by pegging them to the dollar/gold. Indeed, from this perspective the Bretton Woods era proper was remarkably ephemeral.

The Nation-State in the Post-Bretton Woods Era

By the late 1960s inflation in the United States had begun to have an impact on the dollar's purchasing power. The latter's location at the center of the world financial system meant that US economic problems soon began to be transmitted globally, particularly given the significant and growing quantities of dollars held overseas. Although actions by the US government and overseas central banks and governments allowed the Bretton Woods system to "stagger on for longer than it would have otherwise" (Eichengreen 2000: 191), by the early 1970s it had become increasingly untenable to value the dollar at $35 per ounce of gold.[11] In August 1971, to prevent an international run on US gold stocks, the dollar's convertibility was suspended. Whilst several nations sought to develop a replacement system wherein exchange rates would continue to be adjusted by government, this was abandoned in 1974 and a system emerged in which some currencies' values were determined through a "free float" (i.e., solely through market forces), some through a "dirty float" (wherein central banks intervene in currency markets to guide exchange rates), and others continued currency pegging (usually to the dollar but other currencies too). For many, this represents the end of the nation-state's ability to manage domestic economic affairs and the growing independence of international currency markets – 1974 marked, as Bryan and Farrell (1996: 19) have put it, the conclusion of an era in which "the world's financial marketplace was segmented into local national markets [and each] local government had effective control over its own market and used that control to achieve policy objectives."

The Bretton Woods era's putative end has seen many of its institutions and arrangements continue to operate as they always had, a transformation in the activities of others, and the creation in some instances of a whole new set of institutions and arrangements. For example, although it is being challenged in some regions by the euro, the dollar remains the world's de facto reserve currency, although whereas previously it was backed by gold it is now essentially backed by oil, the result of an agreement between the US and Saudi Arabia to continue pricing oil in dollars.[12] Indeed, the global demand for dollars created by periodic oil price increases – as with the "oil shocks" of 1973, 1979, and more recently – has kept the US economy afloat, even as trade and budgetary deficits have burgeoned, whilst the dramatic increase in "petrodollars" circulating planetarily has provided some governments with large quantities of funds by which to purchase US securities in a self-reinforcing cycle of credit extension.[13] For their part, the IMF and World Bank Group have continued to operate in the post-Bretton Woods era, as has the GATT, though they have changed their missions to greater or lesser degrees. In the case of the World Bank, a significant change took place in its modus operandi towards the end of the 1960s under the direction of its new president (and former US Defense Secretary) Robert McNamara. In particular, whereas the Bank had long defined development in terms of the degree to which per capita GNP was increased, McNamara advocated a broader conception, focusing upon matters like nutrition, employment, and literacy, and significantly expanded the Bank's lending and research capacities (Woods 2006: 44–5). Undoubtedly, some of this new direction was a genuine desire to aid developing countries – the Bank worked to eradicate river blindness (*onchocerciasis*), for instance – but it must also be placed within the context of Cold War geopolitics and a concern amongst US policy makers that failure to address issues of underdevelopment would encourage communism's spread. Thus, although the Bank continued to believe in the fundamental advantages of "the free market," it also "took a more equivocal view of ownership, believing that managerial competence was more important than private entrepreneurship, so that loans could be made under public ownership systems within an overall conception of greater governmental intervention in the development process" (Peet 2003: 119). By the early 1980s, though, as neoliberal governments came to power in the US, Britain, and West Germany, it increasingly assumed a policy of inducing "reform" in nations receiving funding, with reform meaning the adoption of policies liberalizing trade and encouraging export-oriented industrialization to correct what it saw as deeper structural problems in such countries' economies (Kapur et al. 1997).

The IMF also has transmogrified from the organization envisioned in 1944. Until the mid-1970s it primarily sought to stabilize the international monetary system by making short-term loans to countries (particularly the industrialized core) experiencing balance-of-payments problems. However, as the world economy began changing in character – especially as the quantity of dollars circulating beyond the US ("eurodollars") grew – the IMF transformed "from an institution regulating money as a means of circulation to one guaranteeing money as a means of payment and defending precarious international credit relations" (Altvater 1993: 121). Beginning with its 1976–7 forcing of the British government to adopt a program of structural adjustment (Harmon 1997) and the growing debt crisis in LDCs caused by their inability to repay loans made with the petrodollars which had been recycled through the world economy in the 1970s, the IMF increasingly took on the role of ensuring global fiscal discipline.[14] Consequently, it switched its attentions from the industrialized world to the LDCs and its role "effectively changed from being a means of collaboration on exchange rates and payments, mainly among industrial countries, to being a means of First World control over Third World economic policy" (Peet 2003: 71) through various structural adjustment policies (typically including foreign exchange and import control liberalization, currency devaluation, wage controls/reductions, dismantling of price controls, reduced government expenditure on social welfare, and implementation of policies encouraging FDI) which governments would have to accept to be eligible for "debt rescheduling." The Fund also encouraged borrower nations to abandon import-substitution policies protecting nascent industries and to develop export-oriented industries, with such industries' products destined primarily for global North countries. In this regard, the IMF scripts itself as an important modernizer of economies, ensuring that "economically correct" policies are implemented (Popke 1994). Furthermore, although the IMF had been fairly uninvolved in Cold War politics until the late 1970s, in the 1980s it became more interlaced with US geopolitical interests through the US's ability to shape its lending policies (Woods 2006: 36).[15]

The GATT's purview has also changed. Whereas negotiators in the 1950s focused on reducing tariffs through bilateral arrangements which were then adopted by other nations, in the 1970s the focus became non-tariff barriers (like quotas and subsidies). The 1973–9 negotiating round focused specifically on such matters, which were considered more difficult to address than tariffs because they were frequently inextricably tied to domestic policy and were often harder to identify as trade barriers.[16] It also sought to address agricultural issues (previously largely ignored), though no pact was ultimately reached because of the US's and Europeans' inability to come to agreement – US negotiators wanted agriculture included under the same rubric as industrial sectors, the Europeans did not. Accordingly, the 1986–94 round sought to tighten up the original 1947 GATT rules (resulting in the "GATT 1994"), to address the growing use of non-tariff barriers, and to widen significantly GATT's ambit to include agricultural trade and subsidies, services, intellectual property, and textiles. As a way to do so it established three new regulatory agreements: the General Agreement on Trade in Services (GATS); the Agreement on Trade-Related Aspects of Intellectual Property Rights (TRIP); and the Agreement on Trade-Related Aspects of Investment Measures (TRIM). These entities reflected changes that had occurred in the world economy since GATT's inauguration – the growth of transnational services (as with telemedicine and e-commerce) and the growing privatization of basic commodities like water (often resulting from World Bank and IMF structural adjustment programs) – and complaints from the publishing, film-making, music, and fashion industries about their products being illegally copied in places like China. Significantly, GATT 1994 did not merely affect cross-border activities but also some activities carried on entirely within nations, as with limiting government regulation of FDI.

The post-Bretton Woods era, however, has also seen the creation of new institutions and arrangements, probably the most significant of which has been the WTO, headquartered in Geneva. The successor to the GATT, the WTO formally came into existence on January 1, 1995.[17] There are several differences between GATT 1994 and the WTO, including that the side-agreements frequently concluded by certain GATT-contracting parties are now brought under the unified umbrella of a single organization (the WTO); that opt-out agreements (like those covering clothing and textiles) and "grey area" measures like voluntary export restraints would gradually be eliminated; and that WTO members cannot block decisions arrived at under the dispute settlement mechanism in the way in which they could many GATT dispute panel findings (Adamantopoulos 1997: 29–30). The WTO has also included rules prohibiting using health and safety standards to "unfairly" restrict trade and against local content laws which require manufacturers either to source a given proportion of their components from a particular country or to produce certain components domestically. The result is that virtually all transnational trade in goods and services is now subject to WTO rules. Meanwhile, the Organization's Dispute Settlement Body is empowered to adjudicate disputes over alleged rule infringements and to direct the non-compliant party to bring its laws into compliance, whilst authorizing the aggrieved party to take retaliatory measures should this not occur. Whereas all members are expected to follow WTO rules, "developing countries" are provided additional time to meet them and may also enjoy the benefits of the Generalized System of Preferences, through which industrialized nations can waive tariffs on goods coming from certain LDCs.

The WTO, though, has been controversial, for several reasons. One has been the manner in which it came about, with its basic rules largely crafted by global North nations and with the direct input of myriad TNCs. Transparency, both internal and external, has also been an issue and many members – especially LDCs – have complained that they are only brought into discussions on particular topics late in the day, with issues presented as faits accomplis. As a result, at the 1999 Seattle WTO meeting numerous African and Caribbean nations declared they would not support any agreements made, although the talks' collapse due to widespread public demonstration in that city left their resolve untested (Davey 2005). Equally, despite WTO protestations that it is a democratic institution, many documents are not made public until well after the fact and the public is generally barred from observing meetings. Others have argued that WTO rules may appear fair on the surface by treating everyone equally but are actually structurally biased against LDCs because of the latter's greater reliance on exporting agricultural products – WTO anti-dumping regulations affect agricultural producers (for whom selling products at below cost is a rational business decision at certain times of the year) to a greater degree than they do manufacturers (Hartigan 2000).

The WTO has perhaps been most castigated, however, for its rules' impacts on environmental and labor issues. Thus, whereas prohibitions on using health and safety requirements to discriminate unfairly against imports may appear reasonable, they raise questions concerning who gets to decide if these requirements are being used in such a fashion, with that decision frequently resting on competing scientific opinions. In 1999, for instance, WTO regulators authorized the US and Canada to impose sanctions upon the EU for banning imports of beef treated with Bovine Growth Hormone, the use of which was long prohibited for public health reasons in European livestock. Significantly, in selecting targets for retaliation, the US and Canada were allowed to choose industries which were not necessarily beef-related but which were sufficiently profitable to promote EU compliance with the WTO's ruling – a provision which encourages powerful industries within individual countries to push for "free trade" for fear of being caught up in such retaliatory measures. Similar complaints have been leveled by environmentalists and labor rights activists, who claim WTO rules undermine environmental protections and limit nations' ability to regulate imports made under poor working conditions (Wallach & Woodall 2004).

Globalization versus the Nation-State or Globalization of the Nation-State?

So, what does this all mean for arguments concerning the nation-state's contemporary status? Above I hope to have provided a context within which to discuss the "post-Bretton Woods" era by suggesting both that the nation-state was not as omnipotent in earlier times as some imply but also that the contemporary nation-state's location within a network of international arrangements and organizations which can exercise power over it is both neither unprecedented nor necessarily disabling. This perspective provides an appropriate backdrop to explore two particular matters: how are forces of globalization impacting the nation-state's power?; and what does this mean for distinctions between govern*ment* of the world economy and govern*ance* of it? This latter question is particularly important, for even if the nation-state's power *is* being eroded in some areas, this does not necessarily imply emergence of a world economy without rules. Indeed, it is possible to see two apparently contradictory tendencies occurring simultaneously: even as the nation-state may enjoy less control over the economy, the volume of rules regulating economies may actually increase. Put another way, we should not simply assume that any withering of the nation-state will augur the rise of a more anarchic economy free of rules (I return to this in the following section).

With regard to the first of these matters, we can address at least four issues. The first questions to what degree the growth of supranational entities like the WTO and of increasingly integrated global markets are, in fact, undermining the nation-state. Thus, despite redounding claims of its passing, there are myriad examples of nation-states continuing to exert significant regulatory power – as anyone who has tried to cross an international border without a passport can perhaps attest. To give but one example, despite complaints by the European Commission (2007) that it represents an unfair barrier to trade, the US 1992 American Automobile Labeling Act requiring the labeling of auto parts' country of origin remains in force and US regulatory requirements continue to be a source of Commission complaint in numerous other areas, including pharmaceuticals, textiles, wines and spirits, and electrical equipment. For example, the Commission points out that it is not uncommon for electrical equipment to be "subject to U.S. Department of Labor certification, a county authority's electrical equipment standards, specific regulations imposed by large municipalities, and other product safety requirements as determined by insurance companies," a situation that is "aggravated by the lack of clear distinction between essential safety regulations and optional requirements for quality, which is due in part to the role of some private organisations as providers of assessment and certification in both areas" (2007: 8). Equally, despite its global reach, the World Bank cannot dictate policy to individual governments, who can adopt policies which run counter to Bank desires; although they run the risk of losing Bank aid, the point is that this is something about which governments make their own calculated decisions. For its part, the WTO cannot force nations to change their tariff and other regulations, though it can authorize governments to levy retaliatory penalties against those not toeing the line. Furthermore, some governments have exerted considerable influence over such international organizations to change how they operate. In the late 1970s, for instance, the US threatened to withhold contributions to the World Bank's International Development Association arm unless it refused loans to Vietnam (the Bank complied), whereas in 1977 the US secured a provision allowing all countries to reduce by a prorated amount their Bank contributions should the US choose to reduce its own – a provision that greatly magnifies the effect of any US threat to decrease its contributions should the Bank pursue policies the US government feels are not in its own best interests (Woods 2006: 28–9).

The second issue concerns whether globalization is weakening the nation-state *in toto* or whether it is, instead, weakening it in some regards but not in others or even, perhaps, allowing it to augment its powers in some areas. For instance, fears of "terrorism" and concerns that globalization is making populations more vulnerable to attack through greater cross-border movements of cargo in which nuclear or biological agents can be hidden has led many governments to tighten freight inspection protocols and requirements for individuals sending money internationally via wire services – thus, the 2001 USA Patriot Act allows the President to restrict foreign governments' and financial institutions' access to the US financial system if they lack adequate money-laundering controls. At the same time, various US states have enacted laws making it more difficult for undocumented immigrants to remain within their domain by requiring police officers to check for citizen status during routine traffic stops. Equally, the rise of "Fortress Europe," in which migration within the EU has been made easier but in which entry into it has been made more difficult, also represents an extension of various nation-states' power to isolate their domestic territories within a world of growing flows of goods, money, information, and people. Such developments have raised the question of whether we are seeing an undermining of government's regulatory powers or simply their redistribution to different levels (from national to local) or areas of the state (reduced trade barriers but increased regulation of immigration). Meanwhile, in the case of China, the government has successfully dictated where TNCs can invest (Liu & Dicken 2006) and has extracted from firms like Boeing agreements to provide technology and establish manufacturing facilities in exchange for being allowed to sell their products there (Greider 1997).

The third issue concerns a universalization in much contemporary discourse, specifically globalization's alleged impact upon *the* (singular) nation-state – Ohmae's *The End of the Nation State* is an exemplar in this regard. However, such universalization – "singularization" is possibly a better term – hides more than it reveals, perhaps for good reason. Thus, representing as a universal phenomenon any undermining of nation-states' power in the face of transformations in the world economy avoids the appearance that some (more powerful) states are actively engaged in promulgating such transformations and/or using their influence over international organizations like the WTO and World Bank to pursue their own national aims. Rather, such a discourse makes any pressures upon various nation-states – say, to reduce their controls over international capital flows – appear inevitable, something which *all* states must face together. Yet, in reality, it is highly unlikely that whatever globalization processes may be occurring will have equal impacts upon all nation-states – the US's capacity to retain degrees of economic sovereignty is much greater than Mali's, for instance. In essence, then, promulgating a discourse in which all are seen to be equally affected gives nation-states (like the US) accused of having any degree of authorship over the processes currently reshaping the world economy "plausible deniability." It allows, in Harvey's (2003) words, for what used to be presented in the highly charged terms of "neo-imperialism" to be rendered in the seemingly more politically neutral terms of "globalization." Similarly, whereas in theory all countries are equally free to withdraw from supranational organizations like the IMF, the consequences of so doing are not equal, for some countries are more capable of weathering withdrawal's costs (perhaps loan suspensions or trade boycotts) than are others.

This raises the fourth and conceivably most oxymoronic issue, namely that whereas "globalization" and "the nation-state" are frequently presented as opposites – the more globalization there is, the less powerful will be the nation-state; the more sovereignty the nation-state exerts, the less will processes of globalization progress – in fact certain nation-states have had a significant role in pushing the laissez-faire policies which have epitomized neoliberalism, whereas others may benefit from the growth of the supranational entities which are seen as the hallmarks of the new global age. Thus, with regard to the latter, and despite the negatives that membership may bring for their inability to defend nascent industries through protective tariffs, participation in the WTO may allow countries like Mali to appeal to an entity which has the capacity to force economically powerful nations to play (at least nominally) by the same set of rules – as the WTO (2007) puts it, "[w]ithout a multilateral regime such as the WTO's system, the more powerful countries would be freer to impose their will unilaterally on their smaller trading partners." With regard to the former, it is imperative to recognize that nation-states have played critically important roles in facilitating "deregulation," some more than others. Hence, the British government's decision in the 1960s to allow the euromarket to function within its territory effectively without regulation stimulated transnational integration of currency markets by encouraging

more individuals and firms to dabble in this market (Helleiner 1994).[18] It also provided the competitive spark for the US government's 1974 removal of various capital
controls which, in turn, led the British government to remove its own capital controls
in 1979 and later engage in a sweeping deregulation of the City of London that mirrored the New York Stock Exchange's 1975 deregulation (Plender 1987). Meanwhile,
in 1981 the US government introduced International Banking Facilities allowing
depository institutions to offer services to foreign residents and institutions free of
certain Federal Reserve requirements and some state and local income taxes. Largely
in response, in 1984 the West German government abolished its own tax on foreign
holdings of German securities, thereby making them more attractive to foreign
investors (Helleiner 1994). Meanwhile, in the early 1980s the US sought to export the
deregulation wave by making Japanese economic liberalization a centerpiece of trade
negotiations with Tokyo in the hope of encouraging Japan's domestic, rather than export-
led, growth and thus shrinking the US's growing trade deficit (Plender 1987; Moran
1991). Consequently, several laws concerning currency markets were liberalized and
the Tokyo stock market underwent its own deregulation in 1998.

In considering how the UK and US governments helped fuel the transnational
integration of capital markets, it is important to recognize that despite their predilec-
tion towards deregulating the financial arena, in neither country were governments
anywhere near as enthusiastic for pushing the deregulation of manufacturing trade.
In fact, both passed and/or continued a number of trade restrictions, like the Reagan
administration's 1981 cap on Japanese car imports and the Thatcher government's
continuance of limits on shoes and other goods coming from Eastern Europe. Indeed,
Helleiner (1994: 299) argues that financial liberalization and manufacturing protectionism were in point of fact closely linked, for historically it has "proven difficult
for states to maintain liberal practices in finance and trade at the same time." This
highlights two important issues. First, any process of undermining a nation-state's
economic sovereignty is highly complex and even contradictory, perhaps involving
a simultaneous undermining in one realm yet augmentation in another. Second, the
primary reason the US and UK pushed financial deregulation whilst retaining
protectionism in other regulatory areas relates to their pursuit of distinct national
interests, as both had significant advantages in the financial sector – in the US's case,
policy makers recognized that, "if given the freedom to invest internationally, private
investors around the world would be attracted to the unmatched depth and liquidity
of US financial markets which . . . would help to fund US external and internal
deficits," whilst for Britain making London the center for the global euromarket "provided a means by which [its] leading position in international finance could be
reestablished" (Helleiner 1994: 308–9).

Given its national interests, then, it was primarily the US government that pushed
the floating exchange rates that have come to epitomize the post-Bretton Woods era.
Whereas European and Japanese officials advocated continued currency pegging to "prevent reserve currency countries [like the US] from living beyond their means . . .
[, thereby limiting] America's exorbitant privilege of financing its external liabilities
with dollars" (Eichengreen 1996: 140), US Treasury Secretary George Shultz preferred to

float the dollar as a way to free the US from fiscal constraints which had emerged in the late 1960s – doing so would provide flexibility to pursue full employment policies at home whilst maintaining a trade deficit (Wray 2006). Although Schultz and his undersecretary Paul Volcker were prepared to contemplate a pegging system if the bands within which currencies could float were sufficiently wide that US domestic policy would not be constrained, in the end a compromise was agreed wherein currencies would float but stability would be maintained through central bank and/or IMF intervention.[19] Whereas, then, the system of floating exchange rates established by international agreement in 1975 and mandated by the IMF in 1978 has been taken as emblematic of global financial markets' power and read to represent a loss of national economic sovereignty because nations can no longer fix their currency's value à la Bretton Woods, in fact it is equally possible to read such a development as actually increasing governments' ability to exert control over domestic economic policies by freeing them from the strictures of fixed exchange rates. Indeed, even such a neoliberal as Milton Friedman (1968) has argued that moving to floating exchange rates opens greater room for domestic economic policy making, a fact also acknowledged by Bank of Canada Governor David Dodge (2005), who has stated that Canada's embrace of floating exchange rates was part of a deliberate strategy to avoid domestic inflationary pressures.

Parallel developments have taken place in other economic sectors. For instance, in 1971 the US government's Overseas Private Investment Corporation began providing insurance to investors should their overseas assets be seized and guaranteed bank loans and bonds to encourage FDI in countries favorable to US geopolitical interests, especially LDCs. Perhaps the most significant role the US government played in helping kindle manufacturing's transnationalization, though, was through what President Eisenhower called the "military-industrial complex" and what Melman (1970) provocatively called "Pentagon capitalism," wherein the US increasingly supplied weapons to nations like Zaire in exchange for minerals like cobalt, which is essential for making jet engines (Congressional Budget Office 1983; Hartung 2000). Similarly, the Marshall Plan, which Mee (1984) suggests marked the Pax Americana's launch, provided money through which post-war Europe could pay for US imports and encouraged US firms to establish Western European operations, as well as mandated that recipients pursue free trade policies. The UK Department of Trade and Industry has similarly played an important role helping UK firms transnationalize, whilst Japan's Ministry of International Trade and Industry (MITI) has been central to Japanese firms' growth – to the point of actively coordinating which would develop particular products – and their overseas investment strategies (Okimoto 1989). Most other nation-states have also played roles helping their own firms develop overseas markets and so shaping FDI flows.

Taking this issue a step further, though, not only have nation-states – at least a select number – played significant roles in spawning capital transnationalization, but they have done so because such transnationalization generally increases their power, as does maintaining strong regulatory regimes. Indeed, as Wolf (2001: 190) has argued, nation-states have a vested interest in maintaining regulatory vigor as they compete for globally circulating capital, because the alternative is likely financial red-lining: "Failed states, disorderly states, weak states, and corrupt states are shunned as the black holes of the global economic system." Furthermore, active reinforcement of nationalist feelings by various state actors can be a powerful way of extending the nation-state's authority within the context of growing transnational flows of goods, information, money, and people, especially if such flows are perceived to be undermining its cultural heritage. Thus, Barber (2001) has argued that as globalization unfolds so can nationalism, indeed that globalization typically spawns such responses as part of a dialectical process – people and their representatives who feel their national cultures being undermined often seek to defend them more vigorously than before (as with "English-only" language campaigns in the US). Similarly, newly independent states engaged in nation-building projects against the backdrop of globalization often flex their regulatory muscles by extending their reach into the daily lives of their citizens to a much greater degree than previously – as with laws establishing Lithuanian, Latvian, and Estonian as the only official languages of the Baltic states and, in the case of the latter two, linking this to national citizenship to de-Russify them (Ozolins 1999), and the decision by the East Timor government, newly independent from Indonesia, to abandon Indonesian and adopt Portuguese as the language of the court and school systems (*New York Times* July 31, 2007).

Moreover, transnationalization can give states both greater regulatory power and greater geographical range over which to exert that power. For instance, the growth of fish processing plants on the shores of Lake Victoria in Tanzania, with their output consumed in Europe, has allowed the EU to insist that each producer follow European Commission directives concerning health and safety. Certainly, this represents an impingement upon Tanzanian national sovereignty but it also represents an extension of EU sovereignty beyond Europe's geographical boundaries. Likewise, although already approved by US regulators, in 2001 the EU rejected General Electric's proposed $42 billion acquisition of Honeywell International Inc. on the basis that it would limit too greatly competition in the aerospace industry, with this being the first time a proposed merger between two US companies was blocked solely by European regulators (who could do so because both companies operate within the EU). Although in this particular case the regulatory difference of opinion was between two very powerful entities (the US and the EU), we can imagine other situations in which a less powerful state – again, we might consider Mali – may feel itself unable to stand up to, say, a US company's desires but, as a result of colonial linkages to France, may find the EU (perhaps under French pressure) willing to do so for it.

Membership in international organizations can also give nations – some more than others – added power to shape globally flows of money and goods. For example, because the US has been the largest lender to the World Bank and IMF, US interests have greatly shaped these organizations' activities and allowed US governments an avenue by which to mold the planet's economic and political geography in ways that having to deal with other nations bilaterally (or through militarily intervention) would not so easily allow – the voting structure at both, for instance, allows the US to veto single-handedly many decisions (Woods 2006: 27–34). Similarly, in the mid-1970s the Ford administration used its power within the IMF to force change in British government policy in ways

not politically feasible using more direct pressure (Harmon 1997), whilst more generally the US has used its influence in the IMF to push countries to accept US FDI and
military bases (Calleo & Rowland 1973).[20] Participating in international organizations,
then, may be more about domestic policy than international policy, whereas in
some cases the pursuit of particular domestic policies may have more to do with posi-
tioning oneself internationally than with domestic goals. The bright line between
international and domestic concerns as drivers of policy, in other words, is murkier
than often presented. Consequently, arguments that present the global and national
scales as discrete entities, with "the global" undermining "the national" but with no
recognition of how "the national" shapes "the global" or may even be strengthened
by it, are highly problematic.

Sovereignty and Governance

In examining nation-states within the contemporary world economy, there are two
additional matters to explore. The first relates to the issue of "sovereignty." Just as
it is important to recognize that we should not speak of *the* nation-state when
considering contemporary developments, so is it important to recognize that there are
different types of sovereignty to consider when discussing how various nation-states
are purportedly being undermined by such developments. In this regard, Krasner (1999)
has distinguished four: "recognition sovereignty" (a state's acceptance by others as
part of the international community); "Westphalian sovereignty" (the principle that
one nation-state should not interfere in the internal workings of another); "border
sovereignty" (the capacity, ability, and willingness to regulate flows of information,
commodities, money, and people into, or out of, a country); and "domestic sovereignty" (the power of a nation-state to decide upon and execute specific policies within
its borders and beyond). Clearly, much as we can imagine how nation-states like the
US and Mali have dissimilar structural capacities to shape contemporary economic
and political processes and may be affected by them differently, so can we imagine
that separate elements of nations' sovereignty may be affected differently, depending
upon which nation we are considering and in what context. Hence, a nation like North
Korea may not enjoy much "recognition sovereignty" but may exert a great deal of
"border sovereignty," even in a globalizing world, whereas the US may enjoy both "recognition sovereignty" and "domestic sovereignty" but have difficulties with its "border
sovereignty" in limiting immigrant flows from Central America. Furthermore, even
here there are analytical complications. Thus, whilst US governments may have the
capacity and willingness to limit such immigration, they may not have the ability due
to immigrants' skill in evading the Border Patrol, whereas in the case of other types
of cross-border flows (e.g., capital) the government may have the capacity and
ability but not the willingness to exercise its sovereignty. Finally, whereas a particular
nation-state may see a loss of sovereignty in one area it may see an increase in
another. Hence, whilst the British nation-state has undoubtedly increased its economic
power vis-à-vis the US through EU membership, it has also lost some of its legal

sovereignty because it has been forced to align British law with EU law. These examples highlight the multifaceted nature of sovereignty and how different interests may pull in opposite directions on separate matters, all of which is to say that the sovereignty issue is much more complicated and historically and geographically conditioned than somewhat simplistic "globalization is undermining nation-states' sovereignty" arguments generally acknowledge.

The second matter to consider is that of the relationship between government's ability to exercise economic and political power and governance of the world economy. Thus, neoliberal narratives typically represent any withdrawal of the nation-state in the economic realm and its replacement by "the market" as a process of "deregulation" – implicitly, the retreat of the state is assumed to lead us from a world with more regulations to one with fewer. However, it is important not to be blinded by zero-sum game arguments which suggest either that more of the market means less of the nation-state (as we have seen, certain nation-states have been actively involved in extending markets' reach) or that more economic liberalization means less regulation. Thus, even in situations where nation-states *are* lifting restrictions on, say, capital holdings, it would be a mistake – or at least a highly ideological act – to interpret this simply as a process of "deregulation," for two reasons. First, "deregulation," in which some government regulations are reduced or eliminated, frequently results in no overall reduction in the rules by which governments shape how economies operate and may actually increase their quantity significantly. Hence, in the US, federal "deregulation" has often simply been about allowing the states to play a greater role in regulating particular industries – all that has changed is the geographical level at which regulation takes place. In other instances "deregulation" – as in the City of London's 1986 "Big Bang" – has actually been followed by new laws with significantly more rules, designed to ensure deregulation is enacted, like the Financial Services Act which quickly followed the Big Bang. Deregulation, then, is regularly more about *re*regulation by the state in a different form than it is about encouraging greater market autonomy. In fact, the liberalization of markets generally requires *re*regulation. In this context, Vogel (1996: 2) argues that the rhetoric of deregulation espoused by neoliberals largely "serves only to obscure what is going on," as *de*regulation is commonly simply code for the privatization either of assets or of regulatory capacities – electric and water utilities that have been "deregulated" still have very complex government rules concerning how they should be run, except that now they answer to private owners rather than public ones. Deregulation in the neoliberal imagination, then, is a highly qualified concept.

Second, even when "deregulation" does result in fewer government rules overall, the retreat of government from the market does not necessarily result in a concomitant absolute decrease in rules shaping how it is regulated. Whereas globalization is regularly represented as a process whereby markets free themselves from regulation, as often as not its drivers prefer the security of global rules-based ways of managing the world economy over market anarchy. Arguably, this is nowhere clearer than in the case of the WTO, which, through its regulations, claims to make the trading system "more secure and predictable" (WTO 2007). The growth of transnational organizations and arrangements, then, raises the question of how non-governmental and quasi-governmental organizations play roles in regulating the world economy. In this regard there are myriad non-elected entities which help govern the world economy, but discussion of a few should suffice to make the point. Thus, the Bank for International Settlements has functioned since the 1930s to promote cooperation between various nations' central banks and its decisions significantly mold capital flows. Equally, organizations like the Asia-Pacific Economic Cooperation Forum and the Association of Southeast Asian Nations play major roles in shaping how the world economy is regulated through the rules they impose on their members. At the same time, numerous other entities have established global rules for everything from accounting practices to industrial and commercial standards – hence, the International Accounting Standards Board (2007), "an independent, privately-funded accounting standard-setter based in London," is "committed to developing . . . a single set of . . . global accounting standards," whilst the International Organization for Standardization (2006) has instituted global manufacturing and service standards to "make trade between countries easier and fairer." The fact that scores of such organizations have been created by governments and non-governmental entities raises questions about both how globalization is impacting governments' behavior and the nature of economic governance, but also how private entities' rules in matters like accounting practices structure the world economy. Put another way, a world without nation-states – were that ever possible – would not be a world without rules and regulations.

Questions for Reflection

- How does globalization affect different nation-states and their constituent elements in different ways?
- What roles have nation-states played in shaping globalization?
- How does portraying the nation-state as being powerful in the past and weak today shape how we think about globalization?

Notes

1. In all fairness, such accounts are not confined to the neoliberal imagination: many on the left have likewise portrayed globalization as a process of the economy's and its regulatory entities' "denationalization" (Sassen 2003).
2. "Dollarization" occurs when one nation's currency is used within another's territory and can be done unofficially (a black market) or officially. Presently, an estimated 40–60 percent of US currency is held abroad (Feige et al. 2003: 47). Although its nomenclature references US currency, others also serve the function.
3. Recent calls for currency transnationalism have come from the Canadian Fraser Institute (a conservative think-tank), which advocates a North American Monetary Union and currency, "the Amero" (Grubel 1999), and Nobel economist Robert Mundell (2005), who

has argued for a "world currency." There have also been moves towards establishing a West African "Eco" and re-establishing an East African shilling.

4. In 1807 Napoléon wrote to his brother, the King of Naples: "Brother, when you issue coins I would like you to adopt the same valuations as in French money. In this way there will be monetary uniformity all over Europe, which will be a great advantage for trade" (quoted in Fuller 2001: 1). The 1867 Conference planned to standardize coinage to achieve a single currency, based on the fact that the British sovereign (worth £1) and American half-eagle (worth $5) contained virtually identical amounts of gold, whilst French currency laws would have allowed a 25 franc piece to be minted containing a similar quantity (Cohen 1998: 69). Had it been successful, the gold quantity in the three nations' coins would have been standardized. Several coins with dual values were actually circulated – for instance, the Austrian 8 florin piece was stamped as worth 20 francs. Even Britain minted prototypes – like the 1 franc/10 pence piece – though these were never issued (Einaudi 2001: xiv provides illustrations).

5. In Ecuador the local currency is still used for small change.

6. The Liberian government stopped using the dollar in 1982, though has considered returning to it. Panama mints its own coins but uses the dollar for all paper transactions.

7. Britain was fortunate in this regard, with large gold finds in Australia in the 1840s, gold from British West Africa in the 1880s and South Africa and Canada in the 1890s, and inflows into India from other parts of Asia.

8. Franklin Roosevelt advocated an international currency unit – the "unitam" – for the Americas as early as 1939 (Helleiner 2006: fn. 5). For their part, the bancor and unitas schemes would operate very differently. Under the bancor plan, a central Clearing Union would provide each nation overdraft facilities, such that whenever it had a negative balance of trade a bancor credit would appear in another's account. Countries could also acquire bancor credits by depositing gold with the Clearing Union, but could never withdraw bancors from the system – bancors could only be transferred from one account to another. Under the unitas plan, each country would make an initial gold deposit in a central fund and place an amount of domestic currency at its disposal for sale to other nations, such that all transactions would take the form of the purchase/sale of currency for currency. Whereas the British scheme, then, meant that the central financial institution of the new order – the IMF – would operate like a bank, with countries in surplus funding those with balance-of-payments deficits, thereby sharing the burden of economic adjustment with deficit nations, the US's favorable balance of payments outlook led it to prefer a system which shifted the burden of deficits onto the deficit countries alone (Körner et al. 1986; Popke 1994).

9. Over time inflation would reduce the real value of dollars used to pay for imports, leaving foreigners holding devalued dollars and Americans holding tangible goods. This was particularly so in the 1960s when the government increased significantly the quantity of dollars circulating to pay for the Vietnam War. Many of these advantages have continued into the present. As Sandra Pianalto (2007), President and CEO of the Federal Reserve Bank of Cleveland, has reported: "Between 1980 and 2005 . . . the income paid to foreigners who owned U.S. assets was 4.9 percent on average. However, the income paid to Americans who owned foreign assets was 6.3 percent. In other words, the United States effectively borrowed at a discount of $1^{1}/_{2}$ percent."

10. As McMahon (2005: 189) has put it, although "in theory GATT covered trade in agricultural products, the Contracting Parties were unwilling to subject their domestic

agricultural policies to the same disciplines as industrial products" and only one of the
original GATT provisions (Article XI: 2) referred specifically to agriculture.

11.   US government actions included threats to retaliate against foreign governments seeking
      to convert their dollars into gold at the official rate, whilst overseas central banks and
      governments used their reserves to defend US currency, largely to protect the value of
      their own dollar holdings.

12.   Between 1999 and 2007 the proportion of LDCs' allocated reserves held in euros
      increased from 18 percent to 30 percent, whilst that held in dollars declined from 71 per-
      cent to 59 percent (International Monetary Fund 2007). Other countries have likewise
      reduced their dollar holdings – Russia's Central Bank decreased its holdings from almost
      90 percent of its total in 2003 to less than 80 percent in just two years (Radio Free Europe
      2005). The continued pricing of oil in dollars results from an agreement emerging from
      the 1974 United States–Saudi Arabian Joint Commission on Economic Cooperation
      (GAO 1979), wherein the US agreed to provide investment, goods, and technical skills
      to Saudi Arabia if the Saudis agreed to pressure other oil producers to continue selling
      in dollars (Spiro 1999).

13.   In early 2007 64 percent of world reserves were held in dollars, up from 59 percent in
      1995 (International Monetary Fund 2007). The purchase of US securities is self-reinforcing
      in the following sense: for overseas investors, buying US securities represents continued
      confidence in the dollar, which helps maintain the value of their own dollar-denominated
      assets, even as the US continues running significant trade and budgetary imbalances.
      However, should that confidence falter then overseas investors are likely to stop lending
      the US government money through purchasing government securities and may actually
      dump their dollars. Some (e.g., Clark 2005) have even argued that Saddam Hussein's 2000
      decision to sell Iraqi oil in euros rather than dollars was the primary reason for the
      US-led invasion, since dumping the dollar might cause a domino effect and lead to total
      collapse of confidence in the currency (for more on the geopolitics of oil and the
      euro/dollar issue, see Noreng 1999).

14.   The flooding of the world economy in the 1970s with petrodollars reveals an interesting
      geography: large quantities of dollars were paid by industrialized nations to oil producers
      after OPEC quadrupled the price of oil; such producers deposited their revenues
      primarily in European banks and European branches of US banks (to keep them beyond
      the Federal Reserve's regulatory reach); such banks loaned petrodollars to various LDCs
      which were trying to encourage development to overcome colonialism's legacy. In 1982
      Mexico became the first country to default on its loan obligations, paradoxically in part
      because oil prices declined after spiking in 1979. A major oil producer, Mexico had taken
      on obligations when oil's price was high, trusting that this high price would provide sufficient
      revenue to fund its loan repayments; when the price fell, so did Mexico's revenue stream.

15.   Although the World Bank has traditionally been led by an American (three of whom
      have worked at Chase Manhattan Bank) and the IMF by a European, given the US's large
      voting share in both entities, and the fact that US law requires that any assistance given
      international financial organizations meet geopolitical needs, the Bank and IMF have
      frequently acted in concordance with US foreign policy interests – the Bank, for instance,
      provided loans to Iran under the Shah and to Nicaragua as the Somoza regime fought
      the Sandinistas.

16.   Negotiating "rounds" took place in 1947, 1949, 1950–1, 1955–6, 1961–2, and 1964–7.
      Initially, negotiations involved a product's principal supplier and consumer nations

agreeing to tariff reductions, with the benefits extended to other countries through the "most-favored nation status" mechanism. In the 1964–7 round the US pushed a new strategy of offering across-the-board reductions in tariffs, with negotiations concentrating upon which products to include/exclude (Evans 1971). Two further rounds in the "post-Bretton Woods era" took place, in 1973–9 and 1986–94, in which a wider range of trade barriers (like product dumping) and economic sectors (agriculture, intellectual property) were addressed.

17. Macrory et al. (2005) provide an in-depth overview of the WTO. The GATT 1994 continues to exist and covers trade in commodities, but now functions under the WTO's aegis.

18. The euromarket is any market in currencies and securities denominated in non-local currencies, primarily US dollars.

19. The compromise was formalized at the 1975 Rambouillet Conference. In addition to establishing a system of managed floating exchange rates, the attendees agreed to revise Article IV of the IMF's Articles of Agreement to permit a member to choose its own exchange arrangements – including floating. Although this revision allowed for a return to a generalized system of fixed exchange rates at some future point, such a move will require an 85 percent majority vote of the IMF membership – a condition effectively giving the US veto power over it (Pauls 1990).

20. As of 2006, there were 766 US military installations in foreign countries and 77 in US overseas territories (US Department of Defense 2006).

Further Reading

Henderson, J. (1993) The role of the state in the economic transformation of East Asia. In: C. Dixon and D. Drakakis-Smith (eds.), *Economic and Social Development in Pacific Asia*. London: Routledge, pp. 85–114.

Mansfield, B. M. (2005) Beyond rescaling: Reintegrating the "national" as a dimension of scalar relations. *Progress in Human Geography* 29.4: 458–73.

Ó Tuathail, G. and Luke, T. (1994) Present at the (dis)integration: Deterritorialization and re-territorialization in the New Wor(l)d Order. *Annals of the Association of American Geographers* 84: 381–98.

Pitelis, C. (1991) Beyond the nation-state? The transnational firm and the nation-state. *Review of Radical Political Economics* 22.1: 98–114.

Sparke, M. (2006) Political geography: Political geographies of globalization (2) – governance. *Progress in Human Geography* 30.3: 357–72.

Electronic resources

Bank for International Settlements (www.bis.org)

International Monetary Fund (www.imf.org)

United Nations Conference on Trade and Development World Investment Reports (www.unctad.org/Templates/Page.asp?intItemID=1485&lang=1)

World Trade Organisation (www.wto.org)

Chapter 8

Globalizing Labor

Chapter summary: This chapter recounts worker efforts to organize transnationally, stretching back to the early nineteenth century. It also shows how Cold War politics shaped labor transnationalism and how new forms and geographies of transnationalism are being developed in the post-Cold War "globalization" era. Finally, the chapter considers how economic landscapes shape, and are shaped by, labor transnationalism.

- Early Worker Transnationalism
- Worker Transnationalism and the Cold War
- Worker Transnationalism after the Cold War
- Geographies of Labor Organizing in the Global Economy

The owner of capital is already cosmopolitan as regards the use of that capital for the purposes of exploitation. No need for us to complain of this. We don't want any walls built round cities or nations for fear of invasion; what we do now stand in urgent need of, is, AN INTERNATIONAL WORKING ALLIANCE AMONG THE WORKERS OF THE WHOLE WORLD.
(*Tom Mann*, The International Labour Movement, *1897, p. 6*)

We must organise resistance against a financial capitalism vagabonding around the world, looking for victims and destroying work places and workers' resistance. This new capitalism is a wild animal. It must be brought under control. This is our main challenge in the international trade union movement in the months and years ahead.
(*John Monks, General Secretary, European Trade Union Confederation, 2007*)

As I have argued throughout this book, a central neoliberal claim is that globalization is an unstoppable process, one which affects workers but over which they themselves have little control: workers, we are told, must "get on the globalization train" and give up "old fashioned" collectivist notions and behaviors or they will be "left at the station." As former Congressman Newt Gingrich (1995: 63) has put it, evoking an era when labor unions were barely in their infancy: "While the Industrial Revolution herded people into gigantic social institutions – big corporations, big unions, big government – [contemporary transformations are] breaking up these giants and leading us back to something that is – strangely enough – much more like Tocqueville's 1830s America." For Gingrich, then, the globalized twenty-first century will look much like the early nineteenth, with weak central government, non-existent unions and workers individually negotiating with their bosses, and a competitive – rather than monopolistic/oligopolistic – form of capitalism.

Despite the significant ideological work that goes into promulgating and propagating such a fatalist narrative, though, in fact many *are* challenging contemporary neoliberal practices and discourses, from various ideological positions. Thus the "Slow Food" movement, begun in Europe in the 1980s as a counter to the "fast food" of companies like McDonald's, has branches in over a hundred nations and has sought to preserve local food crops and culinary traditions – for instance, it has been active in Spanish efforts to maintain the traditional 9 a.m. to 8 p.m. working day (with a two- or three-hour lunch-break) as a bulwark against its Americanization via adopting the 9 a.m. to 5 p.m. *jornada intensiva* ("intensive day").[1] Similar movements have emerged in other realms, including the "slow travel" and "slow cities" movements designed to challenge the perceived homogenizing and culturally destructive aspects of globalization and their erasure of traditional ways of living (Honoré 2004), with such movements' ethos epitomized by novelist Milan Kundera's (1996: 39) contention that "the degree of slowness is directly proportional to the intensity of memory; the degree of speed is directly proportional to the intensity of forgetting." For its part, in the United States the nationalistic Patriot Movement (Gallaher 2002) emerged to challenge globalization, although with a very different politics. Similarly, environmentalists are opposing the global spread of genetically modified organisms by agribusinesses like Monsanto whilst fair trade advocates are confronting trade relations they perceive as unfair to farmers in LDCs. All of these groups are shaping how the world economy is being structured geographically. However, in this chapter I focus upon organized labor, because of all these groups it is organized labor that has the longest involvement with transnationalism.

The chapter is organized as follows. First, I outline efforts at worker transnationalism stretching back to the early nineteenth century. Second, I recount how Cold War politics shaped the geography of labor transnationalism and what this meant for capitalist transnationalism. Third, I explore a number of recent developments in cross-border labor organizing. Finally, I contemplate several geographical issues concerning the practice of transnational labor solidarity.

Early Worker Transnationalism

Although it is impossible to say exactly when the first labor unions were established, we do know that 2 500 years ago Greek artisans were banding together to exert some control over their employment conditions. Two thousand years later European stonemasons and others began establishing guilds to better their working conditions. However, it was not until the late eighteenth century that labor unionism in its modern guise emerged, as workers formed groups to provide members assistance with things like burial costs. Significantly, almost from their very inception in modern form unions and organizations dedicated to defending workers' rights and labor conditions had appreciable international orientations. The growing synchronization of economic booms and busts across Europe in the early nineteenth century, occasioned by industrialization's geographical spread, soon led British workers to seek transnational linkages to prevent their employers from importing strikebreakers – in 1836 London workers sent a missive to Belgian workers encouraging them to create a federation with workers in the Netherlands and the Provinces of the Rhine to improve their conditions and limit opportunities for their hiring as strikebreakers (van der Linden 1988). Starting in the 1840s, a number of formal structures emerged to promote workers' interests internationally, including the Democratic Friends of all Nations, founded in 1844 in London by British, French, German, and Polish workers and labor advocates, the Fraternal Democrats, and the German Workers' Educational Society, all of which operated amongst the European exile community in Britain (Weisser 1971; Lattek 1988, 2006). The 1840s, in fact, marked the beginning of a century of what van Holthoon and van der Linden (1988: vii) have described as the "classical age" of working-class internationalism, and other international democratic and workers' organizations soon followed, including the International Association (founded in 1855), the Congrès Démocratique International (1862), the Association Fédérative Universelle (1863), the Ligue de la Paix et de la Liberté (1867), and the Alliance de la Démocratie Socialiste (1868) (Devreese 1988). In 1864 the International Workingmen's Association (the "First International") was inaugurated to bring together trade unionists and socialists to discuss augmenting mutual aid internationally between workers – for instance, British unionists worked with French unionists to help Polish workers opposing continued Russian control of their country (Facey et al. 1864).

Whereas organizations like the Democratic Friends of All Nations agitated for workers' rights, they were not, strictly speaking, labor unions. However, as labor movements developed in various countries so did unions become increasingly involved in international agreements with their overseas counterparts. In 1872, for example, the president of the US iron molders' union signed an agreement with his British equivalent to limit the use of immigrants as strikebreakers, and in the 1890s a plan – ultimately not enacted – was hatched to hire British unionists to watch for iron molders leaving the ports of Liverpool and Glasgow and to telegraph US representatives as to their destinations and arrival dates, so that they could be met upon disembarking and persuaded not to break strikes. Unionists also sought to develop an international card allowing workers from one country automatically to join the equivalent union in another, thereby creating cooperation between "molders of all nations" (Stockton 1916: 290). In 1909 an American Federation of Labor (AFL) representative toured Europe, meeting with molders' unions in Denmark, Norway, Sweden, Austria, Germany, and France. Several other US unions concluded similar agreements with overseas unions in the decades before World War I, including the mineworkers, boilermakers, iron shipbuilders, and seamen, whilst US and British unionists began attending each others' conferences as fraternal delegates. Taking this a step further, in 1884 glassworkers from the US, Britain, France, Belgium, and Italy met to discuss the creation of a "Universal Federation of Glass Workers" (Pelling 1956), whilst by 1889 printing unions had formed an international organization through which they could address issues of mutual concern. By century's end, unions were also providing material assistance to workers overseas during strikes, as with Australian waterfront workers' 1890 donation of £30 400 [£25.8m] to the London dockers' strike fund (Portus 1930: 923). Additionally, workers and their organizations began to address overseas matters that were not strictly workplace related: the AFL, for instance, condemned British policies in Ireland and Russian persecution of Jews. Moreover, workers organized transnationally not only as producers but also as consumers – in 1895 members of the Christian socialist movement inaugurated the International Co-operative Alliance (ICA), whose goal was to set up transnational cooperative trading associations (Gurney 1988).[2]

If cooperation between unions in different countries was one element of nineteenth-century labor politics, the actual transnationalization (dare we say "globalization"?) of their organizational structures was another. As early as 1853 the British Amalgamated Society of Engineers union established a branch in Sydney, Australia, and in the next few years founded branches in Canada, the US, New Zealand, Malta, France, South Africa, Turkey, India, and Gibraltar (Southall 1989). Others followed suit. In 1870 the British Steam Engine Makers' Society created a New York branch, whilst in 1889 the London-based Dock, Wharf, Riverside and General Labourer's Union launched overseas branches in Rotterdam and Amsterdam. Meanwhile, by 1910 the Amalgamated Society of Carpenters and Joiners had US, Canadian, Australian, New Zealand, and South African branches. US unions similarly instituted branches overseas. By the mid-1880s, the Knights of Labor had 10 000 members in Britain and made efforts to organize branches in France, Germany, and Belgium (Pelling 1956; Michel 1978). In the early 1890s they expanded into Australia (James 1986). AFL unions likewise chartered foreign branches. Given commonalities of language and culture, these were mostly in Canada but there were also some elsewhere – the machinists founded a branch in Mexico whereas the typographical union chartered a local union in Georgetown, British Guiana, in 1908 (Roberts 1964: 11). Such developments led many unions to adopt the title of "international" – as in the International Longshoremen's Association. For their part, within a few years of their 1905 founding the Industrial Workers of the World (IWW) had established branches in the UK, Australia, Belgium, Germany, Canada, South Africa, Mexico, and Chile (Philips 1978; Burgmann 1995; Caulfield 1995). Significantly, then, much labor transnational activity had occurred prior to the emergence of what van der Linden (1988) has called labor transnationalism's "national phase," by which he means it occurred before labor movements developed robust and formalized national organizational structures through which to engage in transnationalism.[3] However, as both individual unions and national union centers began to strengthen after 1900, it was nationally constituted trade union movements – rather than individual union branches or entities like the Second International (founded in 1889) – which increasingly came to dominate the field of labor transnationalism.

First, beginning in 1889 national unions representing typographers and printers, hatters and milliners, cigar-makers and tobacco workers, shoemakers, transportation workers, miners, metalworkers, potters, and even hairdressers began forming "international trade secretariats" (ITSs) in particular industries or economic sectors as a way to foster transnational cooperation (Price 1945; Segal 1953; Windmuller 1980; Busch 1983). By 1902 17 ITSs had been formed, and by 1914 there were 28, with a combined membership of 6.3 million workers, of which the secretariats for miners, metalworkers, and transport workers were the largest (Dreyfus 2000: 36–9). Although most unions involved in ITSs were influenced by their countries' socialist movements and so had similar ideological outlooks, in practice the secretariats served principally as media through which information on wage rates and working conditions could be shared, union membership transferred when a unionist migrated to another country, and sometimes financial support given. Second, in 1901 a number of European national trade union centers founded the International Secretariat of National Trade Union Centers, through which they could network transnationally. Significantly, whereas ITSs represented workers organized according to what they did, the Secretariat organized workers according to national affiliation. Initially exclusively European in composition, in 1910 the first non-European center (the AFL) joined. AFL prompting soon led the Secretariat to change its name to the International Federation of Trade Unions (IFTU) (Fimmen 1922; van Goethem 2006).[4] Although the IFTU was jurisdictionally separate from the ITSs, it nevertheless had close links with them, since the latter's members made up a significant proportion of the national trade union centers in their respective countries. Indeed, as a way to strengthen linkages between the ITSs and to develop closer links with the IFTU, in 1913 the secretariats participated in the IFTU's Zurich congress, determined to develop greater unity in the face of their employers, many of whom were large firms beginning to diversify their operations both geographically and sectorally (Lowe 1921: xxii). By 1920 the IFTU represented almost 23 million workers (Carew et al. 2000: 565).

Whereas the ITSs and IFTU represented union organizations, the establishment of an international organization to protect all workers – unionized or not – had been discussed periodically since the 1880s. In 1901 the International Association for the Legal Protection of Labor was inaugurated (Lowe 1921), but it was not until World War I that momentum was really achieved. Thus, even as war raged, unions on both sides argued that workers would need protections when peace finally came, and they held a number of separate conferences to plan legislation concerning labor migration, social security, safety, and hours of work. Subsequently, in 1919 the International Labour Organisation was launched under the League of Nations' aegis, whilst the IFTU also reformulated itself after the Great War's tribulations. At about the same time, though, two new international labor entities emerged to challenge the IFTU: the Red Trade Union International (RILU), founded in 1921 by the Bolsheviks and frequently referred to as the Profintern, after its Russian acronym; and the International Federation of Christian Trade Unions (usually known as CISC, after its French name – Confédération Internationale des Syndicats Chrétiens), inaugurated in 1920 to represent the interests largely of Catholic (and some Protestant) trade unionists who had been inspired by Pope Leo XIII's encyclical *Rerum Novarum* "On the Condition of the Working Classes" and who had organized Christian trade unions since the 1890s to counter the growth of socialist and anti-clerical unions. Like the IFTU, the CISC set up international bodies in several industrial and craft sectors, including for workers in the metal trades, transport, and farming. Meanwhile, the RILU – which was more geographically diverse than the European-dominated IFTU and CISC (Lozovsky 1927) – sponsored the formation of the Pan-Pacific Trade Union Secretariat, created in 1927 by various communist-influenced unions and designed to be a regional entity for unions from the USSR, China, Australia, India, Japan, and the Philippines, amongst other countries (Portus 1930).

Multinational entities like the Profintern were not the only organizations to establish regional groupings, though. Individual unions and national federations also developed such organizations. Arguably, one of the most energetic in this regard was the AFL, which has a history of international activities dating back to the late 1800s when it supported anti-Spanish activities in Cuba and Puerto Rico. Publicly, Federation leaders argued they were seeking to further the interests of all workers and to fight colonialism, and individual unions played active roles in developing solidarity with overseas workers – a number of Chicago-based unions protested vigorously the arrest by US military authorities of strikers in Cuba in 1899, for instance (Foner 1975: 431). Underneath the rubric of international solidarity, however, the Federation increasingly came to adopt a rather ambiguous position: although leaders like Samuel Gompers regularly condemned European imperialism, many also believed that by sweeping away European colonialism in the Americas and elsewhere the market for US goods would be expanded, thereby creating jobs for American workers – a position made explicit by the Typographical Union, which proclaimed that US annexation of Hawai'i, Puerto Rico, Guam, and the Philippines would lead to English becoming the language of instruction in schools, resulting in more textbooks being printed in the US for export (Foner 1975: 419). Although, then, most opposed territorial acquisition via military means, they were generally not inclined to challenge US corporations' overseas expansion nor to challenge what were perceived to be radical and/or anti-US unions abroad – a position shaped by an ideological mindset in which many US workers saw their historical mission as one of bringing US-style liberal democracy and "modern" economic systems (i.e., capitalism) to nations left economically and politically "backward" by European colonialism (Herod 2001). Consequently, working with the Confederación Regional Obrera Mexicana (CROM – the Mexican Regional Labor Confederation), Gompers was a central figure in creating in 1918 the Pan-American Federation of Labor (PAFL) as a bulwark against what he saw as radical anti-capitalist labor movements in countries like Argentina, Chile, and Uruguay (Snow 1964). Seeing it as "based upon the spirit of the Monroe Doctrine" (quoted in Simms 1992: 37), for Gompers the PAFL would help prevent Mexico's bolshevization after its 1910 revolution and promote Mexican workers' organization, thereby discouraging the migration to Mexico of US firms seeking cheap labor, which had begun to occur. For the CROM, the PAFL would facilitate the normalization of Mexican–US relations after General Pershing's 1916 invasion.

Other imperial powers' labor movements played similar roles in developing international organizations and shaping the geography of the world economy, either through their support of imperial adventurism or through their decision not to challenge it. In the case of France, although some leftist political parties denounced imperialism, others argued that freedom for colonial workers could only come *after* the French working class's domestic victory, a position colonial workers generally viewed as simply another way to continue French imperialism (Agyeman 2003). Indeed, in 1924 Léon Jouhaux, head of the communist Confédération Générale du Travail (CGT – General Confederation of Work), went to Tunisia with the singular intent of preventing the emergence of an indigenous labor movement that might challenge the CGT's position in the colony (Beling 1964). However, most workers generally adopted the view, inculcated since youth (Ozouf & Ozouf 1964), that whilst the worst excesses and abuses of colonial subjects should be avoided, the geographical spread of French republican ideas would bring enlightenment and civilization to the colonies. In such a view, "democratic universalism tended towards ideological imperialism" and transnational labor solidarity was largely rhetorical, an *internationalisme verbal*, to use Bédarida's (1974: 28–9) felicitous phrase. Consequently, whereas there were 32 French national labor congresses between 1886 and 1914, only two made any mention of colonialism, and both references were to the desire to ensure that French labor laws were upheld in the colonies. Similarly, Dutch unions' support for colonialism in Indonesia was based on the premise that having colonies increased national prestige (Tichelman 1988), whilst the German labor movement was generally of the opinion that colonialism brought civilization to the colonized (Mergner 1988). For their part, although they encouraged unionization in the colonies, British unions generally did not challenge colonialism itself (Roberts 1964). Rather, they followed the same patterns as other European unions: whereas some denounced imperialism's evils, many others either thought little of workers in the colonies other than to ensure labor laws were enforced, viewed colonial subjects as blessed by their exposure to British modernizing tendencies, or actively supported Britain's imperial conquests so as to secure raw materials for domestic factories and markets for the products thereof – in the 1930s, for instance, Lancashire textile unions joined with their employers to limit competition from Indian manufacturers and sought to keep Japanese textiles out of Britain's African colonies (Gupta 1975: 232, 280; also, Hennessy 1954).

Worker Transnationalism and the Cold War

The late 1930s and World War II's subsequent outbreak marked a significant transformation in the world of transnational labor politics. For one thing, it spelt the death knell of both the Profintern and the IFTU, whilst the CISC saw many German, Austrian, and Italian affiliates suppressed under fascism and lost its remaining Eastern European affiliates after the war. In an effort to resurrect the international labor movement, in 1945 unionists from many nations (including the British TUC, the Soviet All-Union Central Council of Trade Unions, and the US Congress of Industrial Organizations [CIO], though not the AFL, which refused to work with the Soviets) founded a new international body to replace the old IFTU – the World Federation of Trade Unions (WFTU). At the same time they revived many ITSs that had languished during the war. Initially, part of the WFTU's goal was to encourage trade unionism in the developing world – the TUC, for instance, arranged for representatives from several British colonies to attend the WFTU's founding conference, and in April 1947 a conference was held in Dakar, French West Africa, for the purposes of collecting information and developing linkages between nascent African unions (Roberts 1964: 157–8). However, rising tensions between communist and non-communist unions quickly caused ideological splits. In 1949 the non-communist national centers, led by the TUC and CIO, withdrew and formed a new organization (along with the AFL), the International Confederation of Free Trade Unions (ICFTU) (Carew et al. 2000). The WFTU remained home mostly to communist unions, primarily those of the Eastern European nations now under Soviet control but also some from Western Europe, like France's CGT and the Italian Confederazione Generale Italiana del Lavoro (CGIL – Italian General Confederation of Work). Many developing countries' unions also affiliated with the WFTU. Maintaining pre-war patterns, the ITSs allied with the ICFTU, whilst the CISC maintained its industry-specific international entities (called International Trade Federations), including in teaching, transport, textiles, and construction, and expanded its influence amongst national union centers in Latin America, French West Africa (where French Catholic unions had founded branches), Canada (Québec), and Asia (especially French Indochina). The WFTU, meanwhile, founded its own industry organizations – the Trade Union Internationals (TUIs). Unsurprisingly, perhaps, domestic ideological divisions were replicated in the colonies. Thus, in French West Africa local unions' loyalties were divided on the basis of their relationship with federations in France: the communist CGT affiliated with the WFTU; the socialist Force Ouvrière (Workers' Force) affiliated with the ICFTU; and the Confédération française des travailleurs chrétiens (CFTC – French Confederation of Christian Workers) affiliated with the CISC (Agyeman 2003: 61–5).

Each of the three international trade union centers also founded regional organizations. The ICFTU created its European Regional Organisation (ERO) in 1950, the Asia and Pacific Regional Organisation (APRO), and the Organización Regional Interamericana de Trabajadores (ORIT – Inter-American Regional Organization of Workers) in 1951, and the African Regional Organisation (AFRO) in 1957. It also lobbied governments to ratify labor standards concerning matters like freedom of association and the abolition of forced labor. The CISC established regional organizations in Africa (L'Organisation démocratique syndicale des travailleurs africains [ODSTA – The Democratic Trade Union Organization of African Workers], based in Lomé, Togo), Latin America (Central Latinoamericana de Trabajadores [CLAT – Latin American

Workers' Union], based in Caracas, Venezuela), Asia (the Brotherhood of Asian Trade Unionists [BATU], based in Manila, Philippines), and North America (where it affiliated the National Alliance of Postal and Federal Employees). Moreover, in 1968 it dropped virtually all mention of Christianity as it transformed itself into the World Confederation of Labour (WCL), a move designed to make it more appealing to workers in LDCs, many of whom were not Christians. The WCL also adopted a more radical program, rejecting capitalism and arguing for workers' control of industry through socializing the means of production. Although it had fewer regional organizations, for its part the WFTU put in place an Asian Liaison Bureau in 1949 and worked closely with the Mexican-based Confederación de Trabajadores de América Latina (CTAL – Latin American Confederation of Workers) (Steinbach 1957). More commonly, though, the WFTU tended to develop close links with pre-existing bodies (some of whom had communist leanings, others of which were simply anti-imperialist), like the Congreso Permanente de Unidad Sindical de los Trabajadores de América Latina (Permanent Congress of Trade Union Unity of Latin American Workers), the International Confederation of Arab Trade Unions, and the All-African Trade Union Federation (Herod 1998a). Whereas the WFTU leadership, based in Prague, generally preferred a fairly top-down relationship with its regional organizations and TUIs, by the mid-1960s some Western members began expressing a desire for greater autonomy. Consequently, in 1966 the Federation adopted a slightly less centralized structure when its General Council instructed the various TUIs to adopt their own constitutions, although these organizations were nevertheless still more politically and financially dependent upon the WFTU than were the ITSs on the ICFTU (Windmuller 1980).

During this period the AFL – which merged with the CIO in 1955 to become the AFL-CIO – maintained a virulently anti-communist political ideology. Although it remained affiliated to the ICFTU until 1969, AFL-CIO leaders routinely charged that the ICFTU was insufficiently vigorous in challenging communism. Consequently, during the 1960s it began establishing its own regional entities, finding the ICFTU's regional organizations (like the ORIT) too independent. In 1962, largely in response to Fidel Castro's seizure of power in Cuba, the AFL-CIO established as its operating arm in Latin America and the Caribbean the American Institute for Free Labor Development (AIFLD), a tripartite entity that included elements from organized labor, businesses with interests in the region, and the federal government. In 1964 it set up the African-American Labor Center to serve a similar purpose in Africa, and in 1968 the Asian-American Free Labor Institute, which initially was supposed to be the AFL-CIO's division in Vietnam but which shortly thereafter expanded its operations to other Asian countries. Subsequently, in 1977, the AFL-CIO created the Free Trade Union Institute (FTUI) to act both as its European regional institute and as a distributor of federal government grants to the three other institutes.[5] Given US interests in Latin America and the Caribbean, AIFLD was generally the most active regional body. Working in combination with President Kennedy's Alliance for Progress initiative, AIFLD was designed both to gather intelligence of use to the State Department and to undermine radical anti-US and/or pro-communist trade unions in the region, and did so through various means (Spalding 1988; Barry & Preusch 1990).

First, using funding from the State Department, the United States Agency for International Development, the CIA, the AFL-CIO, and (later) the National Endowment for Democracy, the Institute funneled money to various political and union groups, many of whom were involved in clandestine activities – AIFLD-trained communications workers were involved, for instance, in the 1964 military coup which overthrew Brazilian President João Goulart (Romualdi 1967; Barry & Preusch 1990). Second, AIFLD trained numerous trade unionists, both in-country and at its US facilities, with unionists taking seminars in political organizing, economics, and history, all with the goal of inculcating the benefits of US-style "bread and butter" (i.e., workplace-oriented) unionism rather than more militant social/political unionism.[6] AIFLD also provided funding for labor colleges, including the Trade Union Education Institute of Jamaica, the Critchlow Labour College in Guyana, and the Caribbean Congress of Labour's education program. Third, through its Social Projects Division, AIFLD helped build numerous schools, union buildings, bridges, and, most notably, housing developments during the 1960s and 1970s, the goal of which was to improve the material conditions of life for workers so they would be less inclined towards revolutionary and/or anti-US politics – as one AIFLD (1964a: iv) publication put it, the projects (about 18 000 housing units constructed in a 15-year period) would show that "free trade unions can produce results, while the Communists produce only slogans." AIFLD also set up credit unions, workers' banks, and producer/consumer cooperatives. Such investments in infrastructure and institutions, AIFLD officials believed, would stabilize the hemisphere and energize national development "both directly through provision of jobs and stimulation of materials industries, and indirectly through improving public health, worker motivation, and social stability" (AIFLD 1964b: vi; Herod 2001 provides a more thorough analysis).

Other countries' unions engaged in similar practices. In the case of Britain, the TUC had long been interested in developing links with workers and unions overseas, especially those in British colonies, and various individual unions had established overseas branches as far back as the mid-nineteenth century. Nevertheless, it was really only after World War I that the TUC became involved in colonial labor affairs, both through its own foreign policy and through working with the Foreign and Colonial Offices (Weiler 1988), with the latter in particular seeing unions in the colonies as important development agents because they could help provide a skilled, reliable, and disciplined workforce. However, with the growth of colonial labor movements in the 1930s British officials began to worry that such entities might link up with independence movements and so pose a threat to British rule. Consequently, the Colonial Office sought TUC help in encouraging colonial unions to develop along "responsible lines." This was to be achieved via the Colonial Labour Advisory Committee, established in 1942 and staffed with representatives from the TUC, the Colonial Office, and businesses with colonial interests.

The TUC's adoption of such a position was the result of many of its own members' internalization of paternalistic discourses concerning how British imperialism brought with it economic and political modernization to the colonies, but it was also the product of efforts to create a more influential role for itself within British society in the 1930s by showing that it had standing with the government. Consequently, the TUC provided both in-country training (largely through correspondence courses) and scholarships so that some trade unionists could be educated in Britain. It also recruited British unionists to serve as officers in various colonial labor departments, and several of these played roles in dissolving unions perceived to be too radical, as in Sierra Leone, Malaya, and Kenya. After World War II the Colonial Office encouraged economic development in the colonies as a way to earn foreign exchange to pay off war debts. Accordingly, the TUC became ever more embroiled in providing support for "responsible" trade unionists – in the early 1950s, for instance, it increased its budget for colonial activities to £37 500 [£3.5 million] per year, an amount which would enable it to continue, in one official's words, "not only to exert a formative and educational influence over the more immature Colonial trade unions but also to be a source of seasoned advice and sage counsel to the most advanced" (quoted in Weiler 1988: 49). Although various TUC leaders undoubtedly had a genuine desire to help colonial workers, as Weiler (1988: 51) has put it, ultimately the TUC's "involvement with colonial trade unions reflected its own acceptance of the legitimacy of British imperial rule (and British capitalism overseas) and rejection of any violent attempts to end it."

Although perhaps not as virulently anti-communist as the AFL-CIO, the TUC nevertheless did participate during the post-war period in efforts to undermine communist influence in various parts of the world. One of the earliest such interventions occurred in Greece. After a civil war between communist and non-communist groups, Greece had emerged from the war politically polarized and communists had made significant inroads into the labor movement. Because of its location in the eastern Mediterranean, Greece was a strategically important country for Britain, and the British government determined that a communist-led government would not prevail there. Consequently, the Foreign Office and the TUC worked hand-in-hand to stymie communist influence within the labor movement and to support conservative – or, at least, non-communist – unions, even as it attempted to pressure the right-wing government which had taken power to accept trade unionism and some minor role for such communist unions, this latter position being a sop to the domestic British labor movement and international opinion. When the United States became involved in 1947, AFL and CIO representatives began working closely with those of the TUC, though they pushed a harder line against communist elements (Kofas 1989). Likewise, in Germany TUC officials played important roles in reconstructing the labor movement after the war, largely as a way to limit Soviet influence over German workers. For instance, fearing that a centralized reconstituted movement would make communist domination more likely, the TUC encouraged a decentralized model, one which tended towards factionalism. Meanwhile, by late 1947 TUC officials began inviting non-communist German union officials to Britain as a way of combating both communist inducements and the influence of the AFL, which had begun making overtures to unionists in the Anglo-American zone (Weiler 1988: 178).

Certainly, then, the Cold War played a central role in shaping the contours of labor transnationalism. However, there were also instances of cooperation between the various international labor federations. For example, in 1962 ICFTU and CISC representatives met in Dakar, Senegal, to establish the African Trade Union Confederation as a counter to the All-African Trade Union Federation, which had close links to the WFTU (McKay 1963). Likewise, in the early 1970s, after the AFL-CIO temporarily left the ICFTU and during the period of détente, the ICFTU and the WFTU began developing collaborative agendas concerning challenges posed by TNCs' geographical spread and growing power. Similarly, trade secretariats affiliated with the ICFTU and WCL worked together within the European Trades Union Congress (founded in 1973 to represent workers within the institutions of what would become the EU). Nevertheless, rivalries persisted as the WCL increasingly positioned itself to the left of the ICFTU and as an alternative to it for non-communist unions. Arguably, this was nowhere clearer than in Latin America, where the WCL's regional organization CLAT drew on nationalism, the *aggiornamento* of Vatican II, and the impetus provided by the 1968 Latin American Bishops' Conference and its "preferential option for the poor" to push an anti-capitalist agenda, which invoked liberation theology and "Christian democracy" to develop an economic system that would address issues of persistent and widespread poverty (Gallin 2006).[7] This focus on Latin America (and Asia) became even greater as the WCL lost European affiliates to the ICFTU and ITSs in the 1970s – although it worked with Poland's Solidarność (Solidarity), which received Vatican support, by the early 1990s the WCL retained just two Western European national centers, a Belgian and a Dutch affiliate.

Although much transnational labor activity during the post-World War II era focused upon Cold War machinations and efforts by various entities to outmaneuver their rivals for geopolitical gain, it is important to recognize that growing capitalist transnationalization also encouraged efforts towards greater labor transnationalism. Some of this has been conducted by the ICFTU, WFTU, and WCL, often working in collaboration with agencies like the ILO (International Labour Organisation), the Organization for Economic Cooperation and Development (OECD), and the United Nations Centre on Transnational Corporations, and has involved the various federations calling upon governments and TNCs to adopt ILO conventions related to workers' rights and codes of conduct. However, it is primarily by and through the various trade secretariats and individual unions that transnational labor activity has been conducted (Uehlein 1989). Arguably, the ITSs have been the most active in this regard.[8] Whilst there have been myriad instances of individual unions developing linkages with confederates working for the same or a different employer overseas, the ITSs' importance is that they represent permanent structures through which labor transnationalism may be carried out. Principally, they have focused upon four main sets of activities.

First, the ITSs have developed a degree of international coordination with regard to collective bargaining, especially that involving a single TNC (for several early case studies, see Conference Board 1975). Probably one of the most well-known of such instances occurred in 1969 when the International Federation of Chemical and General Workers' Unions (ICF) helped coordinate negotiations between the French glass manufacturer Companie de Saint Gobain and unions in France, Italy, West Germany, and the US, thereby limiting the company's ability to play them against each other. Arguably, what is most significant here is that the US union involved pressured the company to bargain on the basis of its global profits, rather than national conditions in each of the four countries, even though the firm's US division was actually losing money (Cox 1971). Furthermore, the unions agreed to provide each other financial resources during strikes and refused to accept production transferred from any struck plants. Likewise, during the mid-1980s the ICF helped coordinate a transnational labor campaign against the chemical TNC BASF, with West German unionists pressuring the company to end a lockout at its Louisiana plant (Bendiner 1987). As a way to enhance such transnationalism, at the US United Auto Workers' urging the International Metalworkers' Federation began establishing automobile "world company councils" (WCCs) to coordinate international bargaining by providing information and material aid during strikes. Initially WCCs were established at Ford, General Motors, Chrysler, and Volkswagen (IMF 1967) but were subsequently extended to other firms (Toyota, Fiat, and Honda) and industries, including mechanical engineering (Caterpillar), electronics (General Electric, Matsushita, Siemens, and Electrolux), petroleum refining, tire manufacturing, food production, tobacco processing, and metalworking (Olle & Schoeller 1977). Cold War divisions, however, meant these Councils generally did not include unions with communist links, which was a problem in France and Italy where unions in these sectors were often affiliated with communist labor federations (Stevis 1998).

Second, and relatedly, whilst divergent living costs around the world realistically preclude establishing uniform wage rates, the ITSs have maintained that they should not prevent workers from enjoying similar levels of health and safety at work, nor rights to unionize. In this regard, in 1971 the auto WCCs began a campaign to harmonize non-wage issues (like break-time length). Although perhaps easier to standardize than wages, such a strategy, however, has not been without its own problems. Varying cultural attitudes towards work and labor–management cooperation have made it difficult to pursue harmonization to the fullest extent, whilst the growing geographical dispersion of automobile production out of the high-wage global industrial core (North America, Europe, Japan) into low-wage countries like Mexico, South Korea, and Brazil has also complicated matters, especially as Latin American and Asian WCC representatives have generally been concerned about ensuring basic trade union rights in the face of authoritarian regimes whilst job security, wages, and shorter working hours have often been North American and European unionists' main concern (Bendiner 1987). Nevertheless, WCC participants have generally viewed synchronization as an effective way of challenging TNCs.

Third, secretariats have encouraged mergers between unions operating in different sectors, especially in countries like Britain and Japan where several unions can operate in a single facility, and have themselves engaged in mergers to mirror the concentration and diversification of TNCs. Although this process has not been without issue – ideological differences and turf protection have arisen – there has been some success as both individual unions and trade secretariats have amalgamated. Finally, the secretariats – helped by the WCCs and ICFTU – have spent time and money developing databases on various TNCs to provide information on ownership structures, investment patterns, contract details, and management practices to their affiliates, such that by the early 1990s the International Metalworkers' Federation, by way of example, had data on more than 500 TNCs (IMF 1991). Simultaneously, the ITSs largely entrusted to the ICFTU the job of representing workers' interests to agencies like the ILO, giving the secretariats more time to focus upon data collection and coordinating contract bargaining (Casserini 1993).

In considering the growth of transnational labor activities during this time period, however, it is important to recognize that whereas most of the earliest labor organizations dedicated towards developing cross-border links – the ICFTU, the WFTU, the WCL, the ITSs and their equivalents, the PAFL – originated in Europe or North America, not all efforts have commenced in these regions. Thus, the Japanese International Labour Federation, largely modeling its activities on the AIFLD, has conducted training programs in East Asia and brought high-level unionists to Japan (Williamson 1994). Equally, although largely state-initiated, organizations like the International Confederation of Arab Trade Unions and the Organization of African Trade Union Unity (OATUU) have also served as arenas for cross-border contacts between workers in the global South. The recent rapid industrialization of parts of the global South, though, has also begun to reshape dramatically the geography of global unionism as powerful labor movements have emerged in countries like South Africa, Brazil, the Philippines, and South Korea. Exhibiting significant anti-neoliberal streaks, these workers not only have developed forms of "social movement unionism" characterized by mass strikes and engaging with non-union organizations for mutual support (a model which contrasts with the bureaucratic unionism frequently seen in the global North) but they have also sought to develop greater linkages between global South unions and to avoid being caught up in the economic and political agendas of those of the global North. In 1990, for instance, the Philippine Kilusang Mayo Uno (KMU – May First Movement) proposed a trilateral conference involving itself, the Congress of South African Trade Unions (COSATU), and Brazil's Central Única dos Trabalhadores (CUT – Central Workers' Federation) to bring together major new and autonomous union centers from South Africa, Asia, and Latin America (Waterman 2004). At the same time, both COSATU and CUT refused affiliation with either the ICFTU or the WFTU, with COSATU preferring to work with the OATUU to develop "South–South solidarity, and challenge the domination of the Northern centres/federations" (quoted in Waterman 1998: 122).

Worker Transnationalism after the Cold War

The Soviet Union's collapse transformed many aspects of the global political economy, not the least of which was the world of organized labor. Importantly, the Cold War's end opened up spaces – both material and metaphorical – for increased transnational labor collaboration, even as its legacy has continued to shape union activities and the geography of organized labor. One of the most significant transformations has been the virtual destruction of the WFTU, which has brought with it a dramatic reworking of the geography of trade unionism, particularly in Eastern Europe. Thus, whilst Eastern Europe was a fairly uniform political space between the late 1940s and 1989, with the region's state-controlled unions affiliated with the WFTU, in the 1990s it became an arena of significant political conflict as the WFTU attempted to hold on to affiliates and the ICFTU and WCL expanded into the area. For instance, in Poland, although the old communist labor federation Ogólnopolskie Porozumienie Związków Zawodowych (OPZZ – All Poland Alliance of Trade Unions) initially remained affiliated with the WFTU, Solidarność joined the ICFTU.[9] This process was repeated across the region. In Czechoslovakia the nation's largest national federation, the Československé Konfederace Odborových Svazů (ČSKOS – Czechoslovakian Confederation of Trade Unions), quickly partnered with the ICFTU whilst the Odborové Sdružení Čech, Moravy, Slezka (OSCMS – Trade Union Association of Bohemia, Moravia, and Silesia), a rival center formed by members of the old Czech Communist Party, affiliated with the WFTU. Similar developments took place in Hungary, Estonia, Lithuania, Romania, and elsewhere (Herod 2001). Many such federations' constituent members also joined with the appropriate ITS – the International Metalworkers' Federation, for instance, affiliated unions in the Czech Republic, Slovakia, Romania, Bulgaria, and Hungary, as did ITSs representing workers in food processing, transport, mining, and agriculture. In 1994, the WCL, though generally less influential, affiliated Romania's second largest national center, the Alfa Cartel (Keil & Keil 2002), and the Czech Republic's Křesťanská Odborová Koalice (KOK – Christian Trade Union Coalition). Complicating the picture, however, is the fact that whereas some national centers affiliated with the ICFTU, a number of their individual unions remained associated with the WFTU's Trade Union Internationals, as with the Podkrepa national federation in Bulgaria and in Russia, where the successor to the old Soviet national labor center, the Federatsia Nezavisimiikh Profsoyutz Rossia (FNPR – Federation of Independent Trade Unions of Russia), affiliated with the ICFTU in 2000.[10] Beyond the Soviet satellite states other national centers also left the WFTU, including the French CGT (in 1995).

Both the WCL and the ICFTU have been active in promoting Western-style economic and political systems in the former Soviet bloc. In December 1990, the ICFTU created a Coordinating Committee on Central and Eastern Europe with representatives from its affiliates in and outside the region, from various ITSs, the European Trades Union Congress, and the ILO, with the goal of discussing information exchange, building up technical expertise in the region's nascent unions, and establishing in-country educational centers to train union personnel. As part of this the ICFTU developed linkages with unions in most of the region's countries, including in those not directly under Soviet control (Albania and the countries of the former Yugoslavia) (Herod 1998a). ICFTU officials also established offices in Moscow, Sofia, Sarajevo, Vilnius, and Zagreb, and have sought to promote a regulated market economy and strong and democratic unions whilst challenging the "narrowly conceived" (ICFTU 1996: 275) economic liberalization policies advocated by institutions like the International Monetary Fund. As a way to do so the ICFTU has run seminars and workshops on topics like the impacts of structural adjustment policies, labor

migration, how to engage in effective collective bargaining, and how to develop policies to counter the negative consequences of privatization and economic deregulation
(like increased unemployment). The WCL has engaged in similar activities, often in
collaboration with the ICFTU and the ILO (ILO 2003). Meanwhile, the WFTU has
attempted to reinvent itself, both in Eastern Europe – where it sees itself as the principal
entity challenging economic neoliberalization – and beyond, presenting itself as an
organization largely of global South unions (Zharikov 1995). It has also sought to develop
closer links with both the ICFTU and WCL on several matters – FDI, the LDCs' debt
crisis, trade, and globalization generally – and to allow a greater degree of autonomy
both to its regional offices in Delhi, Dakar, Havana, and Damascus and to its TUIs,
several of which it encouraged to merge in the 1990s to match the centralization and
concentration of transnational capital (WFTU 1994, 1996; Herod 1998a).[11]

Equally, the various industry sector organizations of both the WCL and ICFTU have
been active in Eastern Europe. For instance, the Dutch Christian Teachers' Union
CNV (Onderwijsbond CNV) has provided material assistance to teachers' unions in
Eastern Europe (as well as to those in Africa, Asia, and Latin America). Meanwhile,
ITSs like the International Metalworkers' Federation and the International Union
of Food, Agricultural, Hotel, Restaurant, Catering, Tobacco, and Allied Workers'
Associations (IUF) have organized training seminars on contract negotiating, labor
law, and running pension and welfare plans for newly emergent unions (Herod
1998b). Simultaneously, several individual unions have established direct links with
Eastern European unions. Hence IG Metall, the German metalworkers' union, helped
Poland's Solidarność negotiate collective bargaining agreements in the mid-1990s
(Senft 1995) whilst the American Federation of State, County and Municipal Employees and the Service Employees' International Union, amongst several other US-
based unions, have organized training sessions in the region. The US labor movement
has also been active through the AFL-CIO's Free Trade Union Institute (FTUI), which
provided secret support to Solidarność in the early 1980s and quickly developed links
with anti-communist elements in the Eastern European labor movements after the 1989
revolutions. The FTUI has been concerned with two principal issues.

First, it has worried that Eastern Europe will become a region of unrestrained free
market capitalism, one which might not only undermine workers' standards of living
in Eastern Europe but which may also undercut wages and working conditions in Western
Europe. This is particularly so now that German, Austrian, and other firms have begun
establishing manufacturing facilities in countries like Poland and the Czech Republic.
For the FTUI (1994: 5), then, a key goal has been to provide "independent trade unions
. . . as much support as possible to ensure their survival and to help strengthen them
as effective democratic institutions so that the tyranny of the market will have no place
in a civilized world." Second, the resurgence of communist and former communist
parties in several countries as a result of popular unhappiness over increased unemployment and worsening living standards led the FTUI to worry that reform would
be "stifled, impeding the development of both a private economy and a civil society,"
whilst the emergence of *raubwirtschaften* ("plunder economies") and "a crony capitalism that relies less on free markets than on mafia-like networks" in several nations

– Russia is an obvious example – have meant the process of privatizing state-owned assets has often been "closed and corrupt . . . lock[ing] out workers and free trade unions from the reform process and leav[ing] them even more vulnerable to restructuring and layoffs" (FTUI 1994: 8–11). Consequently, the FTUI has sought to strengthen non-communist unions through educational and training programs, many of which have been conducted by staff from the Institute's Bucharest, Kiev, Moscow, Sofia, and Warsaw field offices.

Whereas the transition in Eastern Europe has dramatically transformed the world of organized labor, this has not been the only geographical arena in which significant changes have occurred. In 1993 South Africa's COSATU and Brazil's CUT, along with Italy's CGIL, held talks to explore developing a southern transatlantic, anti-neoliberal alliance, whilst in 1997 the WCL significantly expanded into the Islamic world by affiliating the Ketua Umum Konfederasi Serikat Buruh Sejahtera Indonesia (K-SBSI – Confederation of Indonesia Prosperity Trade Union). More recently, both COSATU and CUT have been involved with unions from Argentina, Venezuela, Namibia, Egypt, India, Tunisia, Indonesia, and the Philippines in pressuring industrialized countries to provide greater market access for agricultural goods coming from LDCs (*COSATU Daily News* July 27, 2007). Despite its recent troubles, the WFTU and its associated entities have also continued to operate on a small scale in several global South regions through, for instance, sponsoring conferences and running training sessions.

One of the most imaginative efforts in recent years to develop a new organizational structure within which unions in Africa, Asia, Australasia, and Latin America can network has been the Southern Initiative on Globalisation and Trade Union Rights (SIGTUR) (Lambert & Webster 2001). SIGTUR itself began life as an Indian Ocean network, linking Australian unions (which have a long history of transnationalism, especially amongst maritime workers) with the new social movement unions emerging in South Africa in the 1970s. Such links were reinforced during the mid-1990s when COSATU and the Centre of Indian Trade Unions (CITU) supported unions in Western Australia fighting state government efforts to undermine union protections. Initially SIGTUR was opposed by the ICFTU because it appeared to challenge its Asia and Pacific Regional Organisation (APRO), whereas SIGTUR criticized the ICFTU for its hub-and-spoke organizational structure wherein the Confederation's European office (and those of its associated ITSs) generally coordinated activities within and between other regions whilst allowing little East/West or South/South integration. However, with the Cold War's end and growing integration of the Pacific-rim and Indian Ocean-rim economies, relations between the ICFTU and SIGTUR improved dramatically, and the two organizations began working together on issues, including a campaign against the anti-union Rio Tinto mining multinational wherein SIGTUR worked with environmental, indigenous, and human rights activists. SIGTUR itself has also expanded, and now incorporates COSATU, the Korean Confederation of Trade Unions, the Philippines's KMU, Brazil's CUT, India's CITU and the All-India Trade Union Congress, as well as unions in Malaysia, China, Indonesia, Thailand, Sri Lanka, and Pakistan, who have come together on matters of common interest. Such connections have spawned efforts to further integrate unions from different nations, as with the plan inked by the Maritime Union of Australia and the South African Transport and Allied Workers' Union to create a global union linking the port of Fremantle, Western Australia, with that of Durban, South Africa.[12]

Certainly, these organizational changes mark important developments. However, arguably the biggest transformation in the geography of the international labor movement in the post-Cold War period has been the 2006 merger of the ICFTU and WCL to form the International Trade Union Confederation (ITUC). This merger itself followed the 2002 rebranding of the 10 ITSs as Global Union Federations (GUFs), a taxonomic shift designed to emphasize their "globalness." Principally, the ITUC's goal is to present a more unified front to TNCs. Although the WFTU has remained outside this new "super federation," the ICFTU–WCL merger, together with that of their regional and sectoral bodies, has resulted in an entity which claims to represent (as of June 2007) 168 million unionists in 153 countries and territories (ITUC 2007a).[13] The ITUC has also begun establishing new regional organizations, like the Pan-European Regional Council, which represents 87 national centers and 85 million unionists "from Lisbon to Vladivostok" (ITUC 2007b). Much like a TNC, the ITUC operates several offices across the planet, in Amman, Geneva, Moscow, New York, Sarajevo, Vilnius, and Washington, DC, and has begun to engage in broader campaigns, such as its "Fair Play at the Olympics," which urges the International Olympic Committee and sportswear manufacturers to respect workers' rights (ITUC 2007c).

The recent past, then, has seen numerous union efforts to develop strategies to counter the predations of transnational capital. Some of these have involved one-off campaigns, either of the direct union-to-union type or with unions working through third parties, like the ITSs. One of the most global instances of a union working through an ITS – actually, several – to challenge a TNC involved workers at an aluminum smelting plant in Ravenswood, West Virginia, who worked with the International Metalworkers' Federation and the International Federation of Chemical, Energy and General Workers' Unions (ICEM) to challenge their lockout by the multi-billion-dollar Marc Rich group of companies, a campaign that involved workers in 28 countries on five continents, including the Netherlands, Britain, Canada, France, Venezuela, Romania, Bulgaria, Czechoslovakia, and Switzerland, amongst others (Herod 2001).[14] More recently, in 2001 the German metalworkers' union IG Metall helped Brazilian unionists representing over 3 000 fired workers at Volkswagen's São Paolo plant travel to the company's headquarters in Wolfsburg to pressure it to reconsider – which it not only did but, in the process, also agreed to commit greater investment to the plant (Anner 2003). Whereas such disputes represent examples of workers using well-established formal trade union links to pursue their goals, there have also been growing instances of what Castree (2000: 280) has called "grassroots internationalism," wherein workers network largely outside the formal structures of their unions – as in the 1995–8 strike by Liverpool dockers, who engaged with dockers in the US, Australia, Denmark, Canada, and Japan, amongst other places. Myriad other examples abound.

Workers and unions have also developed new structures, though, through which they can bring a more constant pressure to bear on TNCs, rather than scrambling to put together coalitions once disputes have broken out. One structure is the so-called "framing agreement." The first such agreement was negotiated in 1988 between the IUF trade secretariat and the French food company Danone, and covered matters like workers' right to join the union at Danone's plants worldwide and a non-discrimination policy. Similar in some ways to the World Company Councils developed in the 1960s, though with greater powers of enforcement, by 2007 55 framework agreements had been signed, covering issues from union rights to child labor to health and safety rules to a company's environmental practices (as in the agreement signed between the furniture maker IKEA and the International Federation of Building and Wood Workers ITS). Although framing agreements exhibit a distinct European focus – the vast majority of firms agreeing to them have European origins (Stevis & Boswell 2007a: 112–13) – they have nevertheless enabled workers in other parts of the globe to pressure firms as they have invested in Asia, Africa, Australasia, and the Americas. Hence, the agreement between the IUF and the French global hotel chain Accor has provided important support for unionization efforts in North America, Indonesia, New Zealand, the UK, and Africa (Wills 2002), whilst one between Del Monte and the IUF helped Guatemalan banana workers regain their jobs and back wages (Wick 2003: 20). As of 2005, approximately 4.1 million workers worldwide were covered by such agreements, a number greater than the labor force of Belgium (Carley 2005).

A closely linked development has been the growth of various "works councils," especially in the EU. Unlike the World Company Councils, which primarily involved high-level contacts between union and corporate representatives, resulted from negotiations between union and firm, and owe their birth largely to the US autoworkers' union, the works councils have their origins in Europe and the efforts of, particularly, the European Metalworkers' Federation to develop a European-wide system whereby workers in different plants could meet each other regularly to discuss matters of mutual concern (Rehfeldt 1998). This was formalized with a 1994 EU directive requiring firms with at least 1 000 employees, of which at least 150 work in another EU country, to establish European Works Councils (EWCs) as fora for information sharing and consultation with workers.[15] By the early 2000s, an estimated 1 400 firms employing 15 million workers across the EU had created EWCs (Wills 2001). As EWCs have evolved, a number of important changes have occurred. First, whereas initially their role was one of providing a medium through which information could be disseminated and workers consulted about plant matters, they now appear to be developing as entities through which European-wide collective bargaining can occur, as with Ford and GM's European subsidiaries and several other companies. Second, there have been moves to expand geographically worker representation beyond the EU by creating World Works Councils (WWCs) – auto producers Volkswagen, Renault, and Daimler-Chrysler have all established WWCs and other companies are moving in this direction (Carley 2005; Da Costa & Rehfeldt 2006).[16] Significantly, there is also a growing degree of cross-over between international framing agreements and WWCs. Thus, although to date both framing agreements and WWCs have had a distinct European flavor to them, the fact that the GUFs have played central roles in negotiating them has allowed GUFs the opportunity to bring to the table workers' concerns from other parts of the world and encouraged their geographical spread – in November 2006, for instance, several South and North American unions founded a WWC in the Brazilian steelmaker Gerdau, with the goal of persuading it to sign a framework agreement (Stevis & Boswell 2007b).

In another development unions and workers increasingly use media like the internet to coordinate their own activities and to bring pressure to bear on firms, both transnational and non-transnational (Lee 1997). Many unions have used the internet to draw global attention to particular companies and disputes, primarily to solicit support from other workers or members of the public. Others have used the internet in more sustained fashions. For instance, in 1999 ICEM affiliates in Canada, Guatemala, Columbia, Venezuela, Brazil, Chile, the UK, France, Germany, Slovenia, Turkey, Morocco, South Africa, Malaysia, Japan, and the US met to implement a networked global database on the Goodyear company's worldwide operations, both to provide unions with data on working conditions and wage rates across the company for use in local collective bargaining negotiations and to help unions organize Goodyear's non-union plants. Unions and workers have also used such technology to coordinate simultaneous global protests, as when unions on six continents demonstrated concurrently in 2005 for a total ban on asbestos use (UITBB 2005). Numerous similar actions have taken place in recent years, with the internet's speed and flexibility allowing unions and workers to organize synchronized demonstrations far more readily than in the pre-internet age, when they relied on relatively expensive international telephone calls and telex messages or relatively slow postage.

One of the earliest uses of the internet to bring worldwide action against a company, though, was pioneered by the United Steelworkers of America (USWA) and the ICEM against the Bridgestone/Firestone company (Herod 1998a). In 1994 the company fired 2 300 union workers at five US plants and brought in non-union replacement workers. The USWA and ICEM turned to the internet – then a relatively new entity – to force the company to rehire the workers. Sending emails to Bridgestone's South African subsidiary and posting messages on various labor websites, the strikers and their supporters quickly spread word of their plight. Soon the workers – and, perhaps more importantly, sympathetic members of the public – began bombarding Bridgestone with negative comments via its own customer feedback webpages. ICEM posted on its own website a black flag (auto racing's signal to disqualify a driver for breaking the rules) that supporters could electronically clip and send to the company, other Bridgestone customers, and suppliers and shareholders. They also posted the addresses and telephone numbers of local Bridgestone plant managers around the globe, together with the names of banks and other investors in the company, with calls to file complaints with them. The ICEM even linked its own site to that of the US National Highway Traffic Safety Administration so that supporters could more easily complain about defective tires the replacement workers produced. The USWA's and ICEM's innovative campaign not only paved the way for other unions to develop internet actions but was also key in leading the company in September 1996 to call back to work virtually all the union workers it had fired.

Finally, it is important to consider how unions have begun developing strategic alliances internationally. For example, in 2004 the US-based Service Employees' International Union (SEIU) unveiled a global partnerships unit through which to coordinate campaigns against janitorial TNCs and develop collaborations with unions in other countries – in Australia with the Liquor, Hospitality, and Miscellaneous Union (LHMU), in New Zealand with the Service and Food Workers' Union (SFWU), and in the UK with the Transport and General Workers' Union (Ryan & Herod 2006). Emerging from the SEIU's "Global Strength" campaign, the partnership's goal is to organize "whole markets in every part of a global employer's operations" (Crosby 2005). Whereas historically transnational solidarity efforts have typically involved unions in one country sending financial and other resources to unions overseas, this partnership has involved SEIU assigning staff to Australia to work directly with local unions on campaigns – the hiring of some 30 organizers and five researchers makes the SFWU–LHMU–SEIU campaign "probably the biggest growth campaign run in either country in over a hundred years" (Crosby 2005). SEIU has also sent staff to Poland, the UK, India, France, Switzerland, Germany, the Netherlands, and South America to help run campaigns (Stern 2006: 112). Meanwhile, in Britain the GMB union recently began working with Solidarność to organize Polish workers in Britain, with GMB even chartering Polish-speaking local unions in Southampton and Glasgow (*Guardian*, December 6, 2006). Furthermore, much as TNCs transfer skills or "best practices" abroad, so have unions like SEIU exported new labor practices, including the much talked-about "organizing model" (Heery et al. 2002).[17]

Taking this one step further, some unions have begun to explore merging transnationally. Paradoxically, such developments take us back to the kinds of strategies engaged in by unions in the nineteenth-century "pre-national" era of labor internationalism, when the Knights of Labor and the Amalgamated Society of Engineers established overseas branches, although with a twist – whereas these latter unions were established in one country and subsequently founded branches overseas, the current trend seems to be that of unions independently established in different countries merging. For example, in 2006 the International Association of Fire Fighters, representing firefighters in the US and Canada, met with representatives from the UK, Australia, and New Zealand to lay the groundwork for the International Fire Fighter Unions Alliance, an affiliation bringing some 350 000 firefighters and emergency service workers from around the world – and with plans to invite other countries' unions to join – under the umbrella of a single organization (*Australian Firefighter*, 2007). Equally, the British National Union of Marine, Aviation and Shipping Transport Officers established a federation in 2006 with the Dutch Federatie van Werknemers in de Zeevaart (FWZ – Federation of Merchant Marine Workers), with both unions simultaneously changing their names to, respectively, Nautilus UK and Nautilus NL in readiness for a 2008 merger that will produce a single head and governing council. Likewise, the British union UNITE (the UK's largest private-sector union), IG Metall (Germany's largest industrial union), the International Association of Machinists, and the United Steelworkers of America (the largest industrial union in North America), representing over 6 million workers, have agreed to work together to challenge common employers and explore possibilities for a merger. Such mergers are unlikely to be the last.

Geographies of Labor Organizing in the Global Economy

Having outlined some of the institutional structures – and the histories and geographies thereof – that have shaped transnational labor activities during the past century and a half, in this final section I contemplate several conceptual issues which are important to consider when exploring the unfolding geography of the world economy and labor's role in shaping it. A key issue in this regard is to recognize that workers come together over space for varied reasons, each of which have different implications for how the world economy is networked together via workers' praxis. Van der Linden (1999) identifies four reasons why workers engage in transnational solidarity: (1) out of a sense of ethnic unity, as when Irish-Americans support workers of Irish extraction in other parts of the globe; (2) out of collective historical experience, as when pan-Africanism has shaped worker solidarity; (3) because of a common ideological viewpoint and desire to give political support, as when the Amalgamated Clothing Workers of America sent representatives to the Soviet Union in the 1920s to help modernize clothing and textile plants (Morray 1983); and (4) for economic reasons, as when workers in one country help those in another organize, either to improve the latter's situation out of a sense of class cohesion or so that such workers will not become economic threats by serving as sources of cheap labor.

Understanding the varied reasons workers engage in transnationalism is important, for it allows us to consider some of the geographical hurdles to be overcome and some of the spatial consequences thereof. For instance, transnationalism based on ethnic affiliation may occur as part of a generalized notion of solidarity with workers in another country without those engaging in such activities knowing much about the specific employers for whom overseas workers toil. It is also a reflection of how the globe has been networked together via international migration patterns and the consequences thereof – some of the impetus to purge communists from US unions in the 1950s, for instance, resulted from rank-and-file workers expressing solidarity with Soviet-occupied Eastern Europe, given that early twentieth-century immigration patterns meant that much of the US working class at the time was made up of first- and second-generation Czechs, Poles, and Hungarians. Equally, African-Americans' expressions of solidarity with African workers as a demonstration of pan-African unity reflects particular geographies and histories of diasporic migration. On the other hand, the pan-Africanism of African workers themselves – workers in French West Africa supporting those in British West Africa during the colonial period or Tanzanian workers supporting South African workers during apartheid – is not reflective of migration patterns but, rather, of common colonial experiences. It also shows how colonialism transformed worker identities within countries: for instance, Yorubas, Ibos, and Hausas came to think of themselves (at least in certain instances) as part of a Nigerian working class, such that Yoruba workers in the south might support their "Nigerian brothers and sisters" in Hausaland in the north in ways in which they would not previously have done.

On the other hand, transnationalism engaged in for economic reasons might require a more intimate knowledge of a particular TNC's geographical structure, with workers figuring out how their particular plant fits into the overall corporate con-figuration. In other words, it may require a very different spatial knowledge than does solidarity based upon ethnic affiliation. Furthermore, it is important to recognize that whereas transnationalism on the basis of common ethnic background usually does not foreground questions of class – the key is to link with workers overseas on the basis of their ethnic identity, not their class position – transnationalism emerging out of economic concerns generally does: the central element in the process is to link up with workers overseas on the basis of their being workers, with such workers' ethnicity less part of the equation. However, viewing this type of transnationalism as an attempt to transcend space so that workers may come together on the basis of their common class position reveals complex intersections of class and geography. In particular, the reasons why workers engage in such transnationalism is of great consequence for how the world economy's geography is constructed. Thus, Johns (1998) distinguishes two types of transnational solidarity: "accommodationist" and "trans-formatory." For her, the primary difference between them is their goal. Hence, she argues, when workers engage in "accommodationist" solidarity their principal reason for helping foreign workers raise their wages is that so doing will limit the likelihood that capital will flee overseas. Put another way, whilst on the surface accommodationist solidarity appears to be about uniting workers across space, in fact it is about "reasserting the dominance of a particular group of workers within capitalism's spatial structures" – it is about defending the already privileged spaces in the global economy against the possibility of capital flight. Transformatory solidarity, on the other hand, is a type of solidarity in which "universal class interests dominate over interests that are spatially derived and rooted," with such solidarity designed to "unite workers across space in order to fundamentally transform social relations of production into something more humane" (1998: 256). Workers engage in this type of solidarity, then, without concern for how their actions affect their own position and communities, and do so because they are intent upon challenging the social relations of capitalist production and firms' ability to play workers in one location against those in another. Although empirically it may be difficult to differentiate reasons, analytically this is an important distinction to make, for the outcome will shape how the global economic landscape is made: if workers engaging in accommodationist solidarity are successful, then capital investment is likely to stay in their own communities; if they are not, then it is likely to flow elsewhere. Likewise, if workers engage in successful transformatory solidarity, then firms' ability to play different parts of the globe against each other in a downward wage spiral will be limited.[18]

In considering the emerging geographies of labor transnationalism it is essential to reflect not only upon how workers' actions are shaping the spatiality of the world economy but also how the geography of economic activity molds the possibilities for such praxis. Thus, whereas globalization has been viewed as a process whereby commodity chains are being spatially stretched to link various parts of the planet together, it is important to recognize that different types of commodity chains present diverse opportunities and challenges to labor transnationalism. Specifically, whereas some chains are "producer-driven" (meaning that manufacturers control directly the entire production system, including its forward and backward linkages), others are "buyer-driven" (meaning that large retailers and trading companies play the central role in establishing decentralized production networks in which subcontractors – typically in LDCs – produce goods to order, as with toys, clothing, and footwear). What is critical here for considering worker strategies, then, is how power is exercised across the global economic landscape and what that means for bringing pressure to bear on different employers. Hence, in the case of producer-driven chains a single, central actor – the TNC – directly coordinates across space the activities of various elements of the production process, as when an automobile company has its assembly plant in the US but uses components made at its supplier plant in Mexico. In such an arrangement each part of the network is linked directly to the center, such that focusing pressure on a supplier facility as part of a local dispute can bring pressure to bear directly upon the broader entity. However, in the case of buyer-driven chains, wherein manufacturing firms in, say, China, operate as independent contractors for large retailers and distributors like Liz Claiborne or Nike that do not themselves actually manufacture goods, pressuring the Chinese assembler so as to coerce the retailer/distributor is unlikely to have much impact – the retailer can readily switch to another sub-contractor. Furthermore, the fact that profits in buyer-driven chains derive less from economies of scale and technological advances (the case with producer-driven chains) and more from unique combinations of high-value research, design, sales, marketing, and financial arrangements (Gereffi 1994) means that workers and unions challenging each of these two kinds of firms will have to address situations in which quite different corporate resources can be brought to bear, in which profits are made differently (which may affect what the company is willing to do in response to worker demands), and in which the various parts of the assembly process are connected in different ways (through ownership versus through contractual obligation).

These differences in how commodity production is organized geographically and managerially are important, then, if the strategies workers and unions develop are to be successful. Such sensitivity to commodity chains' spatiality has been shown recently by the US UNITE union and the National Consumers League who, in 1995, initiated a "Stop Sweatshops Campaign." Whereas traditional campaigns to defend workers' rights and improve pay and working conditions had focused upon the factories in which garment workers were employed, UNITE's campaign focused instead on pressuring the retailers who carry products made in sweatshops – it shifted the focus, in other words, from the spaces of production to the spaces of consumption (Johns & Vural 2000). By encouraging consumers to badger retailers not to buy from manufacturers violating labor rights, then, the union avoided the costly and lengthy problem of having to organize each individual sweatshop, a potentially fruitless task given how frequently such production facilities open, close, or move location. Likewise, the SEIU has used street theatre and demonstrations to coerce building owners in turn to compel the cleaning companies with whom they contract to raise janitors' wages. This raises issues of how people acting as consumers and people acting as producers may be able to put greater or lesser pressure on different firms on the basis of the kind of commodity chains within which producers are embroiled.

Furthermore, it suggests that there is a significant geographical complexity to this which those challenging particular firms must recognize: how are commodity chains organized and what does this mean for how they tie various parts of the planet together?; are the people who are protesting as workers and those who are doing so as consumers the same individuals and, if not, where are they each located geographically?; what does this matter for the politics and spatiality of the protest?; and what does this all mean for their ability to come together across space to present a united front?

Equally, new geographical patterns of production have brought some workers together – enticing thousands of peasants from rural regions in LDCs into booming industrial cities, perhaps sparking drives towards unionism – whilst splitting others asunder, as when activities are spatially dispersed through outsourcing. Likewise, new geographies of residence, consumption, and sociability have developed in many industrialized nations, such that workers are more likely than in the past to reside "a considerable distance from fellow-workers, [to] possess a largely 'privatized' domestic life or a circle of friends unconnected with work, and [to] pursue cultural or recreational interests quite different from those of other employees in the same workplace," all with the result that "many of the localized networks which [historically] strengthened the supports of union membership (and in some cases made the local union almost a 'total institution')" have been undermined (Hyman 1999: 3). This is requiring unions to develop new strategies sensitive to the reworked spatial contexts within which workers now find themselves – strategies like "geographical unionism," in which workers organize across a particular geographical area rather than according to craft or employer (Herod 2007b). At the same time, the technologies facilitating telecommuting and TNCs' management of offshore operations are also affecting the geography of labor organizing, for not only have they allowed workers in different parts of the globe to more readily communicate but, as Waterman (1993) has argued, in the process they have had a great impact on the geographical structure of organizing itself. Hence, whereas in the past transnational contacts between workers were generally handled by a union's international office and were typically quite hierarchical in nature – workers in one country contacted their national union, which then requested an ITS contact national union officials in another country, who then contacted local union representatives in the second country – today what he calls "networking activists" (those with access to the internet) can make direct contact with overseas workers in a much more organizationally "flat" process.

As they engage with the unevenly developed geography of the world economy, then, workers' praxis will be shaped by their different organizing traditions and geographical goals. Hence, whereas in the 1970s ITSs representing chemical workers and clerical workers began calling for common contract expiration dates internationally as a way to create a unified front across space, the International Metalworkers' Federation preferred not to do so, feeling that different expiration dates – which implicitly represented a heterogeneous geography of contract negotiations – was a more appropriate strategy because it would not stretch unions' financial resources too much at the same time and would allow metal unions to negotiate with employers singularly and on the basis of local and national conditions in the industry (Weinberg 1978: 54).

By the same token, whereas Continental European unions are more likely to be involved in global agreements with employers (perhaps reflecting the history of corporatism in Continental Europe), US unions are more likely to use global campaigns to achieve their goals (Stevis & Boswell 2007b: 188). Depending upon their immediate or long-term goals, then, different unions may engage with the unevenly developed landscape of global capitalism in quite dissimilar ways, all whilst recognizing that how the world economy is spatially structured presents both challenges but also opportunities. Indeed, unions like the SEIU are beginning to think in geographically strategic terms about confronting global employers, as with exploring how unions with strong financial assets might pay strike benefits to overseas workers of the same employer during a dispute, and how to engage in global protests against the clients of the transnational janitorial companies for whom they work (Aguiar & Herod 2006). Thus, in a version of regulatory arbitrage which is the mirror image to that in which TNCs engage, the SEIU has even begun to ponder outsourcing strikes from countries where they are illegal or highly regulated to those where they are not by getting workers in a firm's subsidiary in the second country to strike on behalf of those in the first (Stern 2006: 113). However, whilst unions like SEIU are cognizant of the need to be spatially flexible and open to wider geographical horizons, such tactics raise other questions about the geography of power and union organization – if unions engage in global bargaining of the kind envisioned by SEIU wherein strikes are outsourced, will weak unions in LDCs become too dependent upon the powerful unions of the parent company's country and too logistically cut off from the unions of their own?

Finally, in contemplating strategies by which workers and unions might confront transnationally organized firms, it is clear that the geographical strategies of workers need not necessarily match those of their employers. In other words, workers and unions may not have to develop transnational linkages and structures when challenging TNCs. Rather, their strategy choices will be shaped by how firms themselves are linked, what kinds of organizational structures they employ, how geographically embedded are both firms and workers, and a host of other issues. For instance, in cases where firms have highly integrated organizational structures – as with automobile firms using just-in-time (JIT) delivery methods – workers' ability to hold up arrival of components allows them to disrupt a firm's national or even global operations by targeting a small number of plants. Such was the case when disputes at just two plants in Flint, Michigan, in 1998 enabled autoworkers to shut down virtually all of General Motors's North American operations, as well as a plant in Singapore (Herod 2000). Instead of "going global," workers were able to focus their efforts on plants in a single community to force GM to provide concessions before ending their strike. Certainly, this was only possible because GM has come to rely upon JIT delivery systems, but it does illustrate that how firms are structured geographically and how workers understand the spatial and organizational relationships between their own plants and those of the rest of the company can dramatically influence the avenues of action open to them: had GM not been reliant upon JIT, workers would likely have had to engage in a more traditional transnational solidarity campaign against the company. What this example shows, then, is that workers may have quite different spatial options, depending upon how they are connected to other places geographically and how they understand the ways in which they are so connected.

In viewing such activities, it is clear that – contra the neoliberal mantra – workers and their organizations are indeed playing important roles in shaping how the geography of the world economy is unfolding in the early twenty-first century, much as they played a role in shaping how that of the nineteenth and twentieth centuries developed. Whereas it is frequently transnational firms or nation-states that are seen to have brought about processes of globalization, workers have also been intimately involved, either through their proactive or their reactive stances. Hence, the AIFLD's activities in Latin America helped open up the region to US FDI, whilst US and British union involvement in securing access to raw materials – as with Jamaican bauxite (Harrod 1973) – shaped TNCs' FDI practices. Likewise, unions from both within and beyond Eastern Europe played significant roles in furthering processes of privatization in the 1990s that have been central to the emergence of market economies and the associated political transformations, whilst transnational campaigns have also shaped flows of capital globally – as when the Ravenswood campaign against Marc Rich prevented him from expanding his operations into Eastern Europe and elsewhere.[19] They have also helped establish various codes of conduct (both voluntary and required), as with a side-agreement to the North American Free Trade Agreement on labor standards, although this has also caused workers to have to face the realities of organizing in an unevenly developed global context – some workers in the global South have accused those in the global North of pursuing such codes simply as a protectionist move. Furthermore, it is important to recognize that unions and workers have operated transnationally both as adjuncts to the wishes of various TNCs and governments but also in pursuit of their own geographical agendas – a fact perhaps no more clearly evidenced than in United Mine Workers of America President John Lewis's 1939 declaration that "Central and South America are capable of absorbing all of our excess and surplus commodities" (quoted in Scott 1978: 201).[20] The key question, then, for understanding how the geography of the world economy is unfolding is to see how the spatial visions of TNCs, governments, unions, and unorganized workers (and sub-groups within each) clash or are in accord, and how each of these groups and sub-groups – whose individual spatial agendas are themselves shaped by the geographical contexts within which they find themselves – are going about implementing their visions, sometimes through building coalitions when their visions are similar, and sometimes by being in conflict when they are not. The geography of the world economy, then, is a social product, not an inevitability. As such, its form is shaped by struggle.

Questions for Reflection

- What factors have shaped labor transnationalism?
- How has labor transnationalism shaped the economic geography of the global economy?

Notes

1. The move towards the *jornada intensiva* is tied to US TNCs' spread and attitudes towards the "proper" workday. As one Spanish commentator has put it: "In a globalized world, we have to have schedules that are more similar to those in the rest of the world so we can be better connected" (quoted in McClean 2005).

2. Representatives from Argentina, Australia, Belgium, Denmark, France, Germany, Hungary, India, Italy, the Netherlands, Russia, Serbia, the UK, and the US attended the ICA's founding conference in London.

3. Although the first lasting national trade union center (the British Trades Union Congress) was founded in 1868, other national centers did not arise until closer to century's end – the Trades and Labour Congress of Canada dates to 1883, the AFL to 1886, Spain's Unión General de Trabajadores (General Union of Workers) to 1888, Germany's Generalkommission der Gewerkschaften Deutschlands (General Committee of German Trade Unions) to 1890, and the Dutch Nationaal Arbeids-Secretariaat (National Secretariat for Labour) and Nederlandsch Verbond van Vakvereenigingen (Dutch Association of Labour Unions) to 1893 and 1906 respectively. Equally, although some individual unions in countries like Britain and the US trace their origins to the mid-nineteenth century, many of these were fairly weakly developed as national – as opposed to local or regional – organizations until the 1880s or so.

4. AFL President Samuel Gompers felt "Secretariat" implied too centralized an organization, whereas "Federation" had a more decentralized ring to it.

5. The FTUI had its origins in the Free Trade Union Committee, which the AFL used to funnel monies – the vast majority supplied by the CIA (Carew 1998) – to anti-communist unions after World War II, especially in France, West Germany, Italy, Austria, and Greece (Radosh 1969: 304–47). Although it fell dormant at the Marshall Plan's conclusion, the Committee was resurrected in the late 1970s to counter socialist movements developing in Spain and Portugal after the end of right-wing military dictatorships there, as well as to combat similar tendencies in Western European labor movements more generally (Simms 1992).

6. Between 1962 and 1990 AIFLD trained 4 300 Latin American and Caribbean trade unionists in the US and 500 000 in-country (Herod 2001).

7. "Christian democracy" emerged as a political movement in the late nineteenth century, largely in response to Pope Leo XIII's encyclical. Originating in the Catholic Church, the movement spread to include Protestant and Orthodox denominations and resulted in various Christian democratic parties' formation, especially in Europe and Latin America. Typically, Christian democrats are conservative on so-called moral issues like abortion but liberal on matters of addressing poverty and social inequity, arguing for constraints on market forces (Mainwaring & Scully 2003).

8. The trade secretariats associated with the WFTU and WCL have also been engaged transnationally. However, the WCL's smaller size means its secretariats have been less extensive than those associated with the ICFTU, whilst the WFTU's international activities were often little more than rhetorical declarations of solidarity (for an account by a former WFTU employee, see Waterman n.d.).

9. In 2006 OPZZ affiliated with the ICFTU's successor organization.

10. At the same time two other Russian national centers – the All-Russian Confederation of Labour (VKT) and the Confederation of Labour of Russia (KTR) – affiliated.

11. The TUIs began organizing themselves into three groupings: public sector work; heavy industry and energy; and agriculture and light manufacturing.

12. As part of this endeavor, in 2000 both unions agreed to exchange officers to get a better understanding of work and social conditions in each other's country.

13. Several WCL affiliates, like the German Christian union Christlicher Gewerkschaftsbund Deutschlands, chose not to join the ITUC.

14. Rich is an international commodities trader with interests in metals and oil, amongst other things. At the time (early 1990s) his empire was worth $30 billion and operated in more than 40 countries.

15. Works councils have a long history in several European countries. They are perhaps most developed in Germany, where they play a central role in "codetermination" in which managers and workers – usually, but not necessarily, members of the plant union – jointly agree upon plant operations (Thelen 1991). Numerous companies had established them prior to the 1994 directive, though typically within an individual national context.

16. Although the EU directive gave impetus to WWCs, a small number of companies have operated them for several decades – for instance, the Swedish ballbearing manufacturer SKF initiated one in 1974 (Hammarström 2003).

17. The "organizing model" is one in which unions focus principally on organizing new members rather than servicing present ones.

18. Neither Johns nor I suggest this is an either/or situation: Johns developed a typology wherein degrees of accommodation and transformation can be evaluated, whereas I would argue that there are myriad examples of workers – often the same ones – adopting accommodationist strategies with regard to certain issues and transformatory ones with regard to others.

19. Rich was prevented from buying an aluminum smelter in Czechoslovakia, for instance (Herod 2001).

20. As Roberts (1964: 154) argues, the TUC's foreign policy, though often coincident with that of British firms and the government, similarly has been developed independently of them: "The reaction of the T.U.C. to the problems of trade union organization and industrial relations has inevitably been heavily influenced by British experience, beliefs and values. Since the labour policy of the Colonial Office has been influenced by the same factors, it is not surprising that the attitude of the T.U.C. has often coincided with that of the British Government. There is no evidence, however, to suggest that the T.U.C. has not arrived at its conclusions independently. When it has believed the British Government to be wrong, or local governments to be acting erroneously, it has attacked them vigorously, in private and public."

Further Reading

Drainville, A. C. (2004) *Contesting Globalization: Space and Place in the World Economy*. New York: Routledge.

Ghigliani, P. (2005) International trade unionism in a globalizing world: A case study of new labour internationalism. *Economic and Industrial Democracy* 26.3: 359–82.

Notes From Nowhere (eds.) (2003) *We are Everywhere: The Irresistible Rise of Global Anticapitalism*. New York: Verso.

Electronic resources

American Federation of Labor-Congress of Industrial Organizations (www.aflcio.org)
Global Union Federations (www.global-unions.org)
International Labour Organisation (www.ilo.org)
International Trade Union Confederation (www.ituc-csi.org)
Trades Union Congress (Britain) (www.tuc.org.uk)

Chapter 9

Conclusion

"Would you tell me, please, which way I ought to go from here?" [asked Alice] "That depends a good deal on where you want to get to," said the Cat. "I don't much care where —" said Alice. "Then it doesn't matter which way you go," said the Cat. "— so long as I get somewhere," Alice added as an explanation. "Oh, you're sure to do that," said the Cat, "if you only walk long enough."
(*Lewis Carroll,* Alice's Adventures in Wonderland)

Human history has been fundamentally shaped by the geographical processes whereby different parts of the globe have been tied together. Whilst this geographical tying together has been ongoing since humans first left East Africa millennia ago, it has been a historically and geographically uneven process – spatial linkages between different parts of the globe have been initiated and deepened more at certain times and between certain places than others. There is, then, a material basis to periodizing human existence by suggesting that particular eras experience more instances of what many have come to call "globalization" than do others – the laying of the transatlantic cable in the nineteenth century, for instance, represented a significant leap forward in global connectivity. However, given that historical and geographical processes are just that – processes – it is often difficult to say with any great precision when a particular process (e.g., globalization) definitively began. Thus, in the case of the transatlantic cable, whereas we can say that its operation in the 1860s marked a significant acceleration in the flow of information and money between Europe and the Americas, it is also important to remember that this itself would not have been possible without a significant number of other processes and inventions which have much longer lineages – experimentations with electricity and metalworking, a knowledge of navigation which allowed the ships carrying the cable to sail accurately between its two endpoints, the colonialism which brought from Southeast Asia to England the gutta-percha gum used to wrap the cable (and which for two centuries had also linked Britain with North America), and, of course, the invention of writing, amongst other things. Put another way, the origins of this singular event which played such an important role in linking the planet together can be read to stretch all the way back to when the first system of writing was established. If we so chose, then, we could plausibly argue that globalization's origins lie at least in the Bronze Age. However, the fact that globalization is typically not seen to have its origins that far back in human history suggests that material events like the laying of the cable are bracketed from the flow of time by the temporal bookends we place around these events for purposes of constructing a narrative about them: we focus on the act of cable-laying rather than those others which were necessary to bring it about, thereby discursively separating the two and allowing the cable's laying to be considered the starting point of a new era. What this shows, then, is that discourse is an important element in understanding the material processes linking parts of the planet together. Although not reducible to one another, material processes and discourse clearly influence each other in profound ways – our periodization of human experience is shaped by material processes, whilst such periodization molds how we conceive of material processes and practices.

So, what does this all have to do with reading critically neoliberal (and other) narratives concerning globalization? Partly it concerns what we actually mean by globalization, for what we mean by it shapes whether we see evidence of it. Thus, as outlined in Chapter 3, if we take any instance of greater transnational linkage to be globalization, then we see more globalization occurring than if we take a narrower view of what is globalization. But it also concerns how discourse is used to frame debates about material situations and thereby to shape such materiality. Thus, the neoliberal rhetoric concerning globalization is much like the Cheshire Cat's response to Alice: we will reach a globalized economy if we walk – or talk – far enough into the future. Hence, if globalization is presented as an inevitable process, then there will be little reason to challenge its material manifestations and globalization – or, rather, the "free trade" and "deregulation" seen as its leitmotif – will become a self-fulfilling prophecy: by being convinced of the futility of opposing free trade and deregulation because of globalization's inevitability, workers, environmentalists, traditionalists, economic nationalists, or anyone else opposed to the neoliberal agenda will not organize and so free trade and deregulation/globalization will unfold as if these had been inevitable all along.

Although material processes whereby places are linked together geographically have been ongoing historically, from Cloncavan Man's use of imported resins to Indian TNCs' FDI in the UK, the representation of such processes as having beginnings and endings – the idea, in other words, that human history has entered a new globalized era, has crossed over some planetary economic event horizon from which there is no escape, one which neatly bisects human history into a "pre-global" era which is now completed and a "global" era in which we shall henceforth exist – raises questions not so much about the material practices whereby the globe is being networked but, rather, about how this process is periodized. In particular, it appears that something of a Whiggish history (Butterfield 1931) is at work here, wherein the narrative suggesting we have recently moved from a pre-global era to an era of globalization tells us more about how particular interest groups divide the past into epochs for purposes of present-day political or analytical considerations than it does about the realities of the past.[1] Such epochal categorization is, in fact, a way of disciplining the past for contemporary purposes. Thus, Ohmae (2005) suggests that 1985 marked the crossing of the global economic event horizon, whereas Bryan and Farrell (1996) date the transition to the end of the Bretton Woods accord. Others have suggested dates like 1973, when OPEC quadrupled oil prices. All, however, are interpreting the past through the lens of the present. Certainly, this is a common – even unavoidable – practice, one not confined to neoliberals. But it does raise questions about how the events of the past have been marshaled to present a narrative which inevitably ends up in a globalized present and future, as if all the events of human history have been leading up to this inevitable outcome.

Constructing such a narrative relies upon particular readings of the past, readings which bear scrutiny. For instance, as we have seen, arguments that we are now in a post-Bretton Woods era of weak nation-states rest upon particular representations of nation-states' waxing and waning powers that themselves are open to question. Hence, a key element in the narrative about global financial capital "breaking free" from nation-states' powerful Bretton Woods grip was the US government's ability to enforce a regime in which other nations' currencies were pegged to gold via the dollar. Yet, despite the centrality of this narrative for neoliberals' arguments about how currency markets have "freed themselves" (Bryan & Farrell 1996: 21), for much of the post-1945 era the US government printed banknotes far in excess of the amount it "should" have were it following Bretton Woods's prescriptions, thereby undermining the dollar's statutorily stipulated value ($1 equaling 1/35th of an ounce of gold) (Table 9.1). This means, then, that a – if not *the* – key element of the Bretton Woods agreement never actually functioned as it was supposed to, a fact which raises profound questions about using the agreement's termination as the jumping-off point for a new era in planetary economic arrangements.

Likewise, whereas the fact that the "Bretton Woods era" is typically given distinctive starting and ending points (1944 and 1971) suggests the era was fairly uniform in

Table 9.1 US currency in circulation and gold stocks at various times during the "Bretton Woods era"

	Currency ($b)	Gold ($b)	Ratio
Oct. 31, 1944	24.4	20.1	1.2:1
Nov. 30, 1945	28.2	20.0	1.4:1
Nov. 30, 1953	30.8	22.0	1.4:1
Nov. 30, 1961	33.5	17.0	2.0:1
Oct. 31, 1970	55.0	11.1	5.0:1
Jan. 31, 1973	64.3	10.4	6.2:1

Sources: Oct. 1944 *Federal Reserve Bulletin* 30.12 (1944: 1203); Nov. 1945 *Federal Reserve Bulletin* 32.1 (1946: 48); Nov. 1953 *Federal Reserve Bulletin* 40.1 (1954: 45); Nov. 1961 *Federal Reserve Bulletin* 48.1 (1962: 50); Oct. 1970 *Federal Reserve Bulletin* 56.12 (1970: A16); Jan. 1973 *Federal Reserve Bulletin* 59. 2 (1973: A16).

its operation – that it was more internally similar in nature than were the times before and, particularly, after it – in actuality these temporal bookends are more porous than the appellation "Bretton Woods era" suggests. Thus, several of the period's hallmark institutions – the IMF, the World Bank, GATT – continue to function in the "post-Bretton Woods era," thereby transcending the chronological bracketing which gives coherence to the very claims that a post-Bretton Woods era has emerged, whilst within the "Bretton Woods era" many of its key planks (e.g., major currencies' convertibility into dollars) did not operate throughout the length of its existence – currency convertibility was only really achieved after 1958, for instance. Such porosity is analytically important, for the more permeable the historical dividing line between the "Bretton Woods" and the "post-Bretton Woods" epochs is seen to be, the more difficult is it to sustain the distinction between an earlier period of strong nation-states and the latter one of nation-states weakened by globalization. Certainly, this is not to say that the practices engaged in between, in this case, 1944 and 1971 were not very real, nor that there have not been real material changes in how the world is interconnected. But it *is* to say that how we interpret them to come up with a periodization of history such that we can say we have transitioned from a pre-global to a globalized state is very much tied to discursive categories by which history has been periodized. Whether the decoupling of the dollar from gold and the emergence of floating exchange rates is taken to be the emergence point of globalization, or whether some other sets of events is – perhaps Ohmae's (2005) four key events of 1985 (see p. 60 above) – the same argument can be made: periodizing history is to place a temporal dividing line within its flow, to artificially stop the passage of time so as to make distinctions between one era and another.

Putting all of this together, then, several issues arise. The first is that any evaluation of whether or not we are moving towards – or have already achieved – a globalized world depends entirely upon how we define globalization. Second, no matter how we define globalization, there are nevertheless important material processes which are indeed tying the planet together in new and different ways – in other words, regardless of definitional quibbles, new geographical linkages are being forged between some places as older ones are continued and/or broken. Third, such processes are playing out in a decidedly historically and geographically uneven manner, as data on FDI, trade, and migration clearly illustrate. This raises questions about just how global anything viewed as "globalization" actually is. Fourth, how history is periodized and disciplined will shape how we understand events and processes and whether we posit that, in fact, a new "global" era has emerged or that we are merely seeing the continuation of processes millennia in the making. Fifth, we should consider how particular terms become popular in different contexts, how they have a social and spatial life of their own (Williams 1976), and what this says both about the linguistic and ideological currents which drive their use and how the fact that they come to the surface at particular times in particular places shapes material and discursive life. Thus, whilst the term "globalization" was first used in English in the 1960s, it did not become widespread in public discourse until the 1990s. What, then, does this say about both the material changes occurring in the world economy and the discursive changes through which such material changes are represented?

Perhaps the most important issue to consider, though, is how where one stands shapes what one sees. Thus, it is no surprise, I think, that much of the rhetoric concerning globalization emerged in those places – particularly the manufacturing economies of the global North – which experienced significant deindustrialization and disruptions in the 1970s and 1980s in how their economies had historically been ordered and which saw the rise of right-wing governments enamored of "free markets." Given such countries' economic, political, military, educational, and cultural power – the latter manifested in control of global media through which visions concerning economic organization can be readily spread planetarily – it takes little imagination to ponder how ideas about neoliberalism and globalization could diffuse quickly and widely from their locations of origin.[2] We should, though, be wary of taking the part for the whole, for as the data in the chapters above show, processes of integration as measured by, for example, FDI patterns or international migration are decidedly not global in geographical spread – just because processes that we take to be evidence of globalization may be affecting people in the US or Europe does not mean that they are necessarily affecting the rest of humanity. Thus, for instance, as of the early twenty-first century the United States was the only major economy to have a higher trade to GDP ratio than in 1910 and is the only country with a significant level of immigration, mostly from LDCs (Wolf 2001: 180). It seems a little too coincidental, then, that much of the globalization talk has also originated in the US. Although representations of globalization wherein commodities, information, capital, and people are viewed as moving across the surface of the Earth ever more quickly and easily may speak to millions of us in the global North who sit in front of our computers or televisions to learn instantaneously of events on the other side of the planet whilst eating our imported foods and wearing our imported clothes, they do not really speak much to the life experiences of hundreds of millions of peasants in rural Africa who are still struggling to get by on less than a dollar a day. Indeed, for such peasants the "globalized" world economy does not look that much different from its pre-global ancestor.

Having reached the end of this book, some readers, no doubt, will be frustrated that its conclusion is not more declarative, that I do not state either that "Yes, globalization is occurring and we are entering/have entered a globalized world economy," or "No, it is not occurring." However, this is not – and never was – my goal. Rather, my goal has been to encourage thinking about how the world economy is structured spatially, how the linkages between its constituent parts are unevenly made in time and space, and how different actors shape their making. It has also been to explore how the ways in which we define "globalization" affect what we perceive to be going on, and how, through deconstructing critically the neoliberal rhetoric around contemporary economic processes, we can see how such discourse is used to pursue a particular agenda – continuing the extraction of surplus value which is at the heart of the capitalist mode of production. Such pursuits might seem of little consequence, an intellectual luxury. But they are deadly serious, for as we have seen with the spread of neoliberal thinking and economic orthodoxy around the world, how knowledge is produced and diffused – and what types of knowledge are produced and diffused – has signal impacts upon material processes of economic restructuring. Ideas, rhetorics, and material practices all have real consequences for real people. A critical analysis of the geographies of the contemporary world economy and how these geographies are made and represented is thus central to any liberatory project aimed at promoting global justice. Through such activities there is, quite literally, a world to be won.

Notes

1. "Whiggish history" is a teleological view that interprets past events from the perspective of the present, imposing an historical order on them so that each event is seen as part of an inevitable transition from the past to the present.
2. The significance of the educational system for transmitting ideas from the global North to the rest of the planet is illustrated by José Piñera (n.d.), one of the architects of Chilean neoliberalism. Piñera received graduate degrees in economics at Harvard, having received his undergraduate degree in economics from the Universidad Católica de Chile in Santiago in 1970, which, he avers, "from an intellectual point of view, was a 'wholly owned subsidiary' of The University of Chicago." Indeed, Piñera notes that due to an exchange between the Schools of Economics at Chicago and the Catholic University, "hundreds of students such as myself . . . learn[t] rigorous economics and discover[ed] public policy ideas based on individual freedom and private enterprise . . . Soon there was a critical mass of free market economists, with a common diagnosis of [Chile's] economic problems and similar views on the needed solutions. Since ideas have consequences, this group began to influence the public debate and began to be referred to as the 'Chicago Boys'."

References

Abu-Lughod, J. L. (1989) *Before European Hegemony: The World System AD 1250–1350.* Oxford: Oxford University Press.

Adamantopoulos, K. (1997) *An Anatomy of the World Trade Organization.* London: Kluwer Law International.

Adelstein, R. P. (1991) "The nation as an economic unit": Keynes, Roosevelt, and the managerial ideal. *The Journal of American History* 78.1: 160–87.

Agger, B. (1989) *Fast Capitalism.* Urbana: University of Illinois Press.

Aguiar, L. M. and Herod, A. (eds.) (2006) *The Dirty Work of Neoliberalism: Cleaners in the Global Economy.* Oxford: Blackwell.

Agyeman, O. (2003) *The Failure of Grassroots Pan-Africanism: The Case of the All-African Trade Union Federation.* Lanham, MD: Lexington.

AIFLD (American Institute for Free Labor Development) (1964a) Country Plan for Mexico. In: AIFLD (ed.), *Country Plans for Latin America.* Washington, DC: AIFLD, Social Projects Department.

AIFLD (American Institute for Free Labor Development) (1964b) Country Plan for Chile. In: AIFLD (ed.), *Country Plans for Latin America.* Washington, DC: AIFLD, Social Projects Department.

Aliber, R. Z. and Click, R. W. (1993) *Readings in International Business: A Decision Approach.* Cambridge, MA: MIT Press.

Altvater, E. (1993) *The Future of the Market: An Essay on the Regulation of Money and Nature after the Collapse of "Actually Existing Socialism."* London: Verso.

Amariglio, J. (1988) The body, economic discourse, and power: An economist's introduction to Foucault. *History of Political Economy* 20.4: 583–613.

Andersen, H. C. (1913) *Creation of a World Centre of Communication.* Paris: Philippe Renouard.

Anderson, S. and Cavanagh, J. (2000) *Top 200: The Rise of Corporate Global Power.* Report for Institute for Policy Studies, Washington, DC.

de Andrade, M. C. (1989) *Historía das usinas de açúcar de Pernambuco.* Recife: Fundaçao Joaquim Nabuco.

Anner, M. (2003) Industrial structure, the state, and ideology: Shaping labor transnationalism in the Brazilian auto industry. *Social Science History* 27.4: 603–34.

Anon. (1994) The world car: Enter the McFord. *The Economist*, July 23, p. 69 (US printed edition).

Asimov, I. (1970) The fourth revolution. *Saturday Review*, October 24, pp. 17–20.

Aston, T. and Philpin, C. (eds.) (1985) *The Brenner Debate: Agrarian Class Structure and Economic Development in Pre-Industrial Europe*. Cambridge: Cambridge University Press.

Australian Firefighter, The (2007) One world. Issue 44.2, pp. 14–16.

Axtell, R. E. (1995) *Do's and Taboos of Using English Around the World*. New York: Wiley.

Backer, B. (1993) *The Care and Feeding of Ideas*. New York: Times Books.

Bagchi, A. K. (1982) *The Political Economy of Underdevelopment*. Cambridge: Cambridge University Press.

Bairoch, P. (1974) Geographical structure and trade balance of European foreign trade from 1800 to 1970. *The Journal of European Economic History* 3.3: 557–608.

Bairoch, P. (1982) International industrialization levels from 1750 to 1980. *The Journal of European Economic History* 11.2: 269–333.

Bairoch, P. (1993) *Economics and World History: Myths and Paradoxes*. Chicago: University of Chicago Press.

Bank for International Settlements (BIS) (2005) *Triennial Central Bank Survey: Foreign Exchange and Derivatives Market Activity in 2004*. Basel: BIS.

Bank of England (2007) A brief history of banknotes. Available at: www.bankofengland.co.uk/banknotes/about/history.htm (accessed July 5, 2007).

Barber, B. R. (2001) *Jihad vs. McWorld: How Globalism and Tribalism are Reshaping the World*. New York: Ballantine Books.

Barrell, R. and Pain, N. (1997) Foreign direct investment, technological change, and economic growth within Europe. *The Economic Journal* 107.445: 1770–86.

Barrett, W. (1990) World bullion flows, 1450–1800. In: J. D. Tracy (ed.), *The Rise of Merchant Empires: Long-Distance Trade in the Early Modern World, 1350–1750*. Cambridge: Cambridge University Press, pp. 224–54.

Barry, T. and Preusch, D. (1990) *AIFLD in Central America: Agents as Organizers*. Albuquerque: Inter-Hemispheric Education Resource Center.

Bartel, R. J. (1974) International monetary unions: The XIXth century experience. *Journal of European Economic History* 3.3: 689–704.

Bartelson, J. (2000) Three concepts of globalization. *International Sociology* 15.2: 180–96.

Bassett, T. J. (1994) Cartography and empire building in nineteenth-century West Africa. *Geographical Review* 84.3: 316–35.

Battilossi, S. (2006) The determinants of multinational banking during the first globalization, 1880–1914. *European Review of Economic History* 10: 361–88.

Bédarida, F. (1974) Perspectives sur le mouvement ouvrier et l'impérialism en France au temps de la conquête coloniale. *Le Mouvement Social* 86: 25–42.

Beling, W. A. (1964) W.F.T.U. and decolonisation: A Tunisian case study. *The Journal of Modern African Studies* 2.4: 551–64.

Bendiner, B. (1987) *International Labour Affairs: The World Trade Unions and the Multinational Companies*. Oxford: Clarendon Press.

Benyus, J. M. (1997) *Biomimicry: Innovation Inspired by Nature*. New York: W. W. Morrow.

Berger, S. (2003) *Notre première mondialisation – Leçons d'un échec oublié*. Paris: Seuil.

Berkis, A. V. (1969) *The History of the Duchy of Courland (1561–1795)*. Towson, MD: Paul M. Harrod.

Bethell, L. (1970) George Canning and the independence of Latin America. Lecture delivered at Canning House by Professor Leslie Bethell to mark the Bicentenary of George Canning, April 15, 1970.

Betts, R. F. (1960) *Assimilation and Association in French Colonial Theory, 1890–1914*. New York: Columbia University Press.

Bissell, D. (1999) *The First Conglomerate: 145 Years of the Singer Sewing Machine Company*. Brunswick, ME: Audenreed Press.

Blackwell, R. D., Ajami, R., and Stephan, K. (1994) Winning the global advertising race: Planning globally, acting locally. In: E. Kaynak and S. S. Hassan (eds.), *Globalization of Consumer Markets: Structures and Strategies*. Binghamton, NY: Haworth Press, pp. 209–32.

Blaise, C. (2000) *Time Lord: Sir Sandford Fleming and the Creation of Standard Time*. New York: Pantheon Books.

Blaut, J. M. (1992) *1492: The Debate on Colonialism, Eurocentrism, and History*. Trenton, NJ: Africa World Press.

Blaut, J. M. (1994) Robert Brenner in the tunnel of time. *Antipode* 26.4: 351–74.

Bordo, M. D. and Rockoff, H. (1996) The gold standard as a "Good Housekeeping Seal of Approval." *Journal of Economic History* 56.2: 389–428.

Bornschier, V. (2000) Western Europe's move toward political union. In: V. Bornschier (ed.), *State-building in Europe: The Revitalization of Western European Integration*. Cambridge: Cambridge University Press, pp. 3–37.

Brayer, H. O. (1949) The influence of British capital on the Western range-cattle industry. *Journal of Economic History* 9 (Supplement): 85–98.

Brenner, R. (1977) The origins of capitalist development: A critique of neo-Smithian Marxism. *New Left Review* 104: 25–92.

Briones, I. and Villela, A. (2006) European bank penetration during the first wave of globalisation: Lessons from Brazil and Chile, 1878–1913. *European Review of Economic History* 10: 329–59.

Broadberry, S., Földvári, P., and van Leeuwen, B. (2006) British economic growth and the business cycle, 1700–1850: Annual estimates. Unpublished paper, posted at: www2.warwick.ac.uk/fac/soc/economics/staff/faculty/broadberry/wp/annualgdp3a.pdf (accessed October 29, 2007).

Bronson, B. (1982) Medieval India: An industrial miracle in a golden age: The 17th-century cloth exports of India. Informational item written for the exhibition *Master Dyers to the World: Early Fabrics from India*, Field Museum, Chicago, January 29, 1982. Copy posted at: http://iref.homestead.com/Textile.html (accessed October 29, 2007).

Brooks, G. E. (1975) Peanuts and colonialism: Consequences of the commercialization of peanuts in West Africa, 1830–70. *The Journal of African History* 16.1: 29–54.

Bryan, L. and Farrell, D. (1996) *Market Unbound: Unleashing Global Capitalism*. New York: Wiley.

Buchanan, P. J. (1998) Patriotism in the boardroom. Available at: www.buchanan.org/pa-98-0630.html (accessed October 29, 2007).

Buck-Morss, S. (1995) Envisioning capital: Political economy on display. *Critical Inquiry* 21.2: 434–67.

Burawoy, M. (1985) *The Politics of Production: Factory Regimes under Capitalism and Socialism*. London: Verso.

Burawoy, M. (1992) The end of Sovietology and the renaissance of modernization theory. *Contemporary Sociology* 21.6: 744–85.

Burgmann, V. (1995) *Revolutionary Industrial Unionism: The Industrial Workers of the World in Australia*. Melbourne: Cambridge University Press.

Burgmann, V. (2003) *Power, Profit and Protest: Australian Social Movements and Globalisation*. Crows Nest, NSW: Allen & Unwin.

Burns, R. W. (1993) Alexander Bain, a most ingenious and meritorious inventor. *Engineering Science and Education Journal* (April).

Busch, G. K. (1983) *The Political Role of International Trades Unions.* New York: St. Martin's Press.

Bush, G. W. (2005a) Remarks by the President upon signing The Dominican Republic–Central American Free Trade Agreement Implementation Act. The White House Office of the Press Secretary, August 2, 2005.

Bush, G. W. (2005b) Speech made July 21 by President George W. Bush to the Organization of American States, Washington, DC.

Butler, J. (1990) *Gender Trouble: Feminism and the Subversion of Identity.* New York: Routledge.

Butterfield, H. (1931) *The Whig Interpretation of History.* London: G. Bell and Sons.

Buzzell, R. D. (1968) Can you standardize multinational marketing? *Harvard Business Review* 46 (November/December): 102–13.

Cairncross, F. (2001) *The Death of Distance: How the Communications Revolution is Changing Our Lives.* Boston, MA: Harvard Business School Press.

Calleo, D. P. and Rowland, B. M. (1973) *America and the World Political Economy: Atlantic Dreams and National Realities.* Bloomington: Indiana University Press.

Cameron, J. and Gibson-Graham, J. K. (2003) Feminising the economy: Metaphors, strategies, politics. *Gender, Place and Culture* 10.2: 145–57.

Carew, A. (1998) The American labor movement in Fizzland: the Free Trade Union Committee and the CIA. *Labor History* 39.1: 25–42.

Carew, A., Dreyfus, M., van Goethem, G., Gumbrell-McCormick, R., and van der Linden, M. (eds.) (2000) *The International Confederation of Free Trade Unions.* Bern: Peter Lang.

Carley, M. (2005) Global agreements – State of play. *European Industrial Relations Review* 381: 14–18.

Carlos, A. M. and Nicholas, S. (1988) "Giants of an earlier capitalism": The chartered trading companies as modern multinationals. *The Business History Review* 62.3: 398–419.

Carpenter, E. and McLuhan, M. (eds.) (1960) *Explorations in Communication: An Anthology.* Boston, MA: Beacon Press.

Carson, C. S. (1975) The history of the United States national income and product accounts: The development of an analytical tool. *Review of Income and Wealth* 21.2: 153–81.

Carstensen, F. V. (1984) *American Enterprise in Foreign Markets: Studies of Singer and International Harvester in Imperial Russia.* Chapel Hill: University of North Carolina Press.

Carter, S. B., Gartner, S. S., Haines, M. R., Olmstead, A. L., Sutch, R., and Wright, G. (eds.) (2006) *Historical Statistics of the United States, Millennial Edition.* Cambridge: Cambridge University Press.

Casserini, K. (1993) *International Metalworkers' Federation, 1893–1993: The First Hundred Years.* Geneva: International Metalworkers' Federation.

Cassing, J. and Husted, S. (2004) Trade pattern persistence. In: M. G. Plummer (ed.), *Empirical Methods in International Trade: Essays in Honor of Mordechai Kreinin.* Cheltenham, UK: Edward Elgar.

Castellino, J. (1999) Territoriality and identity in international law: The struggle for self-determination in the Western Sahara. *Millennium: Journal of International Studies* 28.3: 523–52.

Castree, N. (2000) Geographical scale and grass-roots internationalism: The Liverpool dock dispute, 1995–1998. *Economic Geography* 76.3: 272–92.

Castree, N. (2004) Economy and culture are dead! Long live economy and culture! *Progress in Human Geography* 28.2: 204–26.

Castree, N., Featherstone, D., and Herod, A. (2007) Contrapuntal geographies: The politics of organising across socio-spatial difference. In: K. Cox, M. Low, and J. Robinson (eds.), *Handbook of Political Geography*. London: Sage, pp. 305–21.

Caulfield, N. (1995) Wobblies and Mexican workers in mining and petroleum, 1905–1924. *International Review of Social History* 40: 51–76.

de Cecco, M. (1984) *The International Gold Standard*. New York: St. Martin's Press.

Cell, G. T. (ed.) (1982) *Newfoundland Discovered: English Attempts at Colonisation, 1610–1630*. London: Hakluyt Society.

Chami, F. A. (1999) Roman beads from the Rufiji Delta, Tanzania: First incontrovertible archaeological link with the *Periplus*. *Current Anthropology* 40.2: 237–41.

Chapman, S. D. (1992) *Merchant Enterprise in Britain from the Industrial Revolution to World War I*. Cambridge: Cambridge University Press.

Charlesworth, N. (1982) *British Rule and the Indian Economy, 1800–1914*. London: Macmillan.

Chaudhuri, K. N. (1966) India's foreign trade and the cessation of the East India Company's trading activities, 1828–40. *The Economic History Review, new series*, 19.2: 345–63.

Chaudhuri, K. N. (1978) *The Trading World of Asia and the English East India Company, 1660–1760*. Cambridge: Cambridge University Press.

Chomko, S. A. and Crawford, G. W. (1978) Plant husbandry in prehistoric Eastern North America: New evidence for its development. *American Antiquity* 43.3: 405–8.

Chomsky, N. (1993) *Year 501: The Conquest Continues*. Boston, MA: South End Press.

CIA (2006) *The World Factbook*. Washington, DC: Central Intelligence Agency.

Clark, W. (2005) *Petrodollar Warfare: Oil, Iraq and the Future of the Dollar*. Gabriola Island, British Columbia: New Society Publishers.

Cliff, T. (1988) *State Capitalism in Russia*. London: Bookmarks. (Originally published in 1955 as *Stalinist Russia: A Marxist Analysis*.)

Clinton, B. (1996) *Between Hope and History: Meeting America's Challenges for the 21st Century*. New York: Random House.

Coca-Cola (2006) Email correspondence between author and Department of Industry and Consumer Affairs, The Coca-Cola Company, Atlanta, Georgia. September 27, 2006.

Cohen, B. J. (1998) *The Geography of Money*. Ithaca, NY: Cornell University Press.

Conference Board, The (1975) *The Multinational Union Challenges the Multinational Company*. New York: The Conference Board.

Congressional Budget Office (1983) *Strategic and Critical Nonfuel Minerals: Problems and Policy Alternatives*. Washington, DC: US Congress.

Conrad, J. (1947/1907) *The Secret Agent: A Simple Tale*. London: J. M. Dent and Sons (1965 printing).

COSATU Daily News (July 27, 2007). NAMA talks time bomb.

Cosgrove, D. (1994) Contested global visions: *One-World, Whole-Earth*, and the Apollo space photographs. *Annals of the Association of American Geographers* 84.2: 270–94.

Cosgrove, D. (2001) *Apollo's Eye: A Cartographic Genealogy of the Earth in the Western Imagination*. Baltimore, MD: Johns Hopkins University Press.

Cosgrove, D. (2003) Globalism and tolerance in early modern geography. *Annals of the Association of American Geographers* 93.4: 852–70.

Cox, R. W. (1971) Labor and transnational relations. *International Organization* 25.3: 554–84.

Crang, P. (1997) Cultural turns and the (re)constitution of economic geography. In: R. Lee and J. Wills (eds.), *Geographies of Economies*. London: Edward Arnold, pp. 3–15.

Craven, W. F. (1957) *The Virginia Company of London, 1606–1624*. Charlottesville: University Press of Virginia.

Crick, W. F. (1948) *Origin and Development of the Sterling Area*. London: University of London and the Institute of Bankers.

Crosby, M. (2005) Email correspondence with SEIU representative Michael Crosby, Sydney, Australia, December 9 and 19.

Curzon, G. N. (1909) *The Place of India in the Empire, Being an Address Delivered Before the Philosophical Institute of Edinburgh by Lord Curzon of Kedleston on October 19, 1909*. London: John Murray.

Da Costa, I. and Rehfeldt, U. (2006) European unions and American automobile firms: From European works councils to world councils? Proceedings, 58th Annual Meeting of the Labor and Employment Relations Association, Boston, January 5–8.

Daily Times (Lahore, Pakistan) (October 21, 2006) Tata seeks global footprint with Corus takeover, p. 5.

Dampier, W. (1697) *A New Voyage Round the World*. Printed for James Knapton, at the *Crown* in St. Paul's Church-yard, London.

Darling, A. (2006) Speech to the First Annual UK/India Investment Summit, Lancaster House, London, October 10 by the Rt. Hon. Alistair Darling MP, Secretary of State for Trade and Industry.

Das, D. (1969) *India: From Curzon to Nehru and After*. London: Collins.

Davey, W. J. (2005) Institutional framework. In: P. F. J. Macrory, A. E. Appleton, and M. G. Plummer (eds.), *The World Trade Organization: Legal, Economic and Political Analysis* (3 volumes). New York: Springer, pp. 53–87.

Davies, R. B. (1976) *Peacefully Working to Conquer the World: Singer Sewing Machines in Foreign Markets, 1854–1920*. New York: Arno Press.

Davis, C. T. (1965) The crucifixion of Jesus Christ: The Passion of Christ from a medical point of view. *Arizona Medicine* 22.3: 183–7.

Davis, M. (2001) *Late Victorian Holocausts: El Niño Famines and the Making of the Third World*. New York and London: Verso.

Denis, C., McMorrow, K., and Röger, W. (2006) Globalisation: Trends, issues and macro implications for the EU. No. 254, Economic Papers of the European Commission Directorate-General for Economic and Financial Affairs, Brussels.

Devreese, D. E. (1988) An inquiry into the causes and nature of organization: Some observations on the International Working Men's Association, 1864–1872/1876. In: F. van Holthoon and M. van der Linden (eds.), *Internationalism in the Labour Movement, 1830–1940, Volume 1*. London: E. J. Brill, pp. 283–303.

Dicken, P. (1988) The changing geography of Japanese foreign direct investment in manufacturing industry: A global perspective. *Environment and Planning A* 20.5: 633–53.

Dicken, P. (2003) *Global Shift: Reshaping the Global Economic Map in the 21st Century*. London: Sage; New York: Guilford. 4th edition.

Dicken, P., Peck, J., and Tickell, A. (1997) Unpacking the global. In: R. Lee and J. Wills (eds.), *Geographies of Economies*. London: Edward Arnold, pp. 158–66.

Diebold, J. (1973) Multinational corporations . . . Why be scared of them? *Foreign Policy* 12: 79–95.

Dirlik, A. (2001) Place-based imagination: Globalism and the politics of place. In: R. Prazniak and A. Dirlik (eds.), *Places and Politics in an Age of Globalization*. Lanham, MD: Rowman & Littlefield, pp. 15–51.

Dodge, D. (2005) Monetary policy and the exchange rate in Canada. Remarks by David Dodge, Governor of the Bank of Canada to the Canada China Business Council, Beijing, China, June 2, 2005.

Domenich, B. (1998) Top U.S. corporations reject reciting the Pledge of Allegiance. *Human Events* (July 17 issue).

Dreyfus, M. (2000) The emergence of an international trade union organization (1902–1919). In: A. Carew, M. Dreyfus, G. Van Goethem, R. Gumbrell-McCormick, and M. van der Linden (eds.), *The International Confederation of Free Trade Unions*. Bern: Peter Lang, pp. 25–71.

Dunwoodie, P. (1998) *Writing French Algeria*. Oxford: Clarendon Press.

Economist, The (1986) On the hamburger standard. September 6, p. 77 (US printed edition).

Economist, The (1999) Review of *Natural Capitalism: Creating the Next Industrial Revolution* by P. Hawken, A. Lovins, and L. H. Lovins. November 13, p. 8.

Edgerton, S. Y. (1975) *The Renaissance Rediscovery of Linear Perspective*. New York: Basic Books.

Eichengreen, B. (1996) *Globalizing Capital: A History of the International Monetary System*. Princeton, NJ: Princeton University Press.

Eichengreen, B. (2000) From benign neglect to malignant preoccupation: U.S. balance of payments policy in the 1960s. In: G. L. Perry and J. Tobin (eds.), *Economic Events, Ideas, and Policies: The 1960s and After*. Washington, DC: Brookings Institution, pp. 185–239.

Einaudi, L. L. (2000) From the franc to the "Europe": The attempted transformation of the Latin Monetary Union into a European Monetary Union, 1865–1873. *Economic History Review* 53.2: 284–308.

Einaudi, L. L. (2001) *Money and Politics: European Monetary Unification and the International Gold Standard (1865–1873)*. Oxford: Oxford University Press.

Encyclopedia of Associations International Organizations (1989) 23rd edition, Part 2 Indexes. Detroit: Gale Research Inc.

Encyclopedia of Associations International Organizations (2003) 39th edition, Part 3 Indexes. Detroit: Thomson Gale.

European Commission (2007) *United States Barriers to Trade and Investment Report for 2006*. Brussels: European Commission.

Evans, J. W. (1971) *The Kennedy Round in American Trade Policy: The Twilight of the GATT?* Cambridge, MA: Harvard University Press.

Facey, T. G., Cremer, W., Goddard, C., Eglington, J., and Odgers, G. (1864) To the workmen of France from the working-men of England. Reproduced in W. H. Fraser (ed.) (2007), *British Trade Unions 1707–1918*. London: Pickering & Chatto, pp. 403–4.

Farnie, D. A. (1979) *The English Cotton Industry and the World Market, 1815–1896*. Oxford: Clarendon Press.

Farnie, D. A. (2004) The role of cotton textiles in the economic development of India, 1600–1990. In: D. A. Farnie and D. J. Jeremy (eds.), *The Fibre That Changed the World: The Cotton Industry in International Perspective, 1600–1990s*. Oxford: Oxford University Press, pp. 395–430.

Fatt, A. C. (1967) The danger of "local" international advertising. *Journal of Marketing* 31 (January): 60–2.

Fawcett, C. B. (1938) The question of colonies. *Geographical Review* 28.2: 306–9.

Feige, E. L., Faulend, M., Šonje, V., and Šošić, V. (2003) Unofficial dollarization in Latin America: Currency substitution, network externalities, and irreversibility. In: D. Salvatore, J. W. Dean, and T. D. Willett (eds.), *The Dollarization Debate*. Oxford: Oxford University Press, pp. 46–71.

Feis, H. (1930) *Europe, the World's Banker: 1870–1914*. New Haven, CT: Yale University Press.

Ferguson, N. (2003) *Empire: The Rise and Demise of the British World Order and the Lessons for Global Power.* London: Basic Books.

Ferguson, N. and Schularick, M. (2006) The empire effect: The determinants of country risk in the first age of globalization, 1880–1913. *Journal of Economic History* 66.2: 283–312.

Ferguson, R. W. (2005) Globalisation – evidence and policy implications. Speech by Roger W. Ferguson, Jr., Vice-Chairman of the Board of Governors of the US Federal Reserve System, to the Association for Financial Professionals Global Corporate Treasurers Forum, San Francisco, California, May 12.

Ferns, H. A. (1953) Britain's informal empire in Argentina, 1806–1914. *Past and Present* 4: 60–75.

Fieldhouse, D. K. (1978) *Unilever Overseas: The Anatomy of a Multinational, 1895–1965.* London: Croom Helm.

Filkins, D. (2000) India's office pool. *Los Angeles Times*, April 6, p. A1.

Fimmen, E. (1922) *The International Federation of Trade Unions: Development and Aims.* Amsterdam: International Federation of Trade Unions.

Findlay, R. and O'Rourke, K. H. (2003) Commodity market integration, 1500–2000. In: M. D. Bordo, A. M. Taylor, and J. G. Williamson (eds.), *Globalization in Historical Perspective.* Chicago: University of Chicago Press, pp. 13–64.

Finkelstein, A. (2000) *Harmony and the Balance: An Intellectual History of Seventeenth-Century English Economic Thought.* Ann Arbor: University of Michigan Press.

Fleissner, P. (ed.) (1994) *The Transformation of Slovakia: The Dynamics of Her Economy, Environment, and Demography.* Hamburg: Verlag Dr. Kovač.

Flux, A. W. (1899) The flag and trade: A summary review of the trade of the chief colonial empires. *Journal of the Royal Statistical Society* 62.3: 489–533.

Foner, P. S. (1975) *History of the Labor Movement in the United States,* Vol. 2: *From the Founding of the A.F. of L. to the Emergence of American Imperialism.* New York: International Publishers.

Forbes (2006) *The Forbes Global 2000.* Available at: www.forbes.com/2005/03/30/05f2000land.html (accessed August 10, 2006).

Foreman-Peck, J. (1983) *A History of the World Economy: International Economic Relations Since 1850.* Totowa, NJ: Barnes and Noble.

Foucault, M. (1971) *The Order of Things: An Archaeology of the Human Sciences.* New York: Pantheon.

Foucault, M. (1986) Of other spaces. *Diacritics* 16.1: 22–7.

Frank, T. (2000) *One Market under God: Extreme Capitalism, Market Populism, and the End of Economic Democracy.* New York: Anchor Books.

Franko, L. G. (1975) Patterns in the multinational spread of continental European enterprise. *Journal of International Business Studies* 6.2: 41–53.

Freeman, C. (2001) Is local : global as feminine : masculine? Rethinking the gender of globalization. *Signs: Journal of Women in Culture and Society* 26.4: 1007–37.

Friedman, M. (1968) The role of monetary policy. *The American Economic Review* 58.1: 1–17.

Friedman, T. L. (2000) *The Lexus and the Olive Tree.* New York: Anchor Books.

Friedman, T. L. (2005) *The World is Flat: A Brief History of the Twenty-First Century.* New York: Farrar, Straus & Giroux.

FTUI (Free Trade Union Institute) (1994) *Annual Report of the Free Trade Union Institute.* Washington, DC: AFL-CIO.

Fukazawa, H. (1965) Cotton mill industry. In: V. B. Singh (ed.), *Economic History of India: 1857–1956.* Bombay: Allied Publishers, pp. 223–59.

Fukuyama, F. (1992) *The End of History and the Last Man*. New York: Free Press.

Fuller, T. (2001) Divisiveness doomed previous continental monetary links: The Euro renews a dream of unity. *International Herald Tribune*, December 29, pp. 1 and 4.

Galey, J. (1979) Industrialist in the wilderness: Henry Ford's Amazon venture. *Journal of Interamerican Studies and World Affairs* 21.2: 261–89.

Galison, P. (2003) *Einstein's Clocks, Poincaré's Maps: Empires of Time*. New York: W. W. Norton.

Gallaher, C. (2002) *On the Fault Line: Race, Class, and the American Patriot Movement*. Lanham, MD. Rowman & Littlefield.

Gallin, D. (2006) Transnational pioneers: The international labor movement. In: S. Batliwala and L. D. Brown (eds.), *Transnational Civil Society: An Introduction*. Bloomfield, CT: Kumarian Press, pp. 84–100.

Galloway, J. H. (1968) The sugar industry of Pernambuco during the nineteenth century. *Annals of the Association of American Geographers* 58.2: 285–303.

GAO (General Accounting Office) (1979) The US–Saudi Arabian Joint Commission on Economic Cooperation, March 22, 1979. Report ID-79-7. Washington, DC: General Accounting Office.

Gates, B., with Myhrvold, N. and Rinearson, P. (1995) *The Road Ahead*. New York: Viking.

Gates, D. (2004) The pop prophets. *Newsweek* May 24, pp. 44–50.

Gereffi, G. (1994) The organization of buyer-driven global commodity chains: How U.S. retailers shape overseas production networks. In: G. Gereffi and M. Korzeniewicz (eds.), *Commodity Chains and Global Capitalism*. Westport, CT: Greenwood, pp. 95–122.

Ghose, A. K. (1982) Food supply and starvation: A study of famines with reference to the Indian sub-continent. *Oxford Economic Papers* 34.2: 368–89.

Gibson-Graham, J. K. (1996) *The End of Capitalism (As We Knew it): A Feminist Critique of Political Economy*. Oxford: Blackwell.

Gibson-Graham, J. K. (2002) Beyond global vs. local: Economic politics outside the binary frame. In: A. Herod and M. W. Wright (eds.), *Geographies of Power: Placing Scale*. Oxford: Blackwell, pp. 25–60.

Giddens, A. (1985) *The Nation-State and Violence*. Cambridge: Polity.

Gingrich, N. (1995) *To Renew America*. New York: HarperCollins.

Gleick, J. (1999) *Faster: The Acceleration of Just about Everything*. New York: Pantheon.

Goddard, G. W. (1969) *Overview: A Life-Long Adventure in Aerial Photography*. Garden City, NY: Doubleday.

Godley, A. (2006) Selling the sewing machine around the world: Singer's international marketing strategies, 1850–1920. *Enterprise & Society* 7.2: 266–314.

van Goethem, G. (2006) *The Amsterdam International: The World of the International Federation of Trade Unions (IFTU), 1913–1945*. Aldershot, UK: Ashgate.

de Goey, F. (1999) Dutch overseas investments in the very long run (c.1600–1990). In: R. van Hoesel and R. Narula (eds.), *Multinational Enterprises from the Netherlands*. London: Routledge, pp. 32–60.

Goody, J. (1996) *The East in the West*. Cambridge: Cambridge University Press.

Gordon, L. A. and Klopov, E. V. (1992) The workers' movement in a postsocialist perspective. In: B. Silverman, R. Vogt, and M. Yanowitch (eds.), *Labor and Democracy in the Transition to a Market System*. Armonk, NY: M. E. Sharpe, pp. 27–52.

Goudzwaard, B. (2001) *Globalization and the Kingdom of God*. Grand Rapids, MI: Baker Books.

Graham, E. M. and Marchick, D. M. (2006) *US National Security and Foreign Direct Investment*. Washington, DC: Institute for International Economics.

Gramsci, A. (1971) *Selections from the Prison Notebooks of Antonio Gramsci*. London: Lawrence & Wishart.

Gray, J. (1998) *False Dawn: The Delusions of Global Capitalism*. New York: New Press.

Greaves, I. (1954) The character of British colonial trade. *Journal of Political Economy* 62.1: 1–11.

Greider, W. (1997) *One World, Ready or Not: The Manic Logic of Global Capitalism*. New York: Simon & Schuster.

Griswold, D. T. (2003) Free trade tills the soil of democracy. July 30. Posted at www.cato.org/research/articles/griswold-030730.html (accessed September 26, 2006).

Grubel, H. G. (1999) The case for the Amero: The economics and politics of a North American monetary union. Published as a "Critical Issues Bulletin" by the Fraser Institute, Vancouver, British Columbia, Canada.

Guardian, The (December 6, 2006) Poles are bringing solidarity back into fashion in Britain, p. 32.

Gungwu, W. (1990) Merchants without empire: The Hokkien sojourning communities. In: J. D. Tracy (ed.), *The Rise of Merchant Empires: Long-Distance Trade in the Early Modern World, 1350–1750*. Cambridge: Cambridge University Press, pp. 400–21.

Gupta, P. A. (1975) *Imperialism and the British Labour Movement, 1914–1964*. London: Macmillan.

Gurney, P. (1988) "A higher state of civilisation and happiness": Internationalism in the British co-operative movement between c.1869–1918. In: F. van Holthoon and M. van der Linden (eds.), *Internationalism in the Labour Movement 1830–1940, Volume 2*. London: E. J. Brill, pp. 543–64.

Hagen, J. (2003) Redrawing the imagined map of Europe: The rise and fall of the "center." *Political Geography* 22: 489–517.

Hall, A. R. (1963) *The London Capital Market and Australia, 1870–1914*. Canberra: Australian National University Press.

Hall, S. (1988) New ethnicities. In: K. Mercer (ed.), *Black Film, British Cinema*. London: The Institute of Contemporary Arts, pp. 27–31.

Hambleton, R. (1987) *The Branding of America: From Levi Strauss to Chrysler, From Westinghouse to Gillette, the Forgotten Fathers of America's Best-Known Brands*. Dublin, NH: Yankee Books.

Hamilton, D. S. and Quinlan, J. P. (2004) *Partners in Prosperity: The Changing Geography of the Transatlantic Economy*. Baltimore, MD: Center for Transatlantic Relations, Johns Hopkins University – School of Advanced International Studies.

Hamilton, E. J. (1934) *American Treasure and the Price Revolution in Spain, 1501–1650*. Cambridge, MA: Harvard University Press.

Hammarström, O. (2003) The operation and functioning of European Works Councils: The case of SKF. Working Paper, Institute for Management of Innovation and Technology, Stockholm. Available at: www.imit.se/pdf/reports/2003_134.pdf (accessed September 5, 2007).

Hannah, L. (1996) Multinationality: Size of firm, size of country, and history dependence. *Business and Economic History* 25.2: 144–54.

Haraway, D. J. (1991) *Simians, Cyborgs, and Women: The Reinvention of Nature*. London: Free Association.

Harley, J. B. (1983) Meaning and ambiguity in Tudor cartography. In: S. Tyacke (ed.), *English Map-Making 1500–1650: Historical Essays*. London: The British Library, pp. 22–45.

Harley, J. B. (1988) Silences and secrecy: The hidden agenda of cartography in Early Modern Europe. *Imago Mundi* 40: 57–76.

Harley, J. B. (1989) Deconstructing the map. *Cartographica* 26.2: 1–20.

Harmon, M. D. (1997) *The British Labour Government and the 1976 IMF Crisis.* New York: St. Martin's Press.

Harnetty, P. (1971) Cotton exports and Indian agriculture, 1861–1870. *Economic History Review* 24.3: 414–29.

Harnetty, P. (1972) *Imperialism and Free Trade: Lancashire and India in the Mid-Nineteenth Century.* Vancouver: University of British Columbia Press.

Harris, J. (1971) *The African Presence in Asia: Consequences of the East African Slave Trade.* Evanston, IL: Northwestern University Press.

Harrod, J. (1973) *Trade Union Foreign Policy: A Study of British and American Trade Union Activities in Jamaica.* Garden City, NY: Doubleday.

Hart, J. F. (1982) The highest form of the geographer's art. *Annals of the Association of American Geographers* 72.1: 1–29.

Hartigan, J. C. (2000) Is the GATT/WTO biased against agricultural products in unfair international trade investigations? *Review of International Economics* 8.4: 634–46.

Hartung, W. D. (2000) Military-industrial complex revisited: How weapons makers are shaping US foreign and military policies. In: M. Honey and T. Barry (eds.), *Global Focus: US Foreign Policy at the Turn of the Millennium.* New York: St. Martin's Press, pp. 21–43.

Harvey, D. (1973) *Social Justice and the City.* London: Edward Arnold.

Harvey, D. (1982) *The Limits to Capital.* Oxford: Blackwell.

Harvey, D. (1989) *The Condition of Postmodernity: An Enquiry into the Origins of Cultural Change.* Oxford: Blackwell.

Harvey, D. (2003) *The New Imperialism.* Oxford: Oxford University Press.

Hatch, J. (1975) The decline of British power in Africa. In: T. Smith (ed.), *The End of the European Empire: Decolonization After World War II.* Lexington, KY: D. C. Heath, pp. 72–95.

Haudrère, P. (2006) *Les Compagnies des Indes orientales: Trois siècles de rencontre entre Orientaux et Occidentaux (1600–1858).* Paris: Les Editions Desjonquère.

Hawken, P., Lovins, A., and Lovins, L. H. (1999) *Natural Capitalism: Creating the Next Industrial Revolution.* Boston, MA: Little, Brown.

Hawker, A. (1993) Low unemployment perplexes officials. *Prague Post* August 4–10, p. 5.

Hayduk, R. (2006) *Democracy for All: Restoring Immigrant Voting Rights in the United States.* New York: Routledge.

Heery, E., Delbridge, R., Salmon, J., Simms, M., and Simpson, D. (2002) Global labour? The transfer of the organizing model to the United Kingdom. In: Y. A. Debrah and I. G. Smith (eds.), *Globalization, Employment and the Workplace: Diverse Impacts.* London: Routledge, pp. 41–68.

Helleiner, E. (1994) Freeing money: Why have states been more willing to liberalize capital controls than trade barriers? *Policy Studies* 27.4: 299–318.

Helleiner, E. (2003) Denationalizing money? Economic liberalism and the "national question" in currency affairs. In: M. Flandreau, C.-L. Holtfrerich, and H. James (eds.), *International Financial History in the Twentieth Century: System and Anarchy.* New York: Cambridge University Press, pp. 213–37.

Helleiner, E. (2006) Reinterpreting Bretton Woods: International development and the neglected origins of embedded liberalism. *Development and Change* 37.5: 943–67.

Henderson, G. L. (1999) *California and the Fictions of Capital.* New York: Oxford University Press.

Henderson, W. O. (1938) Germany's trade with her colonies, 1884–1914. *Economic History Review* 9.1: 1–16.

Hennessy, B. C. (1954) British Trade Unions and International Affairs, 1945–1953. Unpublished PhD dissertation, University of Wisconsin.

Herod, A. (1998a) Of blocs, flows and networks: The end of the Cold War, cyberspace, and the geo-economics of organized labor at the *fin de millénaire*. In: A. Herod, G. Ó Tuathail, and S. Roberts (eds.), *An Unruly World? Globalization, Governance and Geography*. New York: Routledge, pp. 162–95.

Herod, A. (1998b) The geostrategics of labor in post-Cold War Eastern Europe: An examination of the activities of the International Metalworkers' Federation. In: A. Herod (ed.), *Organizing the Landscape: Geographical Perspectives on Labor Unionism*. Minneapolis: University of Minnesota Press, pp. 45–74.

Herod, A. (2000) Implications of Just-in-Time production for union strategy: Lessons from the 1998 General Motors-United Auto Workers dispute. *Annals of the Association of American Geographers* 90.3: 521–47.

Herod, A. (2001) *Labor Geographies: Workers and the Landscapes of Capitalism*. New York: Guilford.

Herod, A. (2007a) The agency of labour in global change: Reimagining the spaces and scales of trade union praxis within a global economy. In: J. M. Hobson and L. Seabrooke (eds.), *Everyday Politics of the World Economy*. Cambridge: Cambridge University Press, pp. 27–44.

Herod, A. (2007b) Labour organizing in the New Economy: Examples from the USA and beyond. In: P. Daniels, A. Leyshon, M. Bradshaw, and J. Beaverstock (eds.), *Geographies of the New Economy: Critical Reflections*. London: Routledge, pp. 132–50.

Herod, A. (2008) Scale: The local and the global. In: S. Holloway, S. Rice, G. Valentine, and N. Clifford (eds.), *Key Concepts in Geography*, 2nd edition. London: Sage.

Herod, A., Ó Tuathail, G., and Roberts, S. M. (eds.) (1998) *An Unruly World? Globalization, Governance and Geography*. New York: Routledge.

Hewitson, G. (1999) *Feminist Economics: Interrogating the Masculinity of Rational Economic Man*. Cheltenham, UK: Edward Elgar.

Hillis, K., Petit, M., and Cravey, A. J. (2002) "Adventure travel for the mind®": Analyzing the United States Virtual Trade Mission's promotion of globalization through discourse and corporate media strategies. In: A. Herod and M. W. Wright (eds.), *Geographies of Power: Placing Scale*. Oxford: Blackwell, pp. 154–70.

Hirst, P. and Thompson, G. (1996) *Globalization in Question: The International Economy and the Possibilities of Governance*. Cambridge: Polity.

Hirst, P. and Thompson, G. (1999) *Globalization in Question: The International Economy and the Possibilities of Governance*, 2nd edition. Cambridge: Polity.

Hobsbawm, E. J. (1975) *The Age of Capital, 1848–1875*. New York: Mentor.

Hobsbawm, E. J. (1990) *Nations and Nationalism since 1780: Programme, Myth, Reality*. Cambridge: Cambridge University Press.

Hobson, J. A. (1948 [1902]) *Imperialism: A Study*. London: G. Allen & Unwin.

Hochschild, A. (1998) *King Leopold's Ghost: A Story of Greed, Terror, and Heroism in Colonial Africa*. New York: Houghton Mifflin.

Hodgson, D. (1997) Embodying the contradictions of modernity: Gender and spirit possession among Maasai in Tanzania. In: M. Grosz-Ngate and O. H. Kokole (eds.), *Gendered Encounters: Challenging Cultural Boundaries and Social Hierarchies in Africa*. New York: Routledge, pp. 111–29.

Hoefle, S. W. (2006) Eliminating scale and killing the goose that laid the golden egg? *Transactions of the Institute of British Geographers*, new series, 31: 238–43.

van Holthoon, F. and van der Linden, M. (eds.) (1988) *Internationalism in the Labour Movement 1830–1940*, vols. 1 and 2. London: E. J. Brill.

Honoré, C. (2004) *In Praise of Slowness: How a Worldwide Movement is Challenging the Cult of Speed*. San Francisco: HarperSanFrancisco.

Howitt, R. (1993) "A world in a grain of sand": Towards a reconceptualisation of geographical scale. *Australian Geographer* 24.1: 33–44.

Howse, D. (1980) *Greenwich Time and the Discovery of the Longitude*. Oxford: Oxford University Press.

Hudson, R. (2000) One Europe or many? Reflections on becoming European. *Transactions of the Institute of British Geographers*, new series, 25: 409–26.

Hunt, E. S. (1994) *The Medieval Super-Companies: A Study of the Peruzzi Company of Florence*. Cambridge: Cambridge University Press.

Hyman, R. (1999) *An Emerging Agenda for Trade Unions?* Discussion paper DP/98/1999, Labour and Society Programme, International Labour Organisation, Geneva.

ICFTU (International Confederation of Free Trade Unions) (1996) *Report on Activities of the Confederation and Financial Reports, 1991–1994*. Brussels: ICFTU.

ICFTU (International Confederation of Free Trade Unions) (2003) *Export Processing Zones: Symbols of Exploitation and a Development Dead-End*. Brussels: ICFTU.

ILO (International Labour Organization) (2003) *Training Policies for Vulnerable Groups in Central and Eastern European Countries: Trade Union Seminar Report*. Seminar held in Prague, June, 24–26, 2002.

IMF (International Metalworkers' Federation) (1967) *World Company Councils: Auto Workers' Answer to World Company Power*. Geneva: IMF.

IMF (International Metalworkers' Federation) (1991) *The IMF and the Multinationals: The Role of the IMF World Company Councils* (Report, dated May 23–24). Lisbon: IMF Central Committee.

International Accounting Standards Board (2007) About IASB. Available at: www.iasb.org/About+Us/About+IASB/About+IASB.htm (accessed August 5, 2007).

International Monetary Fund (1944) *Articles of Agreement: International Monetary Fund, United Nations Monetary and Financial Conference, Bretton Woods, N.H., July 1 to 22, 1944*. Washington, DC: International Monetary Fund.

International Monetary Fund (2003) *Foreign Direct Investment Trends and Statistics*. Washington, DC: International Monetary Fund.

International Monetary Fund (2006) *World Economic Outlook: Globalization and Inflation* (April). Washington, DC: International Monetary Fund.

International Monetary Fund (2007) Currency Composition of Official Foreign Exchange Reserves (COFER).

International Organization for Standardization (2006) Overview of the ISO system. Available at: www.iso.org/iso/en/aboutiso/introduction/index.html (accessed August 5, 2007).

Ishemo, S. L. (1995) Forced labour and migration in Portugal's African colonies. In: R. Cohen (ed.), *The Cambridge Survey of World Migration*. Cambridge: Cambridge University Press, pp. 162–5.

Island Packet, The (July 1, 2001) Indians take US consumer calls, p. 8A (newspaper, Hilton Head, SC).

ITUC (International Trade Union Confederation) (2007a) List of affiliated organisations (2nd General Council, June 20–22, Brussels).

ITUC (International Trade Union Confederation) (2007b) Founding of the PERC gives hope to European workers. News release.

ITUC (International Trade Union Confederation) (2007c) Campaigning for the rights of workers.

Jackson, W. (2000) The "Left Behind" series. *Christian Courier* (newspaper). September 25.

James, R. N. (1986) Anarchism and Political Violence in Sydney and Melbourne, 1886–1896. MA thesis, Department of History, La Trobe University, Australia.

Jasper, W. F. (1999) European nightmare. *The New American* (magazine), March 1.

Johns, R. (1998) Bridging the gap between class and space: U.S. worker solidarity with Guatemala. *Economic Geography* 74.3: 252–71.

Johns, R. and Vural, L. (2000) Class, geography, and the consumerist turn: UNITE and the Stop Sweatshops Campaign. *Environment and Planning A* 32, 1193–213.

Jonas, A. E. G. (2006) Pro scale: Further reflections on the "scale debate" in human geography. *Transactions of the Institute of British Geographers*, new series, 31.3: 399–406.

Jones, G. (1984) The growth and performance of British multinational firms before 1939: The case of Dunlop. *Economic History Review*, new series, 37.1: 35–53.

Jones, G. (1984) The expansion of British multinational manufacturing, 1890–1939. In: A. Okochi and T. Inoue (eds.), *Overseas Business Activities: Proceedings of the Fuji Conference*. Tokyo: University of Tokyo Press, pp. 125–53.

Jones, G. (1994) The making of global enterprise. *Business History* 36.1: 1–17.

Jones, G. (1996) *The Evolution of International Business: An Introduction*. London: Routledge.

Jones, G. (2000) *Merchants to Multinationals: British Trading Companies in the Nineteenth and Twentieth Centuries*. Oxford: Oxford University Press.

Jones, G. (2005) *Multinationals and Global Capitalism: From the Nineteenth to the Twenty-first Century*. Oxford: Oxford University Press.

Jones, J. and Wren, C. (2006) *Foreign Direct Investment and the Regional Economy*. Aldershot, UK: Ashgate.

Joseph, A. (1990) Japan's FDI in India's industrial development: A case study of Maruti Udyog Ltd. *The CTC Reporter* 30 (Autumn): 25–7.

Kain, J. (1968) Housing segregation, Negro employment, and metropolitan decentralization. *Quarterly Journal of Economics* 82: 175–97.

Kant, I. (1943 [1781]) *Critique of Pure Reason*. New York: Wiley.

Kanter, R. M. (1995) *World Class: Thriving Locally in the Global Economy*. New York: Simon & Schuster.

Kapur, D., Lewis, J. P., and Webb, R. (1997) *The World Bank: Its First Half Century*. Washington, DC: Brookings Institution.

Keay, J. (1991) *The Honourable Company: A History of the English East India Company*. London: HarperCollins.

Keenan, J. L. (1943) *A Steel Man in India*. New York: Duell, Sloan and Pearce.

Keil, T. J. and Keil, J. M. (2002) The state and labor conflict in postrevolutionary Romania. *Radical History Review* 82: 9–36.

Kern, S. (1983) *The Culture of Time and Space, 1880–1918*. Cambridge, MA: Harvard University Press.

Kim, D.-W. (2005) The British multinational enterprise in Latin America before 1945: The case of J. & P. Coats. *Textile History* 36.1: 69–85.

King, G. (1996) *Mapping Reality: An Exploration of Cultural Cartographies*. New York: St. Martin's Press.

Kirby, A. (2002) Popular culture, academic discourse, and the incongruities of scale. In: A. Herod and M. W. Wright (eds.), *Geographies of Power: Placing Scale*. Oxford: Blackwell, pp. 171–91.

Kirichenko, O. S. and Koudyukin, P. M. (1993) Social partnership in Russia: The first steps. *Economic and Industrial Democracy* 14: 43–54.

Kleiman, E. (1978) Metropolitan exports lost through decolonization – Some conceptual and statistical problems. *Bulletin of the Oxford University Institute of Economics and Statistics* 40.3: 273–8.

Klein, H. S. (1990) Economic aspects of the eighteenth-century Atlantic slave trade. In: J. D. Tracy (ed.), *The Rise of Merchant Empires: Long-Distance Trade in the Early Modern World, 1350–1750*. Cambridge: Cambridge University Press, pp. 287–310.

Klein, N. (2000) *No Space, No Choice, No Jobs, No Logo: Taking Aim at the Brand Bullies*. New York: Picador USA.

Kobayashi, K. (1974) Organizational innovation in the McCormick and International Harvester companies. In: K. Nakagawa (ed.), *Strategy and Structure of Big Business*. Tokyo: University of Tokyo Press, pp. 203–13.

Kofas, J. V. (1989) *Intervention and Underdevelopment: Greece During the Cold War*. University Park, PA: Pennsylvania State University Press.

Koncz, J. L. and Yorgason, D. R. (2006) Direct investment positions for 2005: Country and industry detail. *Survey of Current Business* (July): 20–35. A publication of the US Department of Commerce, Bureau of Economic Analysis, Washington, DC.

Körner, P., Maass, G., Siebold, T., and Tetzlaff, R. (1986) *The IMF and the Debt Crisis: A Guide to the Third World's Dilemmas*. London: Zed Books.

Korten, D. C. (2001) *When Corporations Rule the World* (2nd edition). Bloomfield, CT: Kumarian Press.

Koselleck, R. (1985) Space of experience and horizon of expectation: Two historical categories. In: R. Koselleck (ed.), *Futures Past: On the Semantics of Historical Time*. Boston, MA: MIT Press, pp. 267–88.

Krasner, S. D. (1999) *Sovereignty: Organized Hypocrisy*. Princeton, NJ: Princeton University Press.

Kumar, D. (1983) *The Cambridge Economic History of India, vol. 2: c.1757–c.1970*. Cambridge: Cambridge University Press.

Kundera, M. (1996) *Slowness*. New York: HarperCollins Publishers.

Kurian, G. T. (2001) *Datapedia of the United States, 1790–2005*. Lanham, MD: Bernan.

Kurtz, H. (2002) The politics of environmental justice as the politics of scale: St. James Parish, Louisiana, and the Shintech siting controversy. In: A. Herod and M. W. Wright (eds.), *Geographies of Power: Placing Scale*. Oxford: Blackwell, pp. 249–73.

Lall, R. B. (1986) *Multinationals from the Third World: Indian Firms Investing Abroad*. Delhi: Oxford University Press.

Lamb, H. B. (1955) The "state" and economic development in India. In: S. Kuznets, W. E. Moore, and J. J. Spengler (eds.), *Economic Growth: Brazil, India, Japan*. Durham: Duke University Press, pp. 464–95.

Lambert, R. and Webster, E. (2001) Southern unionism and the new labour internationalism. *Antipode* 33.3: 337–62.

Latour, B. (1993) *We Have Never Been Modern*. Cambridge, MA: Harvard University Press.

Latour, B. (1996) On actor-network theory: A few clarifications. *Soziale Welt* 47: 369–81.

Lattek, K. (1988) The beginnings of socialist internationalism in the 1840s: The "Democratic Friends of all Nations" in London: In: F. van Holthoon and M. van der Linden (eds.), *Internationalism in the Labour Movement 1830–1940*, vol. 1. London: E. J. Brill, pp. 259–82.

Lattek, K. (2006) *Revolutionary Refugees: German Socialism in Britain, 1840–1860*. London: Routledge.

League of Nations (1944) *International Currency Experience: Lessons of the Inter-War Period*. Geneva: League of Nations.

Lee, E. (1997) *The Labour Movement and the Internet: The New Internationalism*. London: Pluto Press.

Lefebvre, H. (1976) *The Survival of Capitalism: Reproduction of the Relations of Production*. London: St. Martin's Press.

Leigh-Browne, F. S. (1942) The International Date Line. *The Geographical Magazine* (April): 302–6.

Lenin, V. I. (1939 [1916]) *Imperialism: The Highest Stage of Capitalism*. New York: International Publishers.

Levitt, T. (1983) The globalization of markets. *Harvard Business Review* 61 (May–June): 92–102.

Lewis, C. (1938) *America's Stake in International Investments*. Washington, DC: Brookings Institution.

van der Linden, M. (1988) The rise and fall of the First International: An interpretation. In: F. van Holthoon and M. van der Linden (eds.), *Internationalism in the Labour Movement 1830–1940*, vol. 1. London: E. J. Brill, pp. 323–35.

van der Linden, M. (1999) Transnationalizing American labor history. *Journal of American History* 86.3: 1078–92.

Liu, W. and Dicken, P. (2006) Transnational corporations and "obligated embeddedness": Foreign direct investment in China's automobile industry. *Environment and Planning A* 38.7: 1229–47.

Logan, F. A. (1958) India – Britain's substitute for American cotton, 1861–1865. *Journal of Southern History* 24.4: 472–80.

Lovelock, J. E. (1979) *Gaia: A New Look at Life on Earth*. New York: Oxford University Press.

Lowe, B. E. (1921) *The International Protection of Labor*. New York: Macmillan.

Lozovsky, S. A. (1927) What is the Red International of Labour Unions? Pamphlet published by The Red International of Labour Unions. Copy available at: www.marx.org/archive/lozovsky/1927/rilu.htm (accessed August 15, 2007).

de Luce, D. (1993) Premier charges extortion by union. *Prague Post* (February17–23): 9.

Lyons, F. S. L. (1963) *Internationalism in Europe 1815–1914*. Leyden: A.W. Sythoff.

Macau, J. (1972) *L'Inde danoise: La première Compagnie, 1616–1670*. Aix-en-Provence: Université de Provence.

Macrory, P. F. J., Appleton, A. E., and Plummer, M. G. (eds.) (2005) *The World Trade Organization: Legal, Economic and Political Analysis* (3 volumes). New York: Springer.

Maddison, A. (1972) *Class Structure and Economic Growth: India and Pakistan since the Moghuls*. London: George Allen.

Madsen, J. B. (2001) Trade barriers and the collapse of world trade during the Great Depression. *Southern Economic Journal* 67.4: 848–68.

Mainwaring, S. and Scully, T. R. (eds.) (2003) *Christian Democracy in Latin America: Electoral Competition and Regime Conflicts*. Stanford, CA: Stanford University Press.

Manchester, A. K. (1964) *British Pre-eminence in Brazil: Its Rise and Decline*. New York: Octagon Books.

Mandel, E. (1951) The theory of "state capitalism." *Fourth International* 12.5: 145–56. (Written as Ernest Germain.)

Mandel, E. (1962a) *Marxist Economic Theory*, vol. 1. New York: Monthly Review Press.

Mandel, E. (1962b) *Marxist Economic Theory*, vol. 2. New York: Monthly Review Press.

Mandelbaum, M. (1993) Introduction. In: S. Islam and M. Mandelbaum (eds.), *Making Markets: Economic Transformation in Eastern Europe and the Post-Soviet States*. New York: Council on Foreign Relations, pp. 1–15.

Mann, T. (1897) *The International Labour Movement*. London: Clarion.

Maquila Portal (2006) "100 Top Maquilas" (as of January 31, 2006). Available at: www. maquilaportal.com/cgi-bin/top100/top100.pl (accessed January 30, 2007).

Markovits, C. (2000) *The Global World of Indian Merchants, 1750–1947: Traders of Sind from Bukhara to Panama*. Cambridge: Cambridge University Press.

Marston, S. A., Jones, J. P. III, and Woodward, K. (2005) Human geography without scale. *Transactions of the Institute of British Geographers*, new series, 30.4: 416–32.

Martin, E. (1991) The egg and the sperm: How science has constructed a romance based on stereotypical male–female roles. *Signs: Journal of Women in Culture and Society* 16.3: 485–501.

Marx, K. (1963 [1852]) *The Eighteenth Brumaire of Louis Bonaparte*. New York: International Publishers.

Marx, K. (1973 [1858]) *Grundrisse: Foundations of the Critique of Political Economy*. London: Penguin (1993 printing).

Marx, K. (1976 [1867]) *Capital: A Critique of Political Economy*, vol. 1. London: Penguin (1990 printing).

Marx, K. (1981 [1894]) *Capital: A Critique of Political Economy*, vol. 3. London: Penguin (1991 printing).

Mason, M. (1994) Historical perspectives on Japanese direct investment in Europe. In: M. Mason and D. Encarnation (eds.), *Does Ownership Matter? Japanese Multinationals in Europe*. Oxford: Clarendon, pp. 3–38.

Mason, M. (1999) The origins and evolution of Japanese direct investment in East Asia. In: D. J. Encarnation (ed.), *Japanese Multinationals in Asia: Regional Operations in Comparative Perspective*. New York: Oxford University Press, pp. 17–45.

Massey, D. (1984) *Spatial Divisions of Labour: Social Structures and the Geography of Production*. London: Macmillan.

Mathers, C. J. (1988) Family partnerships and international trade in early modern Europe: Merchants from Burgos in England and France, 1470–1570. *Business History Review* 62.3: 367–97.

Mayor, F. (1980) Comments by Federico Mayor, Deputy Director-General, UNESCO, on the occasion of a presentation by M. Goldsmith to the Commonwealth Section of the Royal Society on "Technical Change and the Irrelevance of Aid." *The Royal Society for the Encouragement of Arts, Manufactures and Commerce Journal*, no. 5287, vol. 128 (June): 437–49.

McArtor, T. A. (2003) Boycotters beware: "French" products are often American. *European Affairs: A Publication of the European Institute*, Summer/Fall.

McClean, R. (2005) Spaniards dare to question the way the day is ordered. *New York Times*, January 12, p. A4.

McClintock, A. (1995) *Imperial Leather: Race, Gender and Sexuality in the Colonial Conquest*. New York: Routledge.

McKay, V. (1963) Changing external pressures on Africa. In: W. Goldschmidt (ed.), *The United States and Africa*. New York: Praeger, pp. 74–114.

McKinsey & Company (2005) *$118 Trillion and Counting: Taking Stock of the World's Capital Markets*. (February). www.mckinsey.com/mgi/reports/pdfs/gcm/global.pdf (accessed August 2, 2006).

McKinsey & Company (2006a) *Mapping the Global Capital Market 2006 – Second Annual Report*. (January). www.mckinsey.com/mgi/reports/pdfs/GCMAnnualReport/Mapping_the_Global_Capital_Market_2006_Update.pdf (accessed August 2, 2006).

McKinsey & Company (2006b) Growing cross-border capital flows. www.mckinseyquarterly. com/article_page.aspx?ar=1579&L2=7&L3=10&pagenum=11 (accessed August 11, 2006).

McLuhan, M. (1962) *The Gutenberg Galaxy: The Making of Typographic Man*. Toronto: University of Toronto Press.

McLuhan, M. (1964) *Understanding Media: The Extensions of Man*. New York: McGraw-Hill.

McLuhan, M. and Fiore, Q. (1967) *The Medium is the Massage*. New York: Bantam.

McMahon, J. A. (2005) The agreement on agriculture. In: P. F. J. Macrory, A. E. Appleton, and M. G. Plummer (eds.), *The World Trade Organization: Legal, Economic and Political Analysis* (3 volumes). New York: Springer, pp. 187–229.

McNeill, W. H. (1982) *The Pursuit of Power: Technology, Armed Force, and Society since A.D. 1000*. Chicago: University of Chicago Press.

Mee, C. L. (1984) *The Marshall Plan: The Launching of the Pax Americana*. New York: Simon and Schuster.

Melman, S. (1970) *Pentagon Capitalism: The Political Economy of War*. New York: McGraw-Hill.

Merchant, C. (1980) *The Death of Nature: Women, Ecology, and the Scientific Revolution*. San Francisco: Harper and Row.

Mergner, G. (1988) Solidarität mit den "Wilden"? Das Verhältnis der Deutschen Sozialdemokratie zu den Afrikanischen Widerstandskämpfen in den Ehemaligen Deutschen Kolonien um die Jahrhundertwende. In: F. van Holthoon and M. van der Linden (eds.), *Internationalism in the Labour Movement, 1830–1940*, vol. 1. London: E. J. Brill, pp. 68–86.

Michel, J. (1978) La Chevalerie du Travail (1890–1906): Force ou faiblesse du mouvement ouvrier belge? *Revue Belge d'Histoire Contemporaine/Belgisch Tijdschrift voor Nieuwste Geschiedenis* 9.1–2: 117–64.

Milhomme, A. J. (2005) Decolonization and international trade: The Côte d'Ivoire's case. *The Journal of American Academy of Business, Cambridge* 7.1: 20–31.

Miller, C. L. (1998) *Nationalists and Nomads: Essays on Francophone African Literature and Culture*. Chicago: University of Chicago Press.

Miller, R. and Greenhill, R. (2006) The fertilizer commodity chains: Guano and nitrate, 1840–1930. In: S. Topik, C. Marichal, and Z. Frank (eds.), *From Silver to Cocaine: Latin American Commodity Chains and the Building of the World Economy, 1500–2000*. Durham: Duke University Press, pp. 228–70.

Milunovich, G. and Thorp, S. (2006) Information processing and measures of integration: New York, London and Tokyo. Research Paper no. 177, Quantitative Finance Research Centre, University of Technology, Sydney.

Min, P. G. (1992) A comparison of the Korean minorities in China and Japan. *International Migration Review* 26.1: 4–21.

Mitchell, B. R. (1988) *British Historical Statistics*. Cambridge: Cambridge University Press.

Mitchell, B. R. (2003a) *International Historical Statistics: Africa, Asia and Oceania 1750–2000* (4th edition). New York: Palgrave Macmillan.

Mitchell, B. R. (2003b) *International Historical Statistics: Europe 1750–2000* (5th edition). New York: Palgrave Macmillan.

Mitchell, B. R. (2003c) *International Historical Statistics: The Americas 1750–2000* (5th edition). New York: Palgrave Macmillan.

Mitchell, T. (1998) Fixing the economy. *Cultural Studies* 12.1: 82–101.

Mitchell, T. (2002) *Rule of Experts: Egypt, Techno-politics, Modernity*. Berkeley: University of California Press.

Mizoguchi, T. and Yamamoto, Y. (1984) Capital formation in Taiwan and Korea. In: R. H. Myers and M. R. Peattie (eds.), *The Japanese Colonial Empire, 1895–1945*. Princeton, NJ: Princeton University Press, pp. 399–419.

Monks. J. (2007) Speech by John Monks, General Secretary, European Trade Union Confederation, Founding Assembly of the Pan-European Regional Council, March 19. Available at: www.trade-unions.org/a/3489 (accessed September 2, 2007).

Monmonier, M. S. (1977) *Maps, Distortion, and Meaning*. Resource paper 75-4, Association of American Geographers, Washington, DC.

Moody's (1909) *Moody's Manual of Railroads and Corporation Securities*. New York: Moody Manual Corporation.

Moran, M. (1991) *The Politics of the Financial Services Revolution: The USA, UK and Japan*. London: Macmillan.

Morison, T. (1908) Review of *India and the Empire: A Consideration of the Tariff Problem* by M. de P. Webb. *The Economic Journal* 18.71: 424–7.

Morray, J. P. (1983) *Project Kuzbas: American Workers in Siberia (1921–1926)*. New York: International Publishers.

Morrison, W. (2006) China's economic conditions. CRS Issue Brief for Congress, Foreign Affairs, Defense, and Trade Division. Washington, DC: Congressional Research Service, Library of Congress.

Mundell, R. (2005) The case for a world currency. *Journal of Policy Modeling* 27.4: 465–75.

Nader, R. (1999) Transcript of a chat room debate between Patrick Buchanan and Ralph Nader on "The Battle in Seattle" facilitated by *Time* magazine, November 28, 1999. Available at: www.time.com/time/community/transcripts/1999/112899buchanan-nader.html (accessed July 29, 2006).

Nagahara, Y. (2000) Monsieur le capital and Madame la terre do their ghost-dance: Globalization and the nation-state. *The South Atlantic Quarterly* 99.4: 929–61.

Nagano, Y. (1997) Re-examining the foreign trade structure of the colonial Philippines: With special reference to the "Intra-Asian Trade." Discussion paper D 97-28, Asian Historical Statistics Project Office, Institute of Economic Research, Hitotsubashi University, Tokyo, Japan.

Nemmers, E. E. (1956) *Hobson and Underconsumption*. Amsterdam: North-Holland Publishing Company.

New-York Daily Tribune (August 14, 1858) Atlantic Cable [advertisement], p. 7.

New York Times (April 27, 2005) In Japan, time obsession may be culprit, pp. A1 and A6.

New York Times (September 7, 2005) Indian tutors: Miles away but as close as a keyboard, p. A23.

New York Times (December 9, 2005) Ogre to slay? Outsource it to Chinese, pp. A1 and C4.

New York Times (July 31, 2007) A new country's tough non-elective: Portuguese 101, p. A4.

New-York Tribune (November 22, 1883) The war in Tonquin – China and France's Hostile Armies, p. 1.

Newcomer, M. (1944) Contribution of the International Monetary Fund and Bank. In: H. Alfred (ed.), *The Bretton Woods Agreement – And Why It Is Necessary*. New York: Citizens Conference on International Union, pp. 7–13.

Noreng, Ø. (1999) The euro and the oil market: New challenges to the industry. *Journal of Energy Finance and Development* 4.1: 29–68.

Norris, F. (1967 [1901]) *The Octopus: A Story of California*. Port Washington, NY: Kennikat Press.

Northrup, D. (1999) Migration from Africa, Asia, and the South Pacific. In: A. Porter (ed.), *The Oxford History of the British Empire* (vol. III). Oxford: Oxford University Press, pp. 88–100.

Northrup, D. (2002) Freedom and indentured labor in the French Caribbean, 1848–1900. In D. Eltis (ed.), *Coerced and Free Migration: Global Perspectives*. Stanford, CA: Stanford University Press, pp. 204–28.

Nowell, G. P. (2002/03) Imperialism and the era of falling prices. *Journal of Post Keynesian Economics* 25.2: 309–29.

O'Malley, M. (1990) *Keeping Watch: A History of American Time*. New York: Viking Penguin.

O'Rourke, K. H. and Williamson, J. G. (1999) *Globalization and History: The Evolution of a Nineteenth-Century Atlantic Economy*. Cambridge, MA: MIT Press.

Observer, The (October 22, 2006) Empire strikes back: India forges new steel alliance, p. 4.

OECD (Organization for Economic Cooperation and Development) (1999) *International Direct Investment Statistics Yearbook 1999*. Paris: OECD.

OECD (Organization for Economic Cooperation and Development) (2004) *International Direct Investment Statistics Yearbook 1992–2003*. Paris: OECD.

OECD (Organization for Economic Cooperation and Development) (2007) *International Direct Investment By Country* (on-line database). Paris: OECD.

Office of National Statistics [UK] (2006) *United Kingdom Economic Accounts* (Q1).

Officer, L. H. (1996) *Between the Dollar–Sterling Gold Points: Exchange Rates, Parity, and Market Behavior*. Cambridge: Cambridge University Press.

Officer, L. H. (2006) Five ways to compute the relative value of a UK pound amount, 1830–2005. Available at: www.measuringworth.com/calculators/ukcompare/ (accessed December 7, 2006).

Officer, L. H. (2007) What was the exchange rate then? Available at: http://eh.net/hmit/exchangerates/ (accessed January 31, 2007).

Ohmae, K. (1990) *The Borderless World: Power and Strategy in the Interlinked Economy*. New York: HarperBusiness.

Ohmae, K. (1995) *The End of the Nation State: The Rise of Regional Economies*. New York: McKinsey and Company.

Ohmae, K. (2005) *The Next Global Stage: Challenges and Opportunities in Our Borderless World*. Upper Saddle River, NJ: Wharton School Publishing.

Okimoto, D. I. (1989) *Between MITI and the Market: Japanese Industrial Policy for High Technology*. Stanford, CA: Stanford University Press.

Olle, W. and Schoeller, W. (1977) World market competition and restrictions upon international trade-union policies. *Capital and Class* 2: 56–75.

Ormrod, D. (2006) Consuming the Orient in Britain, 1660–1760. Paper presented at the XIVth International Economic History Conference, Helsinki, Finland, August 21–25.

Ost, D. (1989) The transformation of Solidarity and the future of Central Europe. *Telos* 79 (Spring): 69–94.

Ost, D. (1995) Labor, class, and democracy: Shaping political antagonisms in Post-Communist society. In: B. Crawford (ed.), *Markets, States, and Democracy: The Political Economy of Post-Communist Transformation*. Boulder, CO: Westview, pp. 177–203.

Ozolins, U. (1999) Between Russian and European hegemony: Current language policy in the Baltic states. *Current Issues in Language and Society* 6.1: 6–47.

Ozouf, M. and Ozouf, J. (1964) Le thème du patriotisme dans les manuels primaires. *Le Mouvement social* 49: 5–31.

Pares, R. (1956) *Yankees and Creoles: The Trade between North America and the West Indies before the American Revolution*. Cambridge, MA: Harvard University Press.

Parthasarathi, P. (1998) Rethinking wages and competitiveness in the eighteenth century: Britain and South India. *Past and Present* 158: 79–109.

Parthasarathi, P. (2001) *The Transition to a Colonial Economy: Weavers, Merchants and Kings in South India, 1720–1800*. Cambridge: Cambridge University Press.

Parthasarathi, P. (2004) Global trade and textile workers, 1650–2000. Paper presented at the "A global history of textile workers, 1650–2000" Conference organized by the International Institute of Social History, Amsterdam, November 11–13.

Pauls, D. B. (1990) U.S. exchange rate policy: Bretton Woods to the present. *Federal Reserve Bulletin* 76.11: 891–908.

Pedler, F. (1974) *The Lion and the Unicorn in Africa: The United Africa Company, 1787–1931*. London: Heinemann.

Peet, R. (1985) The social origins of environmental determinism. *Annals of the Association of American Geographers* 75.3: 309–33.

Peet, R. (2003) *Unholy Trinity: The IMF, World Bank and WTO*. New York: Zed Books.

Peet, R. and Watts, M. (eds.) (1996) *Liberation Ecologies: Environment, Development, Social Movements*. London: Routledge.

Pelling, H. (1956) The Knights of Labor in Britain, 1880–1901. *The Economic History Review*, new series 9.2: 313–31.

Perlin, F. (1983) Proto-industrialization and pre-colonial South Asia. *Past and Present* 98: 30–95.

Perren, R. (1978) *The Meat Trade in Britain, 1840–1914*. London: Routledge and Kegan Paul.

Philips, J. (1978) The South African Wobblies: The origins of industrial unions in South Africa. *Ufahuma* 8.3: 122–38.

Pianalto, S. (2007) The internationalization of national currencies. Paper presented at the Comenius European Banking and Financial Forum, Czech National Bank, Prague, Czech Republic, March 27.

Picard, L. (1997) *Restoration London*. London: Weidenfeld and Nicolson.

Piñera, J. (n.d.) How the power of ideas can transform a country. Available at: www.josepinera.com/pag/pag_tex_powerideas.htm (accessed October 25, 2007).

Pink, D. H. (2005) Why the world is flat. An interview with Thomas Friedman. May issue. Available at: www.wired.com/wired/archive/13.05/friedman.html (accessed October 31, 2007).

Piore, M. J. (1992) The limits of the market and the transformation of socialism. In: B. Silverman, R. Vogt, and M. Yanowitch (eds.), *Labor and Democracy in the Transition to a Market System*. Armonk, NY: M. E. Sharpe, pp. 171–82.

Plender, J. (1987) London's Big Bang in international context. *International Affairs* 63.1: 39–48.

Polanyi, K. (1957) *The Great Transformation*. Boston, MA: Beacon Press.

Poovey, M. (1996) Accommodating merchants: Accounting, civility, and the natural laws of gender. *Differences: A Journal of Feminist Cultural Studies* 8.3: 1–20.

Popke, E. J. (1994) Recasting geopolitics: The discursive scripting of the International Monetary Fund. *Political Geography* 13.3: 255–69.

Portus, G. V. (1930) The Australian labour movement and the Pacific. *Pacific Affairs* 3.10: 923–32.

Prague Post (1995) Economists question "miracle" of nation's low unemployment rate. December 20–26, p. 1.

Prakash, O. (n.d.) The transformation from a pre-colonial to a colonial order: The case of India. Unpublished paper, available at: www.lse.ac.uk/collections/economicHistory/GEHN/GEHN%20PDF/Transformation%20from%20a%20Pre-Colonial%20-%20Om%20Prakash.pdf (accessed December 7, 2006).

Pratt, M. L. (1992) *Imperial Eyes: Travel Writing and Transculturation*. New York: Routledge.

Price, J. (1945) *The International Labour Movement*. London: Oxford University Press.

PriMetrica (2004) Interregional internet bandwidth, 2004. Available at: www.telegeography.com/ ee/free_resources/gig2005-02.php (accessed July 1, 2005).

Pulley, R. H. (1966) The railroad and Argentine national development, 1852–1914. *The Americas* 23.1: 63–75.

Quah, D. (1999) The weightless economy in economic development. Working Paper, Department of Economics, London School of Economics.

Radice, H. (1984) The national economy – a Keynesian myth? *Capital and Class* 22: 111–40.

Radio Free Europe (2005) World: Signs grow of dollar losing favor as world's reserve currency. February 24. Available at: www.rferl.org/featuresarticle/2005/02/a490ae39-7c94-4193-8648-aac6ec3c9b9d.html (accessed July 28, 2007).

Radosh, R. (1969) *American Labor and United States Foreign Policy*. New York: Random House.

Raghavan, C. (1998) MAI not dead yet. *Third World Economics* 179/180 (February 16–March 15).

Ray, H. (1993) *Trade and Diplomacy in India–China Relations: A Study of Bengal during the Fifteenth Century*. London: Sangam Books.

Reed, P. M. (1958) Standard Oil in Indonesia, 1898–1928. *The Business History Review* 32.3: 311–37.

Rehfeldt, U. (1998) European works councils: An assessment of French initiatives. In: W. Lecherand and H.-W. Platzer (eds.), *European Union – European Industrial Relations? Global Challenges, National Developments and Transnational Dynamics*. New York: Routledge, pp. 207–22.

Resnick, S. A. and Wolff, R. D. (2002) *Class Theory and History: Capitalism and Communism in the U.S.S.R.* New York: Routledge.

Richter, S. (2003) Back to 1913? *The Globalist*, February 6.

Ricks, D. A. (2000) *Blunders in International Business*. Malden, MA: Blackwell.

Rifkin, J. (1987) *Time Wars: The Primary Conflict in Human History*. New York: Henry Holt.

Roberts, B. C. (1964) *Labour in the Tropical Territories of the Commonwealth*. Durham: Duke University Press.

Roberts, J. (1985) *The Triumph of the West*. London: British Broadcasting Corporation.

Roberts, K. (1998) Making New Zealand the hottest destination on the planet. Keynote address to New Zealand Tourism Industry Association conference, Auckland, by Kevin Roberts, Saatchi and Saatchi CEO, August 28.

Robins, N. (2006) *The Corporation That Changed the World: How the East India Company Shaped the Modern Multinational*. London: Pluto.

Rodman, K. A. (1998) Think globally, punish locally: Nonstate actors, multinational corporations and human rights sanctions. *Ethics and International Affairs* 12: 19–41.

Romualdi, S. (1967) *Presidents and Peons: Recollections of a Labor Ambassador in Latin America*. New York: Funk and Wagnalls.

Rothermund, D. (1988) *An Economic History of India: From Pre-Colonial Times to 1986*. London: Croom Helm.

Roy, T. (2000) *The Economic History of India, 1857–1947*. New York: Oxford University Press.

Roy, T. (2002) Economic history and modern India: Redefining the link. *Journal of Economic Perspectives* 16.3: 109–30.

Rudner, D. W. (1994) *Caste and Capitalism in Colonial India: The Nattukottai Chettiars*. Berkeley: University of California Press.

Ruggie, J. G. (1982) International regimes, transactions, and change: Embedded liberalism in the postwar economic order. *International Organization* 36.2: 379–415.

Ruggiero, R. (1997) Charting the trade routes of the future: Towards a borderless economy. Address to the International Industrial Conference (IIC), San Francisco, September 29.

Rushkoff, D. (1994) *Media Virus! Hidden Agendas in Popular Culture.* New York: Ballantine Books.

Rusnok, J. (1993) Statement by Jiří Rusnok, adviser to ČMKOS. Quoted in *Prague Post* August 4–10, p. 5.

Ryan, S. (1996) *The Cartographic Eye: How Explorers Saw Australia.* New York: Cambridge University Press.

Ryan, S. and Herod, A. (2006) Restructuring the architecture of state regulation in the Australian and Aotearoa/New Zealand cleaning industries and the growth of precarious employment. In: L. L. M. Aguiar and A. Herod (eds.), *The Dirty Work of Neoliberalism: Cleaners in the Global Economy.* Oxford: Blackwell, pp. 60–80.

Safire, W. (2004) Outsource – and the urge to insource. *New York Times Magazine*, March 21, p. 30.

Said, E. W. (1978) *Orientalism.* New York: Pantheon Books.

Said, E. W. (1994) *Culture and Imperialism.* London: Random House.

Sainsbury, D. (2005) Speech by Lord Sainsbury of Turville, Parliamentary Under Secretary of State for Science and Innovation, to Investing in Innovation Conference, London, November 10. Available at: www.dti.gov.uk/ministers/speeches/sainsbury101105.html (accessed January 11, 2007).

Samiee, S. and Roth, K. (1992) The influence of global marketing standardization on performance. *Journal of Marketing* 56 (April): 1–17.

Sassen, S. (1988) *The Mobility of Labor and Capital: A Study in International Investment and Labor Flow.* Cambridge: Cambridge University Press.

Sassen, S. (2003) Globalization or denationalization? *Review of International Political Economy* 10.1: 1–22.

Saul, S. B. (1960) *Studies in British Overseas Trade 1870–1914.* Liverpool: Liverpool University Press.

Sauvant, K. P. (1996) International trade and investment trends. Keynote address, First Annual Australian Conference on International Trade, Education and Research, Australian APEC Study Centre, University of Melbourne, Melbourne, December 5–6, 1996.

Scammell, W. M. (1965) The working of the gold standard. *Yorkshire Bulletin of Economic and Social Research* 17.1: 32–45.

Schivelbusch, W. (1986) *The Railway Journey: Industrialization and the Perception of Time and Space.* Berkeley: University of California Press.

Schlote, W. (1952) *British Overseas Trade from 1700 to the 1930s.* Oxford: Blackwell.

Schoonover, T. D. (1998) *Germany in Central America: Competitive Imperialism, 1821–1929.* Tuscaloosa: University of Alabama Press.

Schoonover, T. D. (2000) *The French in Central America: Culture and Commerce, 1820–1930.* Wilmington, DE: Scholarly Resources.

Schröter, H. (1988) Risk and control in multinational enterprise: German businesses in Scandinavia, 1918–1939. *The Business History Review* 62.3: 420–43.

Schuler, K. (2000) Basics of dollarization. Staff Report, Office of the Chairman, Joint Economic Committee, US Congress, January. In: *Achieving Growth and Prosperity through Freedom: A Compilation of 1999–2000 Joint Economic Committee Reports*, Senate Print 106–58, 106th Congress, 2nd Session, Washington, DC: Government Printing Office.

Scott, J. (1978) *Yankee Unions, Go Home! How the AFL Helped the U.S. Build an Empire in Latin America.* Vancouver: New Star Books.

Segal, A. (1993) *An Atlas of International Migration.* London: Hans Zell Publishers.

Segal, M. J. (1953) The international trade secretariats. *Monthly Labor Review* (April): 372–80.

Sene, I. (2004) Colonisation française et main-d'oeuvre carcérale au Sénégal: De l'emploi des détenus des camps penaux sur les chantiers des travaux routiers (1927–1940). *French Colonial History* 5: 153–72.

Senft, P. (1995) Interview by author with Peter Senft, Head, Economic Office, German Metalworkers' Union (IG Metall), Berlin, Germany, September 15.

Seyferth, G. (2000–1) Migração Japonesa e o fenômeno *Dekassegui*: Do país do sol nascente para uma terra cheia de sol. *Com Ciência* 16 (December 2000–January 2001).

Shapiro, H. (1991) Determinants of firm entry into the Brazilian automobile manufacturing industry, 1956–1968. *The Business History Review* 65.4: 876–947.

Shaw, C. (2005) Rothschilds and Brazil: An introduction to sources in the Rothschild archive. *Latin American Research Review* 40.1: 165–85.

Shaw, R. B. (1978) *A History of Railroad Accidents, Safety Precautions and Operating Practices*. Binghamton, NY: Vail–Ballou Press.

Sicking, L. (2004) A colonial echo: France and the colonial dimension of the European Economic Community. *French Colonial History* 5: 207–28.

Siddall, W. R. (1969) Railroad gauges and spatial interaction. *Geographical Review* 59.1: 29–57.

Siddiqi, A. (1981) Money and prices in the earlier stages of empire: India and Britain, 1760 to 1840. *Indian Economic and Social History Review* 18.3–4: 231–62.

Simms, B. (1992) *Workers of the World Undermined: American Labor's Role in U.S. Foreign Policy*. Boston, MA: South End Press.

Slater, D. (2002) Trajectories of development theory: Capitalism, socialism, and beyond. In: R. J. Johnston, P. J. Taylor, and M. J. Watts (eds.), *Geographies of Global Change: Remapping the World* (2nd edition). Oxford: Blackwell, pp. 88–99.

Smith, A. (1961 [1776]) *An Inquiry into the Nature and Causes of the Wealth of Nations*. London: Methuen.

Smith, B. (ed.) (1975) *Documents on Art and Taste in Australia: The Colonial Period 1770–1914*. Melbourne: Oxford University Press.

Smith, N. (1990 [1984]) *Uneven Development: Nature, Capital and the Production of Space* (2nd edition). Oxford: Blackwell.

Smith, N. (2003) *American Empire: Roosevelt's Geographer and the Prelude to Globalization*. Berkeley: University of California Press.

Smith, N. (2005) *The End Game of Globalization*. New York: Routledge.

Smith, R. G. (2003) World city topologies. *Progress in Human Geography* 27.5: 561–82.

Snow, D. A. and Benford, R. D. (1992) Master frames and cycles of protest. In: A. D. Morris and C. M. Mueller (eds.), *Frontiers in Social Movement Theory*. New Haven, CT: Yale University Press, pp. 133–55.

Snow, S. (1964) *The Pan-American Federation of Labor*. Durham, NC: Duke University Press.

Soja, E. (1980) The socio-spatial dialectic. *Annals of the Association of American Geographers* 70.2: 207–25.

SOLIDAR (2001) Export processing zones. *SOLIDAR Briefing*, September. Available at: www.solidar.org/English/pdf/DevEd_EPZ_Briefing_EN.doc (accessed April 11, 2007).

Southall, H. (1989) British artisan unions in the New World. *Journal of Historical Geography* 15.2: 163–82.

Southworth, C. (1931) *The French Colonial Venture*. London: P. S. King & Son.

Spalding Jr., H. A. (1988) US labour intervention in Latin America: The case of the American Institute for Free Labor Development. In: R. Southall (ed.), *Trade Unions and the New Industrialization of the Third World*. London: Zed Books, pp. 259–86.

Spigel, L. (1992) The suburban home companion: Television and the neighborhood ideal in postwar America. In: B. Colomina (ed.), *Sexuality and Space*. New York: Princeton Architectural Press, pp. 185–217.

Spiro, D. E. (1999) *The Hidden Hand of American Hegemony: Petrodollar Recycling and International Markets*. Ithaca, NY: Cornell University Press.

Stamp, L. D. (1937) *Chisholm's Handbook of Commercial Geography* (15th edition). London: Longman's, Green.

Standage, T. (1998) *The Victorian Internet: The Remarkable Story of the Telegraph and the Nineteenth Century's On-Line Pioneers*. New York: Berkley Publishing.

Stark, D. (1996) Recombinant property in East European capitalism. *American Journal of Sociology* 101.4: 993–1027.

Steinbach, A. L. (1957) Regional organizations of international labor. *Annals of the American Academy of Political and Social Science* 310: 12–20.

Stern, A. (2006) *A Country That Works: Getting America Back on Track*. New York: Free Press.

Stevis, D. (1998) International labor organizations, 1864–1997: The weight of history and the challenges of the present. *Journal of World-Systems Research* 4: 52–75.

Stevis, D. and Boswell, T. (2007a) *Globalization and Labor: Democratizing Global Governance*. Lanham, MD: Rowman and Littlefield.

Stevis, D. and Boswell, T. (2007b) International Framework Agreements: Opportunities and challenges for global unionism. In: K. Bronfenbrenner (ed.), *Global Unions: Challenging Transnational Capital through Cross-Border Campaigns*. Ithaca, NY: Cornell University Press, pp. 174–94.

Stockton, F. T. (1916) Agreements between American and European molders' unions. *Journal of Political Economy* 24.3: 284–98.

Stopford, J. M. (1974) The origins of British-based multinational manufacturing enterprises. *The Business History Review* 48.3: 303–35.

Subrahmanyam, S. (1989) The Coromandel trade of the Danish East India Company, 1618–1649. *Scandanavian Economic History Review* 37.1: 41–56.

Tabb, W. K. (2004) *Economic Governance in the Age of Globalization*. New York: Columbia University Press.

Taylor, P. J. (1981) Geographical scales within the world-economy approach. *Review* 5.1: 3–11.

Taylor, P. J. (1982) A materialist framework for political geography. *Transactions of the Institute of British Geographers*, New Series, 7.1: 15–34.

Thelen, K. A. (1991) *Union of Parts: Labor Politics in Postwar Germany*. Ithaca, NY: Cornell University Press.

Thomas, B. (1973) *Migration and Economic Growth: A Study of Great Britain and the Atlantic Economy*. Cambridge: Cambridge University Press.

Thomas, G. and Smith, C. F. (2003) *Flightpaths: Exposing the Myths about Airlines and Airfares*. Perth, WA: Aerospace Technical Publications International Pty Ltd.

Thompstone, S. (1984) Ludwig Knoop, "The Arkwright of Russia." *Textile History* 15.1: 45–73.

Thrift, N. (2000) Pandora's box? Cultural geographies of economies. In: G. Clark, M. Feldmann, and M. Gertler (eds.), *The Oxford Handbook of Economic Geography*. Oxford: Oxford University Press, pp. 689–702.

Tichelman, F. (1988) Socialist "internationalism" and the colonial world: Practical colonial policies of social democracy in Western Europe before 1940 with particular reference to the Dutch SDAP. In: F. van Holthoon and M. van der Linden (eds.), *Internationalism in the Labour Movement 1830–1940*, Volume 1. London: E. J. Brill, pp. 87–108.

Times [London] (August 6, 1858) London, Friday, p. 8.

Times [London] (July 27, 1866) No title, p. 9.

Times [London] (July 30, 1866) No title, p. 8.

Times [London] (September 21, 1866) London, Friday, p. 6.

Todorov, T. (1984) *The Conquest of America: The Question of the Other*. Norman: University of Oklahoma Press.

Tolentino, P. E. (2000) *Multinational Corporations: Emergence and Evolution*. London: Routledge.

Tribe, K. (1981) *Genealogies of Capitalism*. Atlantic Highlands, NJ: Humanities Press.

Triner, G. D. and Wandschneider, K. (2005) The Baring crisis and the Brazilian encilhamento, 1889–1891: An early example of contagion among emerging capital markets. *Financial History Review* 12.2: 199–225.

Trotsky, L. (1970 [1930/1906]) *The Permanent Revolution and Results and Prospects*. New York: Pathfinder.

Twomey, M. (2000) *A Century of Foreign Investment in the Third World*. London: Routledge.

Uche, C. U. (1999) Foreign banks, Africans, and credit in colonial Nigeria, c.1890–1912. *Economic History Review* 52.4: 669–91.

Uehlein, J. (1989) Using labor's trade secretariats. *Labor Research Review* 8.1: 31–41.

UITBB (Trades Union International of Workers in the Building, Wood, Building Materials and Allied Industries) (2005) Asbestos – The momentum from 9 September. *UITBB International* (newsletter), No. 14 (October–December), p. 1.

UK Trade and Investment (2006) *UK Inward Investment 2005/2006*. Report by UK Trade and Investment.

UNCTAD (United Nations Conference on Trade and Development) (1995) *World Investment Report 1995*. Geneva: UNCTAD.

UNCTAD (United Nations Conference on Trade and Development) (2002) *Trade and Development Report, 2002*. Geneva: UNCTAD.

UNCTAD (United Nations Conference on Trade and Development) (2004a) *Occasional Note: Outward FDI from Brazil: Poised to Take Off?* Geneva: UNCTAD.

UNCTAD (United Nations Conference on Trade and Development) (2004b) *Industrialization in Developing Countries: Some Evidence from a New Economic Geography Perspective*. Geneva: UNCTAD.

UNCTAD (United Nations Conference on Trade and Development) (2005) *World Investment Report 2005*. Geneva: UNCTAD.

UNCTAD (United Nations Conference on Trade and Development) (2006) *World Investment Report 2006*. Geneva: UNCTAD.

United Nations (1999) *Human Development Report 1999*. New York: United Nations Development Programme.

United Nations (2000) Globalization: A European perspective. United Nations Economic Commission for Europe Economic Analysis Division Discussion Paper Series No. 1. Geneva: UN Economic Commission for Europe.

United Nations (2006) International migration and development: Report of the Secretary-General. May 18. New York: United Nations.

US Bureau of Economic Analysis (2004) *U.S. Direct Investment Abroad: Final Results from the 1999 Benchmark Survey*. Washington, DC: Department of Commerce.

US Census Bureau (1975) *Historical Statistics of the United States: Colonial Times to 1970, Part 1*. Washington, DC: Department of Commerce.

US Census Bureau (2003) *Statistics of U.S. Businesses: 2003*. Washington, DC: Department of Commerce.

US Census Bureau (2006) *A Profile of U.S. Exporting Companies, 2003–2004*. Washington, DC: Department of Commerce.

US Department of Commerce (1958) *Balance of Payments: Statistical Supplement.* Washington, DC: US Department of Commerce, Office of Business Economics.

US Department of Commerce (1960) *U.S. Business Investments in Foreign Countries.* Washington, DC: US Department of Commerce, Office of Business Economics.

US Department of Defense (2006) *Base Structure Report, Fiscal Year 2006 Baseline.* Washington, DC: Office of the Deputy Under Secretary of Defense (Installations and Environment).

US Senate (1931) *American Branch Factories Abroad.* Senate Document No. 258, 71st Congress, 3rd Session. Washington, DC: United States Senate.

Van Den Bulcke, D., Zhang, H., and do Céu Esteves, M. (2003) *European Union Direct Investment in China: Characteristics, Challenges and Perspectives.* London: Routledge.

Van Dormael, A. (1978) *Bretton Woods: Birth of a Monetary System.* New York: Holmes and Meier.

Varsanyi, M. W. (2005) The rise and fall (and rise?) of non-citizen voting: Immigration and the shifting scales of citizenship and suffrage in the United States. *Space and Polity* 9.2: 113–34.

Vaupel, J. W. and Curhan, J. P. (1969) *The Making of Multinational Enterprise.* Cambridge, MA: Harvard University Press.

Viesti, G. (1988) Size and trends of Italian direct investment abroad: A quantitative assessment. In: F. Onida and G. Viesti (eds.), *The Italian Multinationals.* London: Croom Helm, pp. 30–48.

Virilio, P. (1995) Speed and information: Cyberspace alarm! Originally published in French in *Le Monde Diplomatique,* August 1995. Available at: www.eldespertador.info/despierta/ textdesper/viriliospeed.htm

Virilio, P. (1997) *Open Sky.* New York: Verso.

Vogel, S. K. (1996) *Freer Markets, More Rules: Regulatory Reform in Advanced Industrial Countries.* Ithaca, NY: Cornell University Press.

Wallach, L. and Woodall, P. (2004) *Whose Trade Organization? A Comprehensive Guide to the WTO.* New York: New Press.

Wallerstein, I. (1974) *The Modern World-System: Capitalist Agriculture and the Origins of the European World-Economy in the Sixteenth Century.* New York: Academic Press.

Wallerstein, I. (1976) The three stages of African involvement in the world economy. In: P. C. W. Gutkind and I. Wallerstein (eds.), *The Political Economy of Contemporary Africa.* Beverly Hills, CA: Sage, pp. 30–57.

Waterman, P. (1993) Internationalism is dead! Long live global solidarity? In: J. Brecher, J. Brown Childs, and J. Cutler (eds.), *Global Visions: Beyond the New World Order.* Boston, MA: South End Press, pp. 257–61.

Waterman, P. (1998) *Globalization, Social Movements and the New Internationalisms.* London: Mansell.

Waterman, P. (2004) Adventures of emancipatory labour strategy as the new global movement challenges international unionism. *Journal of World-Systems Research* 10.1: 217–53.

Waterman, P. (n.d.) Prague 1968: The last, late short spring of the World Federation of Trade Unions. Available at: www.global-labour.org/prague_1968_the_last_late_short_spring_ of_the_wftu.htm (accessed August 31, 2007).

de Webb, M. (1908) *India and the Empire: A Consideration of the Tariff Problem.* London: Longmans, Green, and Co.

Weber, R. E. (1992) Seward's other folly: America's first encrypted cable. *Studies in Intelligence* 36.5: 105–9.

Webster, A. (1990) The political economy of trade liberalization: The East India Company Charter Act of 1813. *Economic History Review,* New Series, 43.3: 404–19.

Weil, P. M. (1984) Slavery, groundnuts, and European capitalism in the Wuli kingdom of Senegambia, 1820–1930. *Research in Economic Anthropology* 6: 77–119.

Weiler, P. (1988) *British Labour and the Cold War.* Stanford, CA: Stanford University Press.

Weinberg, P. J. (1978) *European Labor and Multinationals.* New York: Praeger.

Weisser, H. (1971) Chartist internationalism, 1845–1848. *The Historical Journal* 14.1: 49–66.

WFTU (World Federation of Trade Unions) (1994) *Strategies for Unity and Solidarity to Advance the Workers' Interests* (Policy Document, Declarations, Resolutions and Other Documents Adopted by the 13th World Trade Union Congress, Damascus, November 22–26). Prague: WFTU.

WFTU (World Federation of Trade Unions) (1996) *Flashes From the Trade Unions,* No. 5–6. Prague: WFTU.

White, C. H. F. (1942) Laying a submarine cable. *The Geographical Magazine* (January): 142–5.

White, H. D. (1933) *The French International Accounts, 1880–1913.* Cambridge, MA: Harvard University Press.

Wick, I. (2003) *Workers' Tool or PR Ploy?: A Guide to Codes of International Labour Practice* (3rd revised ed.). Bonn: Friedrich-Ebert-Stiftung.

Wilkins, M. (1970) *The Emergence of Multinational Enterprise: American Business Abroad from the Colonial Era to 1914.* Cambridge, MA: Harvard University Press.

Wilkins, M. (1974) *The Maturing of Multinational Enterprise: American Business Abroad from 1914 to 1970.* Cambridge, MA: Harvard University Press.

Wilkins, M. (1988) The free-standing company, 1870–1914: An important type of British foreign direct investment. *Economic History Review,* 2nd series, 41.2: 259–82.

Wilkins, M. (1989) *The History of Foreign Investment in the United States to 1914.* Cambridge, MA: Harvard University Press.

Wilkins, M. (1994) Comparative hosts. *Business History* 36.1: 18–50.

Wilkins, M. and Hill, F. E. (1964) *American Business Abroad: Ford on Six Continents.* Detroit: Wayne State University Press.

Williams, D. (1968) The evolution of the sterling system. In: C. R. Whittlesey and J. S. G. Wilson (eds.), *Essays in Money and Banking in Honour of R. S. Sayers.* Oxford: Oxford University Press, pp. 266–97.

Williams, M. L. (1975) The extent and significance of the nationalization of foreign-owned assets in developing countries, 1956–1972. *Oxford Economic Papers,* New Series, 27.2: 260–73.

Williams, R. (1976) *Keywords.* London: Fontana.

Williamson, H. (1994) *Coping with the Miracle: Japan's Unions Explore New International Relations.* London: Pluto.

Willis, K. (2005) *Theories and Practices of Development.* London: Routledge.

Wills, J. (2001) Uneven geographies of capital and labour: The lessons of European works councils. *Antipode* 33.3: 484–509.

Wills, J. (2002) Bargaining for the space to organise in the global economy: A review of the Accor–IUF trade union rights agreement. *Review of International Political Economy* 9.4: 675–700.

Wilson, C. (1974) The multinational in historical perspective. In: K. Nakagawa (ed.), *Strategy and Structure of Big Business.* Tokyo: University of Tokyo Press, pp. 265–303.

Winder, G. (2006) Webs of enterprise 1850–1914: Applying a broad definition of FDI. *Annals of the Association of American Geographers* 96.4: 788–806.

Windmuller, J. P. (1980) *The International Trade Union Movement.* Deventer, The Netherlands: Kluwer.

Winn, P. (1976) British informal empire in Uruguay in the nineteenth century. *Past and Present* 73: 100–26.

Winpenny, T. R. (1995) Milton S. Hersey ventures into Cuban sugar. *Pennsylvania History* 62.4: 491–502.

Wolf, M. (2001) Will the nation-state survive globalization? *Foreign Affairs* 80.1: 178–90.

Wolff, R. D. and Resnick, S. A. (1987) *Economics: Marxian Versus Neoclassical*. Baltimore, MD: Johns Hopkins University Press.

Woodruff, W. (1966) *Impact of Western Man: A Study of Europe's Role in the World Economy, 1750–1960*. New York: St. Martin's Press.

Woods, N. (2006) *The Globalizers: The IMF, the World Bank, and Their Borrowers*. Ithaca, NY: Cornell University Press.

World Bank (2006) World Development Indicators database. July 1.

World Federation of Exchanges (2006) *Annual Report and Statistics*. Paris: World Federation of Exchanges.

Wray, L. R. (2006) Flexible exchange rates, Fed behavior, and demand constrained growth in the USA. *International Review of Applied Economics* 20.3: 375–89.

Wriston, W. B. (1992) *The Twilight of Sovereignty: How the Information Revolution Is Transforming Our World*. New York: Charles Scribner's Sons.

WTO (World Trade Organization) (2006) *World Trade Report 2006: Exploring the Links Between Subsidies, Trade and the WTO*. Geneva: WTO.

WTO (World Trade Organization) (2007) *10 Benefits of the WTO Trading System*. Geneva: WTO.

Yeats, A. J. (1990) Do African countries pay more for imports? Yes. *The World Bank Economic Review* 4.1: 1–20.

Yonekawa, S. (1974) The strategy and structure of cotton and steel enterprises in Britain, 1900–39. In: K. Nakagawa (ed.), *Strategy and Structure of Big Business*. Tokyo: University of Tokyo Press, pp. 217–57.

Zharikov, A. (1995) Interview by author with Alexander Zharikov, General Secretary, World Federation of Trade Unions, Prague, Czech Republic, September 11.

Index

STEELMAN LIBRARY